Civilizing Mission

Civilizing Mission

Exact Sciences and French Overseas Expansion, 1830–1940

Lewis Pyenson

The Johns Hopkins University Press
Baltimore and London

This book has been brought to publication with the generous assistance of the David M. Stewart Museum, Montreal; the Killam Program of the Canada Council, Ottawa; and the Dudley Observatory, Schenectady, New York.

Printed in the United States of America on acid-free paper

The Johns Hopkins University Press
2715 North Charles Street
Baltimore, Maryland 21218-4319
The Johns Hopkins Press Ltd., London

A catalog record for this book is available from the British Library.

Frontispiece: Detail from the bas-relief on the façade of the Musée national des arts d'Afrique et d'Océanie, Paris, sculpted by Alfred Janniot, 1928–31. Janniot's theme is the contribution of the colonies to France. Originally the Musée des colonies, the building was constructed for the International Colonial Exhibition of 1931. Courtesy of Réunion des musées nationaux, Paris.

Illustrations for part and conclusion openings: Le Septentrion, Le Midi, L'Orient, and *L'Occident,* copper engravings by Nicolas Bonnart (1646–1718); *L'Amérique,* hand-colored copper engraving by Henri Bonnart (1642–1711). All illustrations courtesy of David M. Stewart Museum of Discoveries, Montreal.

Endpapers: The Dream, Henri Rousseau, 1910, Museum of Modern Art, New York.

Library of Congress Cataloging-in-Publication Data

Pyenson, Lewis.
Civilizing mission : exact sciences and French overseas expansion, 1830–1940 / Lewis Pyenson.
p. cm.
Includes bibliographical references and index.
ISBN 0-8018-4421-5
1. Science — France — History — 19th century. 2. Science — France — History — 20th century. I. Title.
Q127.F8P94 1993
500.2'09171'24409034 — dc20 92-22577

Pour les maîtres de ma jeunesse

Contents

Illustrations

Preface

This book completes a trilogy devoted to cultural imperialism and exact sciences.[1] It treats physicists and astronomers who worked in the French overseas empire and in French spheres of influence; it documents how Western learning arrived on distant shores and also how pure knowledge played a role in geopolitical contests. It establishes a framework for a systematic study of exact sciences in nations that achieved political independence during the twentieth century.

My interest in imperialism and science developed after I chose to live and work in French Canada, a nation lacking the resources – agricultural, industrial, and intellectual – that have led poorer cousins to proclaim autonomy. My research has concluded in a climate of indecision and unreason which grips Montreal, especially. I do not pretend to contribute here to debate surrounding the racial destiny of French Canadians, a futile discourse whose persistence I find disturbing.[2] This volume nevertheless reminds us about the limitations of tying knowledge to a particular language or culture. It shows that formal, political independence can mean little in the world of science. It suggests that advantages may accrue to a researcher who writes in a language different from the one spoken by most people in almshouse, marketplace, and bedroom.[3]

In the following pages I concentrate on scientific vectors of cultural imperialism, the men and women who carried out research at permanent, overseas locations.[4] I thus seek to understand how scientific activity served political goals by focusing on the individual researcher, "im-

1. The two earlier volumes are *Cultural Imperialism and Exact Sciences: German Expansion Overseas, 1900–1930* (New York, 1985), and *Empire of Reason: Exact Sciences in Indonesia, 1840–1940* (Leiden, 1989), hereafter *CIES* and *ER*, respectively.

2. See John Grimond, "For Want of Glue," *Economist*, no. 7713, 29 June 1991, centerfold; and Mordecai Richler, "Inside/Outside," *New Yorker*, 23 Sept. 1991, pp. 40–92.

3. See Lewis Pyenson, "Macondo científico: Instituciones científicas en América latina a principios del siglo XX," trans. Lucyna Lastowska, in *La Junta para ampliación de estudios e investigaciones científicas 80 años después*, ed. José Manuel Sánchez Ron, 2 vols. (Madrid, 1988), 2:229–49, related discussion on pp. 240–48.

4. I have not sought systematically to include explorers in my study. On these men see Numa Broc, "Les explorateurs français du XIXe siècle reconsidérés," *Revue française d'histoire d'outre-mer* 69 (1982), 237–73, 323–59. My restriction eliminates a source of confusion raised by Paul van der Velde in the *NRC Handelsblad*, 23 June 1990, Saturday supplement, reviewing *Empire of Reason*.

prisoned within a destiny in which he himself has little hand."[5] In a number of instances, the focus is achieved by refraction through the alidades of ministers of state and their subordinates. Historians try to wrest true feelings from correspondence with state preposters. Because cultural imperialism is about pride, envy, cupidity, and worse—in short, about emotion—personal sentiments find a place in my narrative, which favors correspondence *inter pares* as it probes the social origin and training of overseas scientists. I have only scratched the surface of this kind of documentation. The history of science in modern France awaits a revisionist treatment based on family and institutional archives.

It is remote from my intent to project science as a "malign force" in modern history, and while many moral judgments are offered here, the reader will find no explicitly "political" agenda.[6] I do not engage in debate about the essential causes of French overseas expansion in the nineteenth and twentieth centuries, whether capitalist, mercantilist, or the primitive lust to plunder distant lands and enslave foreign civilizations. Political chieftains and plutocrats through the fall of the Third Republic enter into the following pages, but they play cameo roles.[7] This is a work in the social history of science. It explores the feelings and actions of rank-and-file scientists, many of them retaining at the end of their career only the memory of constant devotion to the search for natural laws. By that devotion, the intellectual historian of science Alexandre Koyré would have emphasized, they are more worthy of our attention than the councillors of state who commanded armies to cross oceans on wars of conquest.[8]

Indigenous scientists in French spheres of influence—those who

5. Fernand Braudel, *The Mediterranean and the Mediterranean World in the Age of Philip II*, trans. Siân Reynolds, 2 vols. (New York, 1975), 2:1244.

6. Quotations from Charles C. Gillispie, in *Isis* 82 (1991), 98, commenting on Lewis Pyenson, "Science and Imperialism," in *Companion to the History of Modern Science*, ed. R. C. Olby et al. (London, 1990), pp. 920–33, and several other contributions to the volume.

7. A recent survey emphasizing French politicians is *Histoire de la France coloniale*, 2 vols. (Paris, 1990), vol. 1 (until 1914) by Jean Meyer, Jean Tarrade, Annie Rey-Goldzeiguer, and Jacques Thobie; vol. 2 (from 1914) by Jacques Thobie, Gilbert Meynier, Catherine Coquery-Vidrovitch, and Charles-Robert Ageron. In nearly fifteen hundred pages science finds almost no place.

8. I. Bernard Cohen, "The Many Faces of the History of Science," in *The Future of History: Essays in the Vanderbilt University Centennial Symposium*, ed. C. F. Delzell (Nashville, 1977), pp. 65–110, Koyré's concern with "lesser figures" on p. 92. On the merits of focusing on small figures from the point of view of social history of science, see Lewis Pyenson, "'Who the Guys Were': Prosopography in the History of Science," *History of Science* 15 (1977), 155–88; Nathan Reingold, "Tales from the Archives," *Proceedings of the American Philosophical Society* 134 (1990), 340–48; and Charles Rosenberg, "Making It in Urban Medicine: A Career in Scientific Medicine," *Bulletin of the History of Medicine* 64 (1990), 163–86.

trace their ancestry to a nation other than France – appear infrequently in the following pages.[9] Only for the New World and to some extent China has it been possible to relate the French imperialist experience to autochthonous developments. One volume could not possibly do justice to the rise of national research communities in such varied settings as Algeria, Madagascar, and Vietnam, to mention some of the places I discuss. This focus also reflects my sources, which are largely based in European languages, but not exclusively so; I cite documents written in French, English, German, Dutch, Spanish, Portuguese, Italian, Czech, Slovak, Polish, and Chinese. One of my aims has been to point the way to future histories of the exact sciences drawn from sources in Vietnamese, Arabic, and Malagasy.

Civilizing Mission probes ties between pure and applied research – between *sciences fondamentales* and *sciences appliquées*. The terms *pure* and *applied* occur repeatedly in nineteenth-century and early twentieth-century writings. By using them I do not mean to imply their reification as "time-invariant types of research activity, of which only the mix is a function of the place, period, or institutional setting."[10] No one less than the astronomer Urbain Jean Joseph Le Verrier devoted years of his life to organizing a national weather service. Examples of this kind of practical activity may be found among German and Dutch theoretical physicists (Hermann von Helmholtz and Hendrik Antoon Lorentz come readily to mind), but institutional structures in France encouraged a union of the pure and the practical. To explore this union, which takes special forms at various overseas locations, it is necessary to discuss research programs and publications, although I have tried to avoid burdening the reader with tedious summaries of scientific articles. *Explication de textes*, while it finds a place in this book, is not my primary goal.[11]

I understand exact sciences to be learned enterprises that make use of mathematical formalisms or generate precise data as part of provisional results or syntheses. Numerical results serving to promote mathematical synthesis lie at the heart of astronomy and physics, especially,

9. It is useful to distinguish, as I do in the present book, between an Algerian geophysical research assistant such as the Nobel laureate Albert Camus and a Chinese astronomer such as the Jesuit-trained Kao Kiun.

10. Paul Forman, "Behind Quantum Electronics: National Security as Basis for Physical Research in the United States, 1940–1960," *Historical Studies in the Physical and Biological Sciences* 18 (1987), 149–229, the caveat on p. 219.

11. The essential fallacy of trying to reconstruct the true path of scientific creation has recently been revealed by Paul Forman, "The Aim of the Historian of Science," *Voprosy istorii estestvoznaniia i tekhniki*, no. 1, 1990, pp. 3–9 (in Russian).

which seek to formulate universal laws. This search lends itself to practical applications, but the exact sciences share little in common with the technologies that we need in our daily lives—the restorative powers of health care and the tinkering of mechanical invention. The considerable prestige of the exact sciences is unrelated to their projection in the agora. However useful the techniques and gadgets they parade before the world, astronomical and physical societies cannot compete in legal authority and popular impact with colleges of physicians and associations of mechanical engineers.

Institutional and personal constraints tie the researcher to his local environment, often preventing him from contributing original results. Scientists, no different from men and women generally, seek a social contract by which they obtain latitude of action in return for fidelity to a political or economic master. And scientists, like other people, seek to seduce their master by appealing to his baser or practical motives without compromising the integrity of the enterprise that is dear to them. The resulting, complicated negotiation and the interpretations ascribed to it are among my central themes.

Since figures and equations speak to all readers directly, the exact sciences provide a means of examining what happens when scientific discourse crosses linguistic and cultural boundaries. If, in this process of transmission, we restrict our examination to modifications and transformations in the discourse itself—most often in scientific publications—we may be disappointed. General prejudices and disciplinary norms originate in a particular social context and have a special character; beyond these prejudices and norms, scientific innovation is open to any interested party.

Because the exact sciences resist ideological contamination, they may serve as a probe for studying how learning supports the extension of political power. Mechanisms of political dominance can certainly be found in public health and social psychology. In the latter fields, however, where the scientist is often a salesman and an entrepreneur as well as a savant and where prejudice suffuses all levels of scientific discourse, it is difficult to identify ideologies that are related, in the first instance, to imperialist expansion. My research has been motivated especially by a desire to understand how abstract knowledge relates to political authority.

The French "civilizing mission" entailed an antinomy that remained unresolved: in the end, envoys of French scientific culture were unable to balance the scientific demands of their specialty against the administrative requirements of a central bureaucracy in the *hexagone* of metropolitan France. The form of that central authority, however, became a prototype for the administration of science later in the twentieth century. During the past two generations, we have witnessed the reassertion of military control over large sectors of scientific activity, the absorption of church-administered scientific institutions by state authorities, and the establishment of major scientific dependencies overseas. France was the first nation to experience these changes. We may dispute the desirability of such developments, but here, as in other spheres, France provided a model for modern civilization.

Abbreviations

ACRACS *Actes et comptes rendus de l'Association colonies-sciences.*

AIP New York, American Institute of Physics.

AIPG *Annales de l'Institut de physique du globe de l'Université de Paris et du Bureau central de magnétisme terrestre.*

Aix Aix-en-Provence, Archives d'Outre-Mer.

APGFOM *Annales de physique du globe de la France d'Outre-Mer.*

Armée France, Ministry of National Defense and War, *Le Service géographique de l'armée: Son histoire, son organisation, ses travaux* (Paris, 1938), under the supervision of Géraud Maurice Marie Joseph de Fontanges.

ASPG Strasbourg, Institut de physique du globe, *Annuaire*, pt. 2, *Seismologie* (containing *Bulletin du Bureau central séismologique français*); the publication is cited by its nominal year.

Astron. Philippe Véron, *Astronomes français, 1850–1950* (Ms.; St-Michel-l'Observatoire, 1989).

BC Comité d'études historiques et scientifiques de l'Afrique orientale française, *Bulletin.*

BEI *Bulletin économique de l'Indochine.*

BEM Colonie de Madagascar et dépendances, *Bulletin économique.*

BUA Shanghai, Université l'Aurore, *Bulletin;* the science section is understood when relevant.

Chant. Chantilly, Cantilliaceusis Domus Sessionum et Exercitiorum, Les Fontaines, Archives jésuites.

CIES Lewis Pyenson, *Cultural Imperialism and Exact Sciences: German Expansion Overseas, 1900–1930* (New York, 1985).

Congrès Paris, Exposition internationale, 1937, *Congrès de la recherche scientifique dans les territoires d'outre-mer* (Paris, 1938).

Cormont. Cormontreuil (Marne), Curtismonasteriensis

Schola Apostolica Sancti, Archives of the former Jesuit Province of Champagne.

CRAS Paris, Académie des sciences, *Comptes-Rendus.*

DBF *Dictionnaire de biographie française*, ed. Jules Balteau et al. (Paris, 1933–).

DSB *Dictionary of Scientific Biography*, ed. Charles C. Gillispie (New York, 1970–80).

diss. Dissertation, variously doctorat ès sciences, Dr. phil., Ph.D. Other kinds of dissertation – such as for French university doctorates, engineering doctorates, medical doctorates, or M.A. memoirs – are explicitly noted: univ. diss., ing. diss., med. diss., M.A. diss. Year of defense takes precedence over year of printing. The title for a French dissertation is taken to be identical with that of the "first thesis" defended by a doctoral candidate.

ER Lewis Pyenson, *Empire of Reason: Exact Sciences in Indonesia, 1840–1940* (Leiden, 1989).

GFOM Raymond Furon, *Histoire de la géologie de la France d'Outre-Mer* (Paris, 1955) [Museum national d'histoire naturelle, Mémoires, ser. C, 5].

GHA *General History of Astronomy*, vol. 4, *Astrophysics and Twentieth-Century Astronomy to 1950, Part A*, ed. Owen Gingerich (Cambridge, 1984).

HJA Hawthorn, Victoria, Australia; Jesuit Archives.

Hydrogr. Maurice Rollet de l'Isle, *Etude historique sur les ingénieurs hydrographes et le Service hydrographique de la Marine (1814–1914)* (Paris, 1951) [*Annales hydrographiques*, 1 bis (1950)].

LO Santa Cruz, University of California, McHenry Library, Mary Lea Shane Archives of the Lick Observatory.

Min. *Minerva: Jahrbuch der gelehrten Welt*, ed. Karl Trübner et al. (Leipzig, 1891–).

MUM Montreal, Université de Montréal, Archives, Fonds de l'Institut scientifique franco-canadien.

PAE Paris, Ministère des affaires étrangères, Archives.

PAN Paris, Archives nationales.

PBN Paris, Bibliothèque nationale.

PIPG Paris, Institut de physique du globe, Archives.

PMN Paris, Météorologie nationale, Bibliothèque.

PO Paris, Observatoire de Paris, Archives.

Pogg. *J. C. Poggendorffs biographisch-literarisches Handwörterbuch für Mathematik, Astronomie, Physik, Chemie und verwandte Wissenschaftsgebiete*, 7 vols. (Leipzig, 1863–).

PSG Paris, Bibliothèque nationale, Archives of the Société de géographie de Paris.

Quipu *Quipu: Revista latinoamericana de historia de las ciencias y la tecnología.*

RC *Relations de Chine.*

SJQ St-Jérôme, Quebec, Archives jésuites.

SMB Shanghai, Shanghai Meteorological Bureau, Archives.

SOM Archives nationales, Section d'Outre-Mer; formerly in the rue Oudinot, since transferred to Aix.

Toul. Toulouse, Tolosana Residentia Maior, Cordis Iesu, rue des Fleurs, Archives jésuites.

Vregille Pierre de Vregille, "Les observatoires de la Compagnie de Jésus au début du XXe siècle," *Revue des questions scientifiques* 19 (1906), 10–72, 493–579.

VAA Vincennes, Service historique de l'armée de l'air.

VAT Vincennes, Service historique de l'armée de la terre.

VM Vincennes, Archives de la marine, "Etat-Major."

Note on Places and Names

The activity of French agents abroad is central to this book, so it seems well to identify places by their French designation whenever common English forms are unavailable. The places considered are so diverse, the languages so varied, that a dissonance may occasionally strike the reader's eye. In particular, Arabic and Chinese names sometimes appear in their early twentieth-century, French transliteration. European family names are alphabetized according to conventions of the time; hyphenations in given names, occurring somewhat capriciously in French documents, have often been retained.

Whenever translation of an institution might generate confusion, the French name remains. The Société de géographie of Paris is not called the Geographical Society, just as the Ecole polytechnique is not referred to as the Polytechnical School. Academic, military, and professional designations — for example, *professeur sans chaire*, *physicien-adjoint*, *capitaine de frégate*, *agrégé* — have generally not been provided with English equivalents. In capitalizing titles and institutions in the Romance languages, I follow North American conventions. For this *lèse-majesté* I ask forgiveness from my friends.

I Stars in the Firmament of French Science

Scientific structures in the *hexagone* of metropolitan France—the vaguely six-sided polygon of the French Republic bounded by mountains and water—played important roles in overseas physics and astronomy. The French state controlled culture and science from the time of Louis XIV. During the nineteenth and twentieth centuries, institutions under the ministry of public instruction naturally extended their authority abroad. Culture did and does transcend educational curricula. The overseas projection of French science, like the prestige enjoyed by French culture generally, originated beyond the grasp of state educational authorities.

In attempting to direct French scientific expansion, the state confronted unruly players in each office responsible for overseas territory. The ministries of the navy and the colonies controlled overseas possessions. The ministry of foreign affairs oversaw protectorates, concessions, and spheres of influence. Special interministerial arrangements existed for Algeria, by the twentieth century an integral part of France. There were various ententes with and injunctions against Catholic institutions. The long arm of the French army reached everywhere. Finally, the ministry of public instruction enjoyed broad powers in cultural matters. Under different circumstances competing interests would have retarded attempts to promote cultural imperialism. Centralization determined France's place in European scientific expansion.

An orchestration of overseas science derived in part from political necessity. By the time of the Third Republic, when ministerial portfolios changed hands with alarming frequency, politicians and administrators sometimes found it useful to view matters from a collegial, not to say cooperative, point of view. Agreement among rival services and offices, however, was seldom harmonious. At times jurisdictional and personal disputes threatened to dissolve principled accord. French overseas expansion derived from more than political compromise.

The prosecution of French science abroad transcended factional squabbles to find broad support among both scientists and functionaries. Their common cause stemmed from a shared belief in France's patronage of and leadership in spreading civilized values. The belief guided every level of administration having authority for overseas activity. We turn now to metropolitan structures at the base of the French civilizing mission, focusing on scientific institutions directed by Jesuits, by the military, and by diverse colonial corps. Especially because of their horizon-spanning vision, these institutions are stars in the firmament of French science.

1

Mafia, Crucifix, and Epee

Civilized Institution

Enlightenment, focused by thinkers who warmed themselves beside state fires, has radiated from France for two centuries. The thinkers — no less than their patron, the state — stood for general principles that did not challenge local traditions in bathing, eating, dying, and loving. They projected the dissolution of the peoples of France into a single race. The nation, race, and language of the kingdom became, in the eyes of the French (and with some agreement elsewhere), the guarantors of civilized values. The terms of the guarantee — formal, French manners — became the rule of civilized intercourse everywhere. In science, just as in learning generally, French intellectuals engaged in a mission to civilize the world.[1]

What were the dimensions of the scientific civilizing mission? In providing an answer, it is useful to begin with abstractions. Motives are obscure in practical discourses, such as agronomy and metallurgy, where the expectation of lucrative gain often overpowers other sensibilities. Intentions are more clearly revealed in enterprises devoted to a dispassionate and disinterested search for general principles and laws. This activity was known in the period before 1940 as *science fondamentale*, or pure science. The present inquiry examines French expansion overseas in the fields of pure science based on mathematics, experiment, and observation.[2]

When France started reconstituting an overseas empire around

1. Theodore Zeldin, *France 1848–1940: Intellect and Pride* (Oxford, 1980), pp. 6–9; Norbert Elias, *The Civilizing Process: The Development of Manners*, trans. Edmund Jephcott (New York, 1978), pp. 38–39.

2. The matter is elaborated in the introduction to *CIES* and in Lewis Pyenson, "Typologie des stratégies d'expansion en sciences exactes," in *Science and Empires*, ed. Patrick Petitjean, Catherine Jami, and Anne Marie Moulin (Dordrecht, 1992), pp. 211–17.

1830, French authority in these exact sciences had arrived at a turning point. During the half-century between Jean le Rond d'Alembert's mature years and the death of Pierre-Simon, marquis de Laplace, Charles Coulston Gillispie has emphasized, "the French community of science predominated in the world to a degree that no other national complex has since done or had ever done. In its eminence, French cultural leadership in Europe reached a climax."[3] The climax came in the form of Laplacian physics, an elaboration of Newtonian mechanics that, despite its extraordinary synthetic accomplishment, was tied to particular hypotheses about the physical world. By the early 1820s Laplace and his physics had declined in tandem. Laplacian notions suffered critical reversals: imponderable fluids came under attack, John Dalton's atomic theory received sympathetic attention, Jean-Baptiste Joseph Fourier's rational mechanics provided a new approach to describing physical processes, André-Marie Ampère's electrodynamics offered a framework for understanding electrical current, and the wave theory of light vanquished Newtonian particle theory.

Proponents of the new order in exact sciences, who allied themselves with humanists under the Bourbon regime, looked forward to a new dawn. But as Robert Fox has argued, "this enthusiasm and spirit of optimism, which was to be so fruitful in literature and the arts, in science came to nothing. . . . What emerged from the ruins of Laplacian orthodoxy was not the new, revivified physical science that the early years of the Restoration had seemed to promise, but only a burst of creativity whose duration was no less brief, and whose decline was even more drastic, than that of Napoleonic physical science."[4] The exact sciences in France disappeared from the cutting edge of the research front. The Laplacian synthesis, ending the program that Newton had begun, defined what became known as *classical physics*. In the hands of those who turned against Laplace, the great instrument of mathematics—analysis—elaborated notions of atoms and continua to account for a large number of physical phenomena. Physics and astronomy came to reside in the mathematical adumbration of precise measurement. The most accomplished of the post-Laplacian French researchers are enshrined in present-day mechanics textbooks, their names attached to physical effects or mathematical expressions.[5] Yet their enterprise opened to limited horizons. One sees no sustained search for fundamental principles or laws.

3. Gillispie, *Science and Polity in France at the End of the Old Regime* (Princeton, 1980), p. 74.

4. Robert Fox, "The Rise and Fall of Laplacian Physics," *Historical Studies in the Physical Sciences* 4 (1974), 89–136; see also *DSB*, s.v. "Laplace," for an extended discussion.

5. For example, the law of Dulong and Petit and the Fourier transformations.

Across the second half of the nineteenth century, new visions of physical reality unanticipated by Laplace moved to center stage — electromagnetism and its resolution in Maxwellian electrodynamics; energetics and its elaboration in the two laws of thermodynamics; and statistical mechanics, providing a new perception of molecular reality. Each vision depended on special hypotheses, the validity of which seemed confirmed by subsequent predictions. But France, after the signal work of Ampère, Fourier, and Nicolas Léonard Sadi Carnot, remained on the periphery of electrodynamics, thermodynamics, and statistical mechanics.

We see, in fact, a rush by French physicists to apply mechanics, especially, to any number of situations. In the pages to come, a number of these polymaths make an appearance during the 1830s and 1840s: Auguste Bravais, Charles Joseph Sainte-Claire Deville, and Georges Aimé. They and dozens of their colleagues divided time between inventorying the natural world, in the spirit of Alexander von Humboldt's *Kosmos*, and analyzing it. They lacked "a single well-defined style of science . . . capable of filling the gap left by the old and of yielding a clear program for the future."[6] They divided their attention among many areas of nature as they raided the storehouse of classical physics for inspiration. And by this activity, French physicists and astronomers came to share authority in their discipline with men and women from other nations.[7]

With the new research style in post-Laplacian France came a general erosion of disciplinary norms, an attenuation that persisted for generations.[8] Publication, advancement, and funding in physics and astronomy depended on negotiating one's way through a hierarchy of interlocking committees, most of which were intentionally cross-disciplinary; the path was invariably marked by a cicerone. The national watchword was not *discipline* but *mafia*. Especially in the fields of physics, astronomy, and geodesy — an ensemble known at the time as the exact sciences — cross-disciplinary mafias policed French scientific life. For this reason, French physicists and astronomers who worked overseas are best understood not with reference to their scientific specialty

6. Fox, "Rise and Fall," pp. 127–28.

7. This is the formulation in Paul Tannery, "Les sciences en Europe depuis le XIVe siècle jusqu'à 1900," in Tannery, *Mémoires scientifiques*, 17, ed. Pierre Louis (Toulouse, 1950), pp. 123–434, p. 354, a new edition of Tannery's late nineteenth-century survey.

8. The failure of disciplinary societies to play a leading role on the research front is suggested in Robert Fox, "Science, the University, and the State in Nineteenth-Century France," in *Professions and the French State, 1700–1900*, ed. Gerald Geison (Philadelphia, 1984), pp. 66–145; on pp. 106–7 Fox discusses the most prosperous disciplinary society, the Société française de physique.

but rather from the viewpoint of the institutions that educated and employed them.[9]

Attenuation of discipline in French science is complemented by the central importance of patronage.[10] One might be tempted to derive the phenomenon of patronage from structural features of scientific life. Advancement in national scientific institutions placed a premium on seniority and administrative acumen rather than on brilliant, original publications. Indeed, publications were not infrequently of two sorts—a brief note communicated by a patron to the Academy of Sciences or a long monograph (worked up by assistants) issued by the director of another learned corporation. It was more prestigious to control the flow of information than to contribute to it. Elite educational institutions recruited by national competition and prepared graduates for a career as schoolteacher or road-builder. No premium was placed on training disciples.

France did have a research diploma, the *doctorat ès sciences*, which corresponded in a formal sense to the doctorate awarded by German philosophy and natural-science faculties. Like their German homologues, many French scientists discussed in the following pages completed a doctorate as a first step toward a career in research. Neither the French nor the German diploma opened doors in the workaday world; each was an indication of purpose and a designation of polite respectability. The French and German diplomas nevertheless diverged in scope and intent. Unlike the German dissertation, the French text did not have to present, strictly speaking, a new result; it could summarize existing knowledge and could even consist of several related essays. Furthermore, there was no formal identification of a supervisor for a French dissertation, like the German *Referent* or *Doktorvater*. The French system required a doctoral candidate to be resourceful and independent, and it diluted responsibility for standards over the entire jury. What professors would assume responsibility for only reluctantly, no admin-

9. This is the general point to Pierre Bourdieu's construction of the *champ scientifique*, which is derived entirely from the French experience. See Fritz Ringer, "The Intellectual Field, Intellectual History, and the Sociology of Knowledge," *Theory and Society* 19 (1990), 269–94, for a sympathetic reading of Bourdieu's work that does not address its cultural specificity; and John L. Davis, "The Influence of Astronomy on the Character of Physics in Mid-Nineteenth Century France," *Historical Studies in the Physical and Biological Sciences* 16 (1986), 59–82, where an argument is made implicitly for considering the exact sciences as an ensemble.

10. Terry Clark, *Prophets and Patrons: The French University and the Emergence of the Social Sciences* (Cambridge, Mass., 1973), pp. 66–92, where Clark points to the importance of patrons and their mafias, which he calls "clusters."

istrator would want to demand of a candidate as a credential. Talented younger researchers were rewarded by prizes from a learned corporation. We shall see, indeed, that *lauréats* of the Paris Academy of Sciences needed no diploma whatever to be named to state scientific corps. When it came to selecting an heir, French scientists preferred mafia allegiance to doctoral research.[11]

Historians must be wary of using the special features of science in France — its attenuated disciplines and centralized mafias — to explain scientific vitality or torpor. Patronage appointments do not in themselves discourage scientific innovation, just as the promotion of research diplomas may not encourage it; disciplines are not necessarily affected by either circumstance. Highly developed systems of patronage have coexisted with articulated specialization, even in the absence of important disciplinary journals, and the formal certification of research potential seems an unnecessary prerequisite for promoting a scientific discipline, much less a school. Nineteenth-century Britain, whose scientific establishment was at least as hierarchical as that of France and whose educational infrastructure lacked provision for research diplomas, offered fertile ground for new disciplines. So did the Netherlands, where despite the absence of strong disciplinary societies, specialized research doctorates underwrote nineteenth-century scientific life.

French scientific institutions manifested themselves in great diversity. *Faculté* and *école*, learned association and proprietary journal, made their presence felt in metropolis and province. And if everything received instruction from Paris, all was not regimented by it. Isolated savants flourished at regional centers, and high-minded provincials set up private institutions for prosecuting (and especially for popularizing)

11. The effective thesis supervisor normally received one of several dedications in the printed dissertation. Sometimes the director was the president of a candidate's examining jury. On the French doctorate, see Terry Shinn, "The French Science Faculty System, 1808–1914: Institutional Change and Research Potential," *Historical Studies in the Physical Sciences* 10 (1979), 271–332, pp. 281–82, 308–9; Fox, "Science, the University, and the State," pp. 74–76, 83; Harry Paul, *From Knowledge to Power: The Rise of the Science Empire in France, 1860–1939* (Cambridge, 1985), pp. 53–55; and for general statistics, Victor Karady, "L'accès aux grades et leurs fonctions universitaires dans les facultés des sciences au 19e siècle: Examen d'une mutation," in *The Making of Frenchmen: Current Directions in the History of Education in France, 1679–1979*, ed. Donald N. Baker and Patrick J. Harrigan (Waterloo, Ont., 1980) [Historical Reflections, Directions, 2], pp. 397–414. On the German doctorate, see Lewis Pyenson and Douglas Skopp, "Educating Physicists in Germany *circa* 1900," *Social Studies of Science* 7 (1977), 329–66; and Lewis Pyenson, *Neohumanism and the Persistence of Pure Mathematics in Wilhelmian Germany* (Philadelphia, 1983) [American Philosophical Society, Memoirs, 150], pp. 17–18.

high science. In this profusion of satellites orbiting a luminous source, the operant discourse by the end of the century was *utility.*[12] The reason is readily apparent.

Well into the twentieth century one could rise to the summit of scientific life — full membership in the Paris Academy — only if one resided in Paris.[13] Excellence in pure science necessarily reflected away from the provincial location that may have engendered it; brilliant contributions to pure learning would naturally pull a scientist toward the capital. Industrial and agricultural innovation, however, could contribute directly to the life of provincial towns. The regeneration of the provincial science faculties late in the nineteenth century and the persistent labors of local scientific associations both derive from a concern with using science to generate wealth. The circumstance is less surprising — although perhaps more confusing to foreigners — when we recall that in French the word *science* refers to everything and nothing.[14]

Understanding the generation of abstract knowledge in French provincial settings balances a traditional emphasis on Parisian discourse, but in the present context it would be misleading to focus on the countryside beyond Paris. Fortuitous alignments of state, religious, and municipal institutions did allow certain provincial settings to supply cadres and matériel for overseas enterprises in the exact sciences. We shall see that organizations in Lyons were prominent in this regard. Other provincial locations with a place in the following pages include Nice, Marseilles, Bordeaux, Toulouse, Cherbourg, Clermont, and Strasbourg. Yet there will be hardly a mention of university towns such as Rennes and Caen, and ports that have been identified with French overseas expan-

12. The central importance of applied and industrial research in late-nineteenth-century provincial settings is revealed in Shinn's seminal study, "The French Science Faculty System"; the theme is elaborated in Robert Fox, "The *Savant* Confronts His Peers: Scientific Societies in France, 1815–1914," in *The Organization of Science and Technology in France, 1808–1914*, ed. Fox and George Weisz (Cambridge, 1980), pp. 241–82. It is a leitmotif in three recent surveys: Mary Jo Nye, *Science in the Provinces: Scientific Communities and Provincial Leadership in France, 1860–1930* (Berkeley and Los Angeles, 1986); George Weisz, *The Emergence of Modern Universities in France, 1863–1914* (Princeton, 1983); and Paul, *Knowledge.* The practical turn taken late in the nineteenth century is the more remarkable for its absence before the 1870s. See Susan Sheets-Pyenson, "Popular Science Periodicals in Paris and London: The Emergence of a Low Scientific Culture, 1820–1875," *Annals of Science* 42 (1985), 549–72, where the author finds reporting about elite science instead of practical or applied discourse in popular Parisian journals.

13. The small, nonresidential section for full members came in 1913 (Nye, *Provinces*, pp. 150–51).

14. Criminology, theology, and astrology have all been taught as *science* over the past several years at the University of Montreal, for example.

sion – Lorient and Le Havre – are notable by their absence.[15] This apparently chaotic situation derives from the political orchestration of physics and astronomy. Every overseas institution in these domains took ultimate instruction not from provincial authorities but from Parisian mandarins. The ebb and flow of intellectual life across the hexagone and the outstretched arm of maritime greed are minor themes in the exact sciences overseas.

As pervasive as the practical or encyclopedic orientation of science may be in the French provincial context, it seems clear that physics and astronomy largely overstep this limited, utilitarian trend, even though it is in evidence. Geophysical prospecting was undertaken in the provinces by scientists situated at such faculties as Montpellier and Clermont-Ferrand; an *école de horlogérie* was set up at Besançon by one of Paris's finest astronomical instrument-makers. Yet the endowment of magnificent astronomical observatories at Lyons and Nice late in the nineteenth century can have little to do with practicality.[16] When institutes of geophysics came to Strasbourg and Clermont-Ferrand in the 1920s, only a small portion of their mandate related to local climatological inventory. These circumstances suggest that in the ecology of learning in France, the exact sciences occupy a special place.[17]

The niche is hardly apparent in the state universities and *grandes écoles*, where disciplinary equality and equanimity dominated governing councils and academic budgets. There an archaeologist was the peer of a physicist. Beyond such environments, which have recently drawn attention from historians, three habitats – religious, military, and colonial scientific institutions – sheltered the exact sciences and also occupied a

15. *Histoire de la France coloniale*, 2 vols. (Paris, 1990), vol. 1 (until 1914) by Jean Meyer, Jean Tarrade, Annie Rey-Goldzeiguer, and Jacques Thobie, pp. 369–76.

16. PO, carton 10, "La Plata," Auguste Fraissinet to François Beuf, 19 Apr. 1893, for A. Fénon (watchmaker to the Paris Observatory) called to Besançon; Harry Paul, "Apollo Courts the Vulcans: The Applied Science Institutes in Nineteenth-Century French Science Faculties," in Fox and Weisz, *Organization of Science and Technology*, pp. 155–81, Besançon horology on p. 173; Charles André, "L'Observatoire universitaire de Lyon (Saint-Genis-Laval)," Lyons, Académie, *Mémoires de la classe des sciences* 23 (1878–79), 53–71; A. Clorennec, R. Feldman, and J-P. Zahn, eds., *Cent ans d'astronomie à l'Observatoire de Nice* (Nice, 1981). On the Besançon astronomical observatory and its relationship with the local watchmaking industry, see René Baillaud, *Souvenirs d'un astronome (1908–1977)* (Rodez, France, 1980), pp. 307–9, 316–21, 361, Fénon on p. 503; the book is the third volume of Baillaud's three-volume *Souvenirs* (Rodez, 1976–80).

17. Geophysics and astronomy are largely absent from the works cited in n. 12, and physics appears only episodically there. The broad and nuanced nature of pure physics in early twentieth-century France is revealed in Dominique Pestre, *Physique et physiciens en France, 1918–1940* (Paris, 1984).

central place in French cultural imperialism. We now turn to their inventory and explication.

Priest and Observer

Daniel J. O'Connell, a director of one of Australia's finest research institutions who rose to head the Vatican Observatory, has emphasized that science played an important part in the activity of the Society of Jesus from the time of its foundation, "both in the training of its own recruits and in its colleges for the laity."[18] The commitment to research by Jesuit physicists, mathematicians, and astronomers during the first phase of the society's existence, from Christopher Clavius and Matteo Ricci to Girolamo Saccheri, has been elaborated by John Heilbron, who points to the key roles of Jesuit colleges and publications in disseminating and encouraging scientific learning.[19] With respect to astronomy, one need only recall, along with O'Connell, that at the time of the suppression of the society, Jesuits ran more than 30 of the world's 130-odd observatories. After the restoration of the society, Jesuits continued to count among their number a large circle of distinguished contributors to the exact sciences. Readily to mind come the astronomers Pietro Angelo Secchi, Stephen John Perry, and Johan Willem Jakob Antoon Stein, the geophysicist James Bernard Macelwane, and the late Father O'Connell himself.

How does one account for this profusion of talent, which in its very enumeration discredits the popular notion of a profound antipathy cleaving the Roman Catholic church from modern scientific research?[20] One might think that the answer simply reflects the circumstance by which the society selects its members in part with an eye to the sharpness of their intellectual talent and then hones this talent, over a long apprenticeship, to a razor-sharp edge. Intellect, however, comes in many forms. For such an argument to be persuasive, one must make the additional assumption that Jesuits are selected for scientific talent. This cannot be the case: a large number of the French Jesuit astronomers and geophysicists who enjoyed distinguished careers beyond Europe entered the order before they gave any evidence of scientific abilities.

The future chancellor of the Université St-Joseph in Beirut, François Dupré La Tour, signed on in 1918, when he was eighteen; he began

18. Daniel J. O'Connell, "Jesuit Men of Science," *Studies* 45 (1956), 307–18.

19. John L. Heilbron, *Electricity in the Seventeenth and Eighteenth Centuries: A Study of Early Modern Physics* (Berkeley and Los Angeles, 1978), pp. 101–14, 180–92.

20. The issue is discussed in Harry Paul, *The Edge of Controversy: French Catholic Reaction to Scientific Change from Darwin to Duhem* (Gainesville, 1979), a work in which Jesuits find scant place.

nuclear physics under Maurice de Broglie and Louis Leprince-Ringuet only in the 1930s. Ernesto Gherzi, a member of the pontifical and Lisbon academies of science and director of the radio weather-prediction service at Zikawei in Shanghai, became a Jesuit in his teens, before his technological studies. Gherzi's predecessor Louis Froc, "Father of the Typhoons," entered at age twenty and only much later learned about advanced physics from Edouard Branly at Paris. Stanislas Chevalier, first director of the major astronomical observatory at Zô-Sè, in a suburb of Shanghai, was essentially self-taught; at the end of his secondary-school studies he failed to obtain the *baccalauréat*, his sole academic degree being a later-acquired *bachelier ès sciences.* The astronomer who supervised erection of the telescope at Zô-Sè around 1900, Robert de Beaurepaire de Louvagny, failed to be named to the corps of engineers upon his leaving the Ecole polytechnique; subsequently placed in the Ecole d'application at Fontainebleau, he promptly resigned his commission to work as a civil engineer. Louis Cattala, specialist in gravimetry on Madagascar, earned his *licence ès sciences* at Lille when he was thirty years old. And Charles Poisson, longtime director of the Ambohidempona Observatory on Madagascar, was a career military officer whose scientific attainments, prior to landing his directorship, were minimal. Distinguished Jesuit scientists did receive academic laurels before they rose to the top of their field — Pierre Lejay coming readily to mind — but this seems to be somewhat epiphenomenal.[21]

To what do we owe scientific talent among men in the Society of Jesus? Intellectual brilliance accounts for a number of scientific achievements, but steadfast character equally underlies important additions to knowledge. And character, expressed in unshakable faith and intellectual honesty, is an attribute highly prized by the society. It is for this reason that the international scientific community has come to have great confidence in the scientific results obtained by Jesuit researchers. Father O'Connell recalls, for example, that the nineteenth-century naval astronomer Simon Newcomb absolved the eighteenth-century

21. For Dupré La Tour, see Henri Jalabert, *Jésuites au Proche-Orient* (Beirut, 1987), pp. 330–31; Dupré La Tour's publications are listed in Bernard A. Fiekers, "Bibliography of Jesuit Publications in Chemistry," *Jesuit Science Bulletin* 34 (1957), 71–94, pp. 73–74, where the French titles are translated into English. On Gherzi, see SJQ, Gherzi, "Autobiographie"; on Froc, André Bonnichon, *Le R. P. Louis Froc de la Compagnie de Jésus, "Le Père des typhons"* (n.p., 1927); on Chevalier, J. de Lapparent, "Le R. P. Stanislas Chevalier (1852–1930)," *Lettres de Jersey* 46 (1937), 323–79; on Beaurepaire de Louvigny, *Le révérend père Robert de Beaurepaire de Louvigny, un vaillant fils de Saint-Ignace* (Chartres, 1917); on Cattala, Bernard Leurent, "Le père Louis Cattala, 1904–1963," *Maduré-Madagascar*, no. 109, Oct. 1963, pp. 359–62; on Poisson, J[ean-Baptiste] C[oze], "Le père Charles Poisson, 1882–1965," ibid., no. 116, July 1965, pp. 218–23; and on Lejay, Pierre Tardi, "Notice nécrologique sur le R. P. Pierre Lejay," *CRAS* 247 (1958), 1265–70.

Jesuit Maximilian Hell from charges that he had altered data taken during observations of the transits of Venus. (It turns out that Hell's principal accuser, Karl Ludwig von Littrow, was led to an erroneous conclusion because he was color-blind.) More recently, the paleontologist Stephen Jay Gould has suggested that the architect of the Piltdown-man hoax was none other than the Jesuit anthropologist Pierre Teilhard de Chardin, a hypothesis that has been discredited. In O'Connell's words, dishonesty in scientific matters is "totally opposed to the whole training of a Jesuit, and to the ideals of the Society." The Jesuits pursued truth in their own way "with constant and devoted labour, often at the cost of great sacrifices."[22]

Closely connected with their devoted scientific labor was Jesuit commitment to education. Since the time of the suppression, Jesuit schools and universities existed in uneasy harmony with many secular authorities, notably those of France. Periodically excluded and tolerated, French Jesuits developed a special interest in establishing overseas institutions. It is not without some irony that foreign educational missions, notably those in Syria and China, came to provide havens for young recruits.[23] In the middle of the nineteenth century, during a period of relative peace with the state, French Jesuits placed a premium on preparing teenagers for the entrance examinations to the Ecole polytechnique, the Ecole militaire at St-Cyr, and the Ecole navale. They succeeded in Christianizing the French military: from 1871 to 1879 the Jesuit Collège Ste-Geneviève (then in the rue des Postes) graduated 316 students who gained entry to the Ecole polytechnique and 669 to St-Cyr. State tolerance led to a Francization of the society generally, which by 1880 counted 30 percent of its members as French citizens.[24]

The anticlerical reorganization of education promulgated by Jules Ferry led to a diaspora of French Jesuits and their institutions. Although

22. O'Connell, "Men of Science," pp. 317–18; John Winslow and Alfred Meyer, "The Perpetrator at Piltdown," *Science 83* [Washington, D.C., American Association for the Advancement of Science], Sept. 1983, pp. 32–43.

23. Paul Duclos, "Introduction," in *Dictionnaire du monde réligieux dans la France contemporaine*, vol. 1, *Les jésuites*, ed. Jean-Marie Mayeur and Yves-Marie Hilaire (Paris, 1985), pp. 5–16. Duclos notes the origin of missionary fervor in uncertainty and suppression at home (pp. 6–7).

24. Ibid., p. 7, statistics on p. 17; John Langdon, "New Light on the Influence of the Jesuit Schools: The Graduates of the Ecole Sainte-Geneviève, Paris, 1854–1913," *Third Republic/Troisième république* 1 (1976), 132–51; idem, "Whither the *postards?* Graduates of the Ecole Sainte-Geneviève, 1914–1954," in Baker and Harrigan, *Making of Frenchmen*, pp. 429–39. As the Society of Jesus flourished, other orders declined (Eugene Weber, *Peasants into Frenchmen: The Modernization of Rural France, 1870–1914* [Stanford, 1976], pp. 371–72).

a number of secondary schools were maintained under nominally secular direction, the Jesuits moved their naval *collège* from Brest to Jersey, just off the French coast; the Jesuit prenaval regime went back to France in 1900. Following a new set of anticlerical laws in 1901, the Jersey collegial buildings were reconstituted as the Collège Notre-Dame de bon-secours. As it was on English soil, the Jersey collège funneled new recruits to the Jesuit authorities at Hales Place, Canterbury; in fact, for a generation a continuing stream of French Jesuits (and some of their cultural allies, for example, the Italian national Ernesto Gherzi) passed through Jersey to matriculate on English soil. On the eve of the First World War there was an established pattern of studying for *navale* or *militaire* at a Jesuit school, joining the Society of Jesus in Canterbury, Enghien (Belgium), or Uclés (Spain), and then becoming an overseas missionary.[25]

Hazard of family and training should not obscure the attraction exerted by the Society of Jesus on a scientifically inclined teenager, especially in the fields of the exact sciences. For the past four centuries Jesuits have been among the most reliable recorders of natural phenomena.[26] Indeed, the impulse behind their search for the laws of nature is usually and incorrectly identified as something special to the Protestant world of the seventeenth and eighteenth centuries. In the present context, it is well to emphasize that a Jesuit scientist who was professed of the three vows had undergone the lengthiest and most exacting course of study in modern France. In addition to completing his scientific work (which often included the usual university diplomas up to and including the doctorate, as well as certification by a grande école), he had satisfied his spiritual masters of accomplishment in a rigorous apprenticeship; the latter included several years of basic philosophy, four years of theological study, and three years as a novice.[27] By virtue of their inner dis-

25. On the Jesuit preparation for French military schools and the subsequent diaspora, see John Cedric H. Aveling, *The Jesuits* (London, 1981), pp. 326–32; and Alain Guichard, *Les jésuites* (Paris, 1974), pp. 109–16. On the key role of Jersey in the French Jesuit establishment, see Pierre Delattre, ed., *Les établissements des jésuites en France depuis quatre siècles*, 5 vols. (Enghien, 1949–57), s.v. "Jersey" and "collèges"; Duclos, "Introduction," p. 8.

26. For a survey by a director of the Vatican Observatory, see Giovanni [Johan] Stein, "La Compagnia di Gesù e le scienze fisiche e matemàtiche," Milan, Seminàrio matemàtico e fisico, *Rendiconti* 15 (1941), 129–46; and Heilbron, *Electricity*, pp. 101–14. A recent bibliography on Jesuits and science, emphasizing the exact sciences over the recent past, is *Bibliographie sur l'histoire de la Compagnie de Jésus, 1901–1980*, ed. Lászlo Polgár, vol. 1 (Rome, 1981), pp. 503–10.

27. See, e.g., the treatments of Jesuit training in *Encyclopaedia Britannica*, 11th ed. (1911), s.v. "Jesuits"; and Gabriel Le Bras, *Les ordres religieux*, 2 vols. (Paris, 1979–80), vol. 2. The normal progression in the French system is: *noviciat*, two years; *juvénat*, two years of literary studies; *régence*, two or three years of teaching; *philosophie*, three years; *théologie*, four years (sacerdotal ordination in the third year); and finally the so-called *troisième an de*

cipline, French Jesuits were as a rule better adept than French civilians or military men at prosecuting science overseas. Near the end of his life Pierre Lejay — the first Jesuit named to full membership in the Academy of Sciences since the eighteenth century (due, as we shall see, to his direction of a French observatory in China) — reflected on just this advantage retained by Jesuit scientists. The scientist-priest

> refused, yesterday, to participate in those hysterical protests that alternatively had science pass from idolatry to bankruptcy; he pursued his task calmly, with a sure, interior enthusiasm of certainty. Tomorrow, if by chance he met with despair, he would provide the example of perseverance, certain that he will be supported by The One whom he in the end seeks and who sent his disciples "to baptize" not only men but "all the nations."[28]

Especially by the middle of the nineteenth century, a Jesuit's world-view was compatible with the ethos that underlines pure scientific research. The two systems of belief reinforced each other in the midst of civilizations far removed from European culture and censure.

Jesuits disseminated the research ethos overseas by example and by epistle at institutions in every kind of imperial administrative regime: the concession of the French Settlement at Shanghai (and several suburbs on land that France had not seized for its own purposes); Madagascar, both as a kingdom and protectorate and as a colony; and the mandate of Lebanon. The spiritual lords accordingly dealt with special kinds of temporal lord in each location: at Shanghai with the navy, the governing councils of the French and International settlements, the ministry of foreign affairs in Paris, and the various imperial, federal, and provincial authorities in China; on Madagascar with the Hova regents, the colonial army, the governor general, and the ministry of colonies; and in Lebanon with the occupying army, the Turkish and then the Lebanese ruling councils, and the ministries of public instruction and foreign affairs in Paris.

It might be thought that each environment hosted special institutions existing as a consequence of unique ententes. In fact all the Jesuit scientific institutions had common traits and evolved according to one pattern. The similarity is certainly due in part to the nature of the enterprise: science done by Jesuits whose immediate superiors resided in France. Yet one feature cuts across the diversity of local potentate and the uniformity of metropolitan, religious bureaucracy: the progressive

---

noviciat, lasting one year, to prepare for the final vows (Mayeur and Hilaire, *Dictionnaire*, p. 17).

28. Pierre Lejay, "La vie intérieure du savant," *Etudes* 252 (1947), 389–98, on p. 398.

assimilation of the institutions by imperial functionaries in Paris. In part 3 we shall see how occasional and exceptional grants awarded by the ministries of foreign affairs, colonies, and public instruction translated into an avalanche of financial support. Having showered funding on overseas Jesuit scientists, the imperial seat exacted its due in fealty to metropolitan conventions and priorities. By around 1930, France wielded an effective veto over appointments at each of the Jesuit scientific operations; metropolitan functionaries invigilated pedagogical standards and formulated research programs.

The Third Republic's assimilation of Jesuit scientific institutions at the outposts of empire finds a reflection in the hexagone, as Jesuit *collèges* and seminaries filtered back from exile during the years after 1900 and as the Catholic faculties achieved a measure of educational credibility.[29] The dénouement of the state's acquiescence to Jesuit scientists late in the 1940s is seen in Pierre Lejay's elevation to the Academy of Sciences and in the chair offered to Pierre Teilhard de Chardin at the Collège de France.[30] It was an inverted Kulturkampf, with the Jesuits wary about state support and resisting, as long as they were able, state encroachment. As time went on they had increasingly little room to maneuver. State cooptation of private initiatives was a consistent trend across the twentieth century, crossing the spectrum from disciplinary societies in meteorology and astronomy to medical laboratories such as the Pasteur institutes.[31]

Sailor and Stargazer

Between the end of the *ancien régime* and the Fourth Republic, astronomy in France was tied to practical surveys of the Earth's surface. The connection provided an excuse for maintaining a central bureaucracy to supervise the exact sciences and a justification for placing talented apprentices in permanent research posts. Measuring the land, both at home and abroad, required military precision, and the military controlled France's finest scientific school, the Ecole polytechnique. The armed forces as a result became an alembic for refining expertise in astronomy and physics.

The meter, called into being by the Constituent Assembly in 1790–91, required for its instantiation a detailed survey of the terrestrial meridian passing from Dunkirk to Barcelona. The surveying work pro-

29. Duclos, "Introduction," pp. 8–13.

30. *DSB*, s.v. "Teilhard"; Lejay is discussed in part 3, below.

31. See, e.g., P-Noël Bernard, ed., *Les instituts Pasteur d'Indochine* (Saigon, 1922); and *DSB*, s.v. "Nicolle, Charles."

ceeded intermittently, finally propelling Jean-Baptiste Biot and Dominique François Jean Arago to the island of Formentera in 1808. The arc they surveyed gave greater precision to the standard meter, and the surveying provided luster to the Bureau of Longitudes, the independent astronomical office set up in 1795 to collect and publish observations and generally to oversee the French observatories.[32] The bureau came to reside at the Paris Observatory, its highly paid staff—an assembly of astronomers, mathematicians, hydrographers, and geographers—charged with organizing astronomical missions and publishing an ephemeris. By the value of its emoluments and an accompanying freedom to undertake research of a general nature, a senior appointment to the bureau was highly prized; and the nature of French learning guaranteed that the bureau's senior staff would be members of the Academy of Sciences.[33]

The bureau functioned as the astronomical research institute of the Paris Academy from its inception, and it figured in every major French astronomical expedition over the succeeding two centuries. Yet it operated no major observatories of its own, except for the portion of the Paris Observatory that it came to control by sufferance.[34] And although it projected a strong presence in distant parts during eclipse or planetary-transit expeditions, it spent very little, if anything at all, on permanent overseas installations. Furthermore, the bureau was by its nature

32. Léon Bassot, *La géodésie française* (Paris, 1891), *separatum*, p. 7. On the meter and its adoption in France, see Ronald Edward Zupko, *Revolution in Measurement: Western European Weights and Measures since the Age of Science* (Philadelphia, 1990), pp. 137–75; and John L. Heilbron, "The Measure of Enlightenment," in *The Quantifying Spirit in the Eighteenth Century*, ed. Tore Frängsmyr, John L. Heilbron, and Robin Rider (Berkeley and Los Angeles, 1990), pp. 207–42.

33. Guillaume Bigourdan, "Le Bureau des longitudes: Son histoire et ses travaux, de l'origine (1795) à ce jour," Bureau des longitudes, *Annuaire pour l'an 1928* (Paris, [1928]), pp. A.1–A.72, continued in each of the next five years of the *Annuaire:* 1929, pp. C.1–C.92; 1930, pp. A.1–A.110; 1931, pp. A.1–A.151; 1932, pp. A.1–A.117; 1933, pp. A.1–A.91. Maurice Crosland, *The Society of Arcueil: A View of French Science at the Time of Napoleon I* (Cambridge, Mass., 1967), pp. 209–13.

34. According to Bigourdan, "Bureau des longitudes," over the first half of the nineteenth century the distinction between bureau initiatives and observatory functions was blurred. The two institutions assumed an independent character with the formation of a civil service at the observatory. In 1875 the Bureau of Longitudes organized an astronomical training school for naval officers and explorers at Montsouris Park in Paris; it shared grounds with a meteorological observatory, to be discussed shortly. Charles Maurain noted about the astronomical part of Montsouris: "This observatory is a kind of school of application for people who are called to carry out astronomical observations, such as explorers, members of geographical missions, for instance missions of boundary delimitations, military or naval officers sent on missions, etc." (PIPG, Maurain, "Rapport à la Commission des subventions aux établissements scientifiques" [1928]).

an interdisciplinary fellowship that transcended the mandate of its administrative superior, the ministry of public instruction. Soldiers and sailors freely entered its ranks. Military and naval assistance was indeed crucial in mounting astronomical expeditions, but multiple allegiances discouraged the emergence of a scientific school that could claim to be attached to the bureau. The bureau's modus operandi facilitated the appointment of Admiral Amadeo Ernesto Bartolomé Mouchez as director of the Paris Observatory in 1878. Mouchez then set the course for French astronomy over the next two generations with an ambitious project to compile a photographical chart of the stars — a project that related directly to French scientific initiatives beyond the hexagone.[35]

A military regimen was precisely what the Paris Observatory required. Its director at mid-century had been Urbain Jean Joseph Le Verrier, perhaps the greatest of French astronomers and the worst of French administrators. He sought to cut off the observatory from its quasi-feudal past by transforming it into a state bureaucracy (the Cassini family and eventually Arago had run the observatory as a private fiefdom, and when Le Verrier inherited it, the staff were bound together by blood and marriage). The annals of early nineteenth-century French science, however, were written in individual accomplishment and discovery. Le Verrier's subordinates would not suffer anonymous submersion in his bureaucracy, and his tenure as director marked the most tempestuous years of French astronomy.[36]

Mouchez's claim to Le Verrier's mantle had nothing to do with the ministry of public instruction and everything to do with the navy's scientific ambitions. Naval engineers were assimilated into a separate and well-paid corps in 1814. By mid-century, under the inspired leadership of Charles-François Beautemps-Beaupré, a member of both the Bureau of Longitudes and the Academy of Sciences, the elite corps was recruiting from graduates of the Ecole polytechnique. Among Beautemps-Beaupré's catches were Magloire-Thomas Daussy, who in 1832 became secretary-librarian of the Bureau of Longitude's observational facilities and who rose in 1855 to take Beautemps-Beaupré's position in the Academy of Sciences; and Louis Urbain Dortet de Tessan, who in 1861 took Daussy's chair at the academy. Other naval hydrographers who supervised major overseas expeditions and enjoyed distinguished scientific careers include Jean Jacques Anatole Bouquet de la Grye, Philippe Eugène Hatt, Maurice Rollet de l'Isle, Henri Roussilhe, and

35. *DSB* and VM, s.v. "Mouchez." Mouchez was born in Madrid, and according to the VM, his birth and marriage registrations bear the Spanish Christian names.

36. *DSB*, s.v. "Le Verrier."

Lazare-Eugène Fichot – all of whom figure in the following pages. The elite hydrographers numbered seventy-three over the century between 1814 and 1914.[37]

Scientific work, whether on astronomical expeditions, with the Bureau of Longitude's *pépinière* at Montsouris (organized by Mouchez), or in the navy's own observatory at Toulon, was also confided to career naval officers outside the corps. Mouchez was in the latter category, which also included Octave Marie Gabriel Joachim de Bernardières, Edouard Jean Pierre Marie Sylvain Perrin, Georges Ernest Fleuriais, Jules-Jean Feillet, Victor François César Beuf, Jean Alexandre Justin Paul Eugène Fitte, Georges Roux, Charles Poisson, and Edmund de la Villemarqué. We shall see that all these men served overseas.[38]

Naval engineers and regular officers constituted the single largest reservoir of astronomical talent in France at mid-century, a circumstance freely exploited by the ministry of public instruction. In 1868, for example, the naval officer and mathematician Jean-Philippe-Ernest de Fauque de Jonquières urged the ministry to lend a spectroscope to Philippe Hatt, then charting the coast of Indochina, so that Hatt could observe the upcoming solar eclipse at the nearby island of Poulo Condore.[39] When France constituted expeditions to observe the transit of Venus in 1874, Mouchez and Bouquet de la Grye were seconded from the navy to the ministry of public instruction; and the navy sent out four teams of its own to observe the 1882 transit.[40] This practice was followed in all French overseas astronomical expeditions during the nineteenth century.

Mouchez was the man of the astronomical hour. After graduating from the Ecole navale, he quickly distinguished himself as an astronomical calculator. He rose through the ranks to become *capitaine de vaisseau* in 1868, member of the Bureau of Longitudes in 1873, member of the

37. *Hydrogr.*, pp. 3–41; *CRAS* 50 (1860), v, gives Daussy's Christian name as Pierre.

38. Feillet went to direct the naval school (and its associated observatory) at Santiago de Chile; Beuf received direction of the La Plata Observatory in Argentina; Fitte was called to direct the Phu-liên Observatory at Haiphong; Roux directed the Moroccan meteorological and geophysical service. After they retired from the navy, Poisson received direction of the Ambohidempona Observatory on Madagascar, and de la Villemarqué the direction of the Zô-Sè Observatory near Shanghai.

39. PAN, F17/3008, E. de Jonquières to Minister of Public Instruction, 14 Mar. 1868; *DSB*, s.v. "de Jonquières," where, however, his Christian names appear in an order different from that given in *CRAS* 130 (1900), 8.

40. PAN, F17/2928, 1, Direction politique, Ministry of Foreign Affairs to Minister of Public Instruction (Jules Simon), 7 May 1873, for the seconding; VM, BB4/1342, Institut de France to Minister of the Navy, 9 May 1883, for the naval teams under Fleuriais, Hatt, Bouquet de la Grye, and de Bernardières.

Amadeo Ernesto Bartolomé Mouchez (Courtesy of Observatoire de Paris)

Academy of Sciences in 1875, and rear admiral in 1878. The veteran of numerous foreign tours of duty (Mouchez spoke Spanish fluently), he was the navy's premier hydrographer and expeditionary. Installed as director, he spent nearly a decade renovating practical astronomy in Paris. Then he committed resources to an ambitious and far-reaching project.

The project turned on advances in photographical astronomy. By the early 1880s, David Gill at Cape Town had demonstrated the feasibility of using celestial photography to inventory the stars. In 1885 he began a map of the southern skies; he took the photographical plates, and Jacobus Cornelius Kapteyn, professor of astronomy at Utrecht, measured them. Mouchez had inherited a star-mapping project of his own, an inventory of the ecliptic started in the 1860s by the brothers Paul and Prosper Henry at the Paris Observatory. The Henrys abandoned their catalogue when they reached the profusion of the Milky Way. They began experimenting with photography, and by the middle 1880s, with a photographical refractor of their own design, they discovered celestial objects that had escaped visual astronomers. Their achievements led Mouchez to correspond with overseas colleagues about the possibility of a photographical map of the heavens. Gill urged an international congress to discuss the proposal. Mouchez was only too happy to offer

Paris as its site.[41] Here was a way to supersede all the star catalogues that had been produced or were under production by foreign rivals such as Friedrich Wilhelm August Argelander at Bonn, Benjamin Apthorp Gould at Córdoba, and of course Gill himself.

The organizing conference of 1887 assembled more than fifty astronomers from nearly a score of countries. They divided the sky into portions assigned to various observatories; they planned for photographical illustrations (in twenty-two thousand sheets showing stars down to the fourteenth magnitude) and tables of some two million brighter stars. Uniformity would be ensured by using identical telescopes and ruled photographical plates. As the project was underwritten by the Paris Observatory, the Academy of Sciences covered the cost of publishing its memoirs and bulletins (although not, of course, all parts of the catalogue). A permanent commission came into being to supervise the project, assign work schedules, direct special studies, and generally handle problems as they arose.[42]

Within several years, the commission established standards and conventions – even for the thorny issue of stellar magnitude.[43] The authorized telescope would conform to the 34-cm refractor designed by the Henrys and built by Gautier in Paris. Photographical plates were to be had from Lumière frères in Lyons. Stellar zones went to eighteen participating observatories. The commission had no money to distribute, however; each participating observatory had to find its own funding. United States astronomers abstained from committing any of their impressive resources to the project (and the Germans committed theirs reluctantly), but Mouchez nevertheless believed that he would be able to set the enterprise on a firm path. French cultural imperialism confirmed his faith.

Among the eighteen observatories taking slices of the sky, three were on French soil: Paris, Bordeaux, and Toulouse. Another, Algiers, was in France's major overseas appendage. France effectively controlled two observatories in the Southern Hemisphere, Santiago de Chile and La Plata in Argentina, as French astronomers had recently arrived to take charge of each operation. A third South American observatory, at Rio de Janeiro, had been managed earlier by a refugee from Le Verrier's

41. *DSB* and *Astron.*, s.v. "Mouchez"; *Astron.*, s.v. "Henry"; Georges Rayet, "Notes sur l'histoire de la photographie astronomique," *Bulletin astronomique* 4 (1887), 165–75, 262–72, 307–20, 344–60, 449–56, on pp. 249–53 for the Henrys and the triumphs of their celestial photography. For the origins of the sky-chart project, see John Lankford, "The Impact of Photography on Astronomy," in *GHA*, pp. 16–39, on pp. 25–29.

42. Lankford, "Impact of Photography."

43. Harold F. Weaver, "The Development of Astronomical Photometry, Period III," *Popular Astronomy* 54 (1946), 287–99, 339–51, 389–404, on pp. 291–94.

tyranny and then had passed into the hands of a native of Belgium. And the Mexican national observatory at Tacubaya was also Francophile. Each of the five overseas observatories would be harnessed to a long-term project, using French astronomical technology, for the greater glory of the imperial seat. To the constellation of eight observatories owing greater or lesser allegiance to Paris, Mouchez added five from the English-speaking world — Greenwich, Oxford, Cape Town, Sydney, and Melbourne — which he thought could be disciplined from the British metropolis. The remaining partners in the project would be pulled along by inertia or national pride.

The tale unwound otherwise. As we shall see, among the French overseas astronomical satrapies only Algiers completed its task within a generation of starting it. Santiago de Chile, La Plata, and Rio de Janeiro withdrew well before the First World War. Tacubaya persisted, completing its assignment during the Second World War. Paris collaborated with the British empire to pick up most of the defaulted areas.

It has been argued that the burden of the project undermined the health of French astronomy by focusing attention on routine operations related to stellar position instead of on exciting new areas in astrophysics.[44] By 1920, with the French portion of the *carte du ciel* still incomplete, the French astronomical community felt fatigued and discouraged. The Academy of Sciences commissioned astrophysicist Henri Deslandres, director of the Meudon Observatory, to undertake a systematic and secret enquiry into the malaise. He concluded that France would do well to set off on a new path by allocating half of its astronomical resources to astrophysics, following the pattern of astronomers in the United States.[45]

We must not imagine that the sky-chart work in itself impeded creative research. Celestial inventory underlay analyses of proper motion and stellar parallax — studies directly related to cosmological issues — and this is precisely how United States and British observers used their catalogues. Furthermore, as Deslandres pointed out in his secret report, the French hexagone had ten major, research observatories; nine belonged to the state, and one (Abbadia) to the Academy of Sciences. Of the six observatories not implicated in the sky-chart project, Lyons and Nice had been created at the end of the nineteenth century and endowed with reasonable facilities; by the early twentieth century, Lyons concentrated on stellar photometry. Meudon, in greater Paris, com-

44. Lankford, "Impact of Photography," p. 32; *Astron.*, "Introduction," citing Jules Baillaud in 1936.

45. Henri Deslandres, "Rapport sur l'inspection des observatoires français décidée par l'ordre ministériel du 29 avril 1922." A copy of this document was kindly lent to me by Philippe Véron, St-Michel-l'Observatoire.

prised a large complex devoted exclusively to astrophysics. None of these observatories had enormous telescopes under ideal observing conditions, but the means were certainly available for significant, new undertakings.

The routine nature of positional astronomy, which absorbed the lion's share of funding at French observatories up to the middle of the twentieth century, was well suited to the social structure of astronomy in France. Observatories affiliated themselves only loosely with local universities when, indeed, they had any university affiliation at all. Astronomers advanced in their career through apprenticeship, generally working their way up in the observatory hierarchy; to obtain a doctorat ès sciences in observational astronomy, one had to be accomplished in the art of observing, and this meant spending years in a junior and poorly paid assistantship. *Anciens élèves* of the Ecole normale supérieure such as Félix Tisserand, Benjamin Baillaud, Ernest Esclangon, and André Danjon did rise to astronomical heights in France, as did graduates of the Ecole polytechnique such as Deslandres, but the recruitment of talented science students remained something of an accident.[46] Among the astronomers who figure in the following pages, many found their calling in the army or navy, and others rose to positions of authority by dogged perseverance. The guild system in astronomy, however, is not special to France. It flourished in the British empire and in the Americas, giving way early in the twentieth century to the German and Dutch model of academically trained astronomers. France, ironically, preceded the English-speaking world in requiring the doctorate of senior observatory staff.[47]

Nor can the attenuation of the astronomical discipline be the cause of the insecurity endemic to French astronomers. It is true that no clear path led to excellence in the field. One could rise to the top by starting in the army, the navy, the provincial universities and lycées, or the observatories, but in each kind of institution a prospective astronomer had to navigate around numerous obstacles and spend a great deal of time on things unrelated to the stars. And if one did discover new stars or mathematical equations, there was no respected, disciplinary journal to

46. The pivotal role in science of an education at the Ecole normale supérieure and the Ecole polytechnique has been elaborated in many studies, e.g., Craig Zwerling, "The Emergence of the Ecole normale supérieure as a Center of Scientific Education in the Nineteenth Century," in Fox and Weisz, *Organization of Science and Technology*, pp. 31–60; and Terry Shinn, *Savoir scientifique et pouvoir social: L'Ecole polytechnique, 1794–1914* (Paris, 1980).

47. Philippe Véron, "Introduction," in *Astron.*, where 1910 is given as the date for requiring a doctorate of all senior observatory staff. Véron argues that astronomy in France sustained great damage from capricious recruitment practices.

receive the discovery, the *Bulletin* of the French Astronomical Society being a house organ that appealed in large measure to amateurs. But with one exception, namely, the strong military presence, these circumstances also applied to Dutch astronomers during the nineteenth century, a community that rose to world prominence in the 1920s.[48]

Tedium and sterility derived from the rigid comportment and inward focus of astronomy in France. With the pervasive presence of military discipline and with all eyes turned toward Paris, few astronomers took the trouble to keep abreast of publications in German and English periodicals. We shall see in part 2 that only with the political convulsions of the twentieth century did foreign-educated astronomers find their way to French observatories — Maurice Loewy and Milan Ratislav Štefánik (who both figure in the following pages) being exceptional, earlier immigrants. Here the contrast with the Netherlands could not have been greater. Most Dutch astronomers would have been literate, if not fluent, in German, English, and French; Dutch astronomers prized individual initiative and outward vision. By the late nineteenth century, French astronomers withdrew into collective work that contributed to the glory of institutions in Paris.

48. *ER*, pp. 1–3, 181–83. On the interaction between the most important Dutch university observatory and the Dutch navy, see Elly Dekker, "Frederik Kaiser en zijn pogingen tot hervorming van 'het sterrekundig deel van onze zeevaart,'" *Tijdschrift voor de geschiedenis der geneeskunde, natuurwetenschappen, wiskunde en techniek* 13 (1990), 23–41.

2

Geophysics, the Military Establishment, and Overseas Expansion

Weatherman

Meteorology, the study of sublunar physical phenomena, received the attention of local observers in France during the first half of the nineteenth century. The focus was on medical climatology and natural-history inventory. The most ardent of the naturalists was Auguste Bravais, a naval lieutenant assigned in the 1830s to the scientific mission in newly conquered Algeria and later to a mission exploring Spitzbergen and Norway. His observations from Algeria led to two theses, for which he obtained a doctorat ès sciences from Lyons; subsequent work on atmospheric physics propelled him to a chair at the Lyons Faculty of Sciences in 1841, and high-altitude studies in the Alps resulted in a call to the Ecole polytechnique four years later. Research into crystal structure gave him, in 1854, a chair in the Academy of Sciences. And at about this time he lent his energy and prestige to the creation of an association for the comprehensive study of the French climate – the Meteorological Society of France. But his health failed in 1856; he retired from the Ecole polytechnique in 1857 and resigned his naval commission.[1] He left the details of meteorological organization to his naturalist colleague Charles Joseph Sainte-Claire Deville.

Deville, a mineralogist pupil and then assistant of Jean-Baptiste-Armand-Louis-Léonce Elie de Beaumont's (he was Elie de Beaumont's *suppléant* at the Collège de France for twenty-one years before succeeding his patron there in 1875), joined Bravais and Antoine d'Abbadie in founding the Meteorological Society of France in 1852. The meteorologists thought to understand the weather as naturalists might, and

1. J-B-A-L-L. Elie de Beaumont, "Memoir of Auguste Bravais," trans. C. A. Alexander, in Smithsonian Institution, Board of Regents, *Annual Report for 1869* (Washington, D.C., 1869), pp. 145–68; *DSB*, s.v. "Bravais"; Edouard Doublet, "La météorologie en France et en Allemagne," *Revue philomathique de Bordeaux et du Sud-Ouest* 14 (1911), 213–32, 250–67, and 15 (1912), 103–28, 169–83, here 14:252.

Charles Sainte-Claire Deville (Courtesy of Météorologie nationale, Paris)

their inductivist outlook required detailed records from all quarters of the world. Deville worked assiduously as the society's secretary, and in 1867 he obtained as his meteorological office a replica of the Bardo Palace (constructed by the bey of Tunis for the Universal Exhibition of 1867), which had been acquired by the city of Paris and transported to Montsouris Park. Laboring to transform his office into a national meteorological service, he began early in 1870 to upgrade it to a meteorological observatory. And by this action he incurred the enmity of Urbain Jean Joseph Le Verrier.[2]

Le Verrier had been running a meteorological operation of his own at the Paris Observatory since 1854, when at the urging of the minister of war, Marshal Jean-Baptiste Philibert, comte de Vaillant, he set up a weather advisory service based on telegraphical despatches exchanged with French and foreign observatories. In 1854 Le Verrier appointed Emmanuel Liais, an ambitious young scientist, to take charge of the international part of the service; after several years, Le Verrier sacked him. *Normalien* Quentin-Paul Desains, a professor of physics at the Sorbonne, replaced Liais in 1858, only to resign himself in 1861. In 1862 Le Verrier called upon Edme Hippolyte Marié-Davy, a normalien who had been a physics professor at the Faculty of Sciences in Mont-

2. Alfred Lacroix, "Charles Sainte-Claire Deville (1814–1876)," in *Figures de savants*, 4 vols. (Paris, 1932–38), 1:45–57.

pellier, to supervise the international service; by 1864 Marié-Davy drew his patron's wrath. Early in 1870 Le Verrier ran afoul of his entire staff, who resigned en masse. The imperial authorities in France finally removed the cantankerous chief, substituting in his place Charles Delaunay, an astronomy professor at the Ecole polytechnique.

To the new regime Deville owed his meteorological observatory, for although his descriptive approach to meteorology offended the mathematician Delaunay, Deville remained on correct terms with Marié-Davy, who had been reinstated as the head of the weather prediction service at the Paris Observatory. Deville's dedication and politicking resulted, in 1872, in the transfer of the weather prediction service from the Paris Observatory to Montsouris. Delaunay nevertheless exerted a measure of control over Deville by absorbing Montsouris as a branch of the Paris Observatory, under the direction of Marié-Davy; he pacified Deville by awarding him the title of *inspecteur général* of meteorology. After engineering the new arrangement, Delaunay drowned in a shipwreck at Cherbourg; Deville once more asserted the independence of his observatory. But Delaunay's successor was Le Verrier, who immediately reappropriated the weather prediction service. The rest of meteorological Montsouris, officially amputated from the Paris Observatory in 1873, passed to the hands of the Department of the Seine, where it was rapidly eclipsed by an observatory set up next to it by the Bureau of Longitudes as an astronomical school for sailors. From 1873 to 1887 Marié-Davy guided the fortunes of meteorology at Montsouris. As for Deville, he kept his title and spent his last years overseeing the establishment of a meteorological service for Algeria, whose weather had intrigued him since the beginning of his career.[3]

Deville died in 1876, and Le Verrier in the following year. The disappearance of the two old antagonists allowed for the creation of a special weather service overseen by an interministerial council in which the

3. Doublet, "Météorologie," 14:258–67, 15:103–20; Lacroix, "Deville," pp. 45–57; *Astron.*, s.v. "Marié-Davy," "Desains," and "Observatoire de Montsouris"; William Napier Shaw, *Manual of Meteorology*, 4 vols. (Cambridge, 1926–31), 1:144, for Desains. See also PAN, F17/13571, administrative notes; and SOM, Généralités, carton 252, dossier 1734, Director of Colonies to Director of Algerian Affairs (Eugène Daumas), 21 Mar. 1853, where Deville's meteorological observations, appearing in the *Revue coloniale*, are recommended. In unraveling the complex institutional changes in mid-nineteenth-century meteorology, I am guided by André Danjon's "Le Verrier, créateur de la météorologie," *La météorologie*, no. 1, 1946, pp. 363–82, a work that takes Le Verrier's side in the disputes; I am also indebted to John L. Davis's "Weather Forecasting and the Development of Meteorological Theory at the Paris Observatory, 1853–1878," *Annals of Science* 41 (1984), 359–82. A useful, recent account is found in Alfred Fierro, *Histoire de la météorologie* (Paris, 1991), pp. 103–7 on the Société météorologique de France, pp. 108–16 on Le Verrier and Delaunay, and pp. 212–13 on Montsouris.

ministry of public instruction had a plurality. The Bureau central météorologique, as it was called, received its own quarters and came on the ledger of the ministry of public instruction. Its direction fell to Eleuthère Elie Nicolas Mascart, professor of physics at the Collège de France. Mascart was a great believer in standardization of information, and under his energetic and inspired leadership the bureau established a meteorological network throughout France.[4] He also set about to extend his reach to all ports of call where France had a presence.

Mascart's international perspective required efficient collaboration between the ministry of public instruction and the ministry of foreign affairs. In 1879 French premier William Henry Waddington, who also retained the foreign affairs portfolio, reported to Jules Ferry, his minister of public instruction, about Mascart's plan to use French consular agents as meteorological observers. The agents were to take observations twice daily with instruments provided by Mascart. Waddington indicated that instructions had been sent or would soon be sent to a score of locations around the Mediterranean basin, and he considered including posts in the West Indies and the Atlantic islands as well. He was willing to help defray the cost of hiring observers at several of the most important sites. Mascart welcomed the help. In some cases (notably Portugal), the consuls would provide more complete local coverage than the official, national weather service. And the occasional, isolated observer at the ends of empire could look forward to Mascart's financial support. Beginning in 1879, the French consul at Khartoum received two thousand francs annually, and twelve hundred francs went to the consul at Pernambuco for working with a local priest. Mascart placed only one foot on the slippery slope of foreign operations. When MacPhee Lee, the Bermuda viceconsul, asked for fifty pounds to install meteorological equipment and one hundred pounds per year to make observations, Mascart announced the principle that would guide his bureau over the next forty years. "This project," Mascart wrote to the ministry of public instruction, "requires a good deal of thought, as the bureau is not able to go in the direction of establishing foreign observatories at its own expense."[5]

If the ministry of foreign affairs could be obliging, Mascart was less successful at working with the navy and the colonial bureaucracy. Until

4. For a description of the organization of the bureau in its early years, see Doublet, "Météorologie," 15:121–22; and Fierro, *Météorologie*, pp. 199–206. See also *DSB*, s.v. "Mascart"; and Charles Maurain, *La météorologie et ses applications* (Paris, 1950), p. 17.

5. PAN, F17/3827, Waddington to Ferry, 11 Apr. 1879; Waddington, circular on establishment of meteorological stations at consular posts, 30 Mar. 1879; Mascart to Ministry of Foreign Affairs, 19 July 1879; Mascart to Ministry of Public Instruction, 30 Aug. 1879 and 26 Sept. 1879 for Khartoum and Pernambuco, 27 Aug. 1879 for quotation.

Eleuthère Mascart (Courtesy of Météorologie nationale, Paris)

1894, when it achieved permanent, ministerial status, the colonial office reported to a secretariat in the ministry of the navy or the ministry of commerce. Throughout the nineteenth century, military and naval medical officers at colonial posts often sent meteorological data to a central desk in France. Sporadic data accumulated over the 1840s, and in 1853 meteorological instructions and instruments worth Fr 2,288 went out to health services in the West Indies, French Guiana, India, St-Pierre and Miquelon, and Central Africa, and a related mailing worth Fr 1,292 went to French colonies in the Indian Ocean.[6] At the same time, the navy's hydrographical service extended beyond coastal and oceanographical mapping to involve itself with ephemeral celestial events (eclipses and planetary transits) and astronomy generally; because knowledge of the weather and compass readings was critical to mariners, the navy accumulated data about meteorology and terrestrial magnetism, some of which found its way into print.[7]

6. SOM, Généralités, carton 252, dossier 1734, "Statistique coloniale" [1849]; Report of Director of Colonies, 29 July 1852, requiring meteorological observations in all colonies to be made by the health service; and "Instructions météorologiques à envoyer aux colonies" [1853].

7. See Fierro, *Météorologie*, pp. 176–80, for the navy's meteorological section in the hydrographical service, originally a separate operation created in 1875; *Hydrogr.*, pp. 349,

Even after epoch-making innovations by Louis Pasteur and others, where disease originating with microbes could be eradicated by chemical means, meteorologists prevailed more upon military medical men than naval officers or diplomats to gather climatical data at distant locations. In the last third of the century, for example, Deville persuaded Alfred-Emile Borius, a naval physician, to take meteorological records at a Catholic mission school in Senegal; Borius's records were complemented by those of the naval pharmacist Emile-Théodore-Marie Morio.[8] Meteorologists in the hexagone had no choice in the matter, for as a functionary in the ministry of commerce observed in 1893, geophysics in the colonies "was excluded from the mandate of the ministry of public instruction."[9]

On the eve of the creation of an independent ministry of colonies, Mascart saw an opportunity to ensure a presence for his bureau in the colonial realm. In 1893 he urged creation in the colonial undersecretariat of a permanent commission for meteorology and geophysics; in exchange, he had his administrative superior Louis Liard, the director of higher education, offer the new ministry of colonies a seat on the council of his bureau. Mascart indicated the colonial sites where data could profitably be taken, and by whom. In Senegal, he suggested the railway company; at Cap Vert, the lighthouse; at Gorée and Dakar, the military hospitals (although in general he signaled that the military health service was quite unreliable in meteorological matters); at Joal and Thiès, Catholic missionaries. He wrote that he had sent two shipments of instruments to Sudan, with a view to outfitting six stations; only scattered reports had come back to Paris, and he had given up hope for accurate reporting from that quarter.[10]

The colonial health service received Mascart's initiative cautiously. It

---

351–52, for a bibliography of the navy's hydrographical service that deals with naval compasses and meteorology; PAN, Marine 3JJ/401, for an example of miscellaneous incoming material in H. Bridet, *Etude sur les ouragans de l'hémisphere austral* (St-Dénis [Réunion], 1861), manuscript (docketed 1879) for the second, revised edition of Bridet's book. Bridet was a retired *capitaine de frégate* and held memberships in the meteorological societies of France and Mauritius, among other affiliations.

8. VM, s.v. "Borius" and "Morio." Both men circulated widely around the French colonial empire, and in the middle 1870s Borius received a special naval commendation for "excellent work in meteorology in Senegal."

9. SOM, Généralités, carton 243, dossier 1720, Ministry of Commerce, Industry, and Colonies, Undersecretariat of Colonies, note for the First Division, 26 Apr. 1893.

10. Ibid., Ministry of Public Instruction to Undersecretariat of Colonies, 24 Feb. 1893; Mascart to Inspector General of Public Works in the Colonies (Fournié), 28 Dec. 1893; Liard to Colonies, 19 Aug. 1893, offering the seat on the bureau's council; Fournié to Colonies, 17 Sept. 1893, offering thanks for appointment to the bureau's council.

affirmed that meteorological observations were taken at all colonial hospitals by pharmacists and published monthly. The data generally did not find their way to Paris, but this omission could be rectified to inform Mascart about colonial weather. Mascart's instructions and his gifts of meteorological instruments would be welcomed at the colonial hospitals.[11] Early in 1894, however, as the proposed commission came into being, the colonial physicians affirmed their complete control over colonial meteorology:

> The work of the commission will be based continually on statistics and observations that have always been and are now collected by officers in the colonial health corps. However, these statistics and observations are the property of the service to which these officers belong, and they must be located at the Conseil supérieure de santé and placed in its archives, where they serve for studies of health and medical climatology, the only scientific end for which they have been and are now instituted. The conseil, nevertheless, will be happy to allow members of the medical corps who are under its authority to participate, from the technical point of view, in analyses undertaken by the commission that shall be instituted, and to provide them with the material that the conseil possesses.[12]

As a result of this arrangement, colonial meteorology suffered under physicians for the next generation. The Central Meteorological Bureau could only lend instruments and request information; it lacked any direct means of disciplining or rewarding meteorologists in the colonial empire.

Mascart's successor in 1907 was his former pupil Charles Alfred Angot, who had worked in Mascart's bureau since 1879. Angot continued and elaborated the program of his teacher. With respect to colonial and foreign stations, Angot depended on the good will of men in distant parts. In 1911 he wrote to the French *résident* on Wallis and Futuna islands to indicate that he had received nothing since the résident's letter of early 1910 except a package of barometrical recordings dating to August 1910. Angot hoped that his diplomat-meteorologist had been able to organize things despite "the difficulties that plague your isolated post in such a primitive country and, furthermore, given that you as the representative of the government must struggle against hostile influences." In 1914 Angot questioned the barometrical recordings of the pharmacist at Brazzaville who was charged with meteorological obser-

11. Ibid., extract from the minutes of the Conseil supérieur de santé des colonies, 11 Jan. 1894.

12. Ibid., President of the Conseil supérieur de santé des colonies, internal note of 31 Jan. 1894.

Alfred Angot (Courtesy of Météorologie nationale, Paris)

vations. The man had systematically failed to make corrections for his instrument and had not taken local gravity anomalies into account. He wrote in 1916 to the director of the health services in the colonial army at Dakar requesting copies of observations for Senegal that he had not received and also alerting the physician to a probable inaccuracy of at least 3 mm in his mercury barometer. In 1920 he congratulated Jacob Dupuis, a supervisor of native affairs at Timbuctu, for "his devoted assistance," and with his praise went a reprint of Dupuis's observations for 1914, which the bureau had published.[13]

The appearance of trench warfare, poison-gas attacks, and military aviation during the First World War brought meteorology to the attention of the high command. After the disastrous Champagne offensive in 1915, the army had the director of its geographical service, General Joseph-Emile-Robert Bourgeois, take effective control of the Central Meteorological Bureau and its network of observing stations. The army also established a meteorological service of its own in the air corps under *lieutenant de vaisseau* Jules Rouch, a graduate of the Ecole navale who had been charged with oceanographical, meteorological, and magnetical measurements on the second Antarctic expedition of Jean-Baptiste-Etienne-Auguste Charcot. In 1917 Commandant Rouch took charge of the navy's meteorological service, the army's being reorga-

13. PMN, Angot to Dr. V. Brochard, résident on Wallis and Futuna islands, 20 Mar. 1911; Angot to Micheli, *pharmacien major* in the colonial army at Brazzaville, 19 Sept. 1914; Angot to Director of the Health Service of the Colonial Army in Dakar, 23 June 1916; Angot to Dupuis, 24 June 1920.

nized under Colonel Emile Delcambre.[14] In 1918 Bourgeois assimilated the army's meteorological service into his civilian network.[15]

The weather was an enemy that would sign no armistice. Jules Rouch wrote during the summer of 1918:

> To use a comparison from current affairs, the synoptic map, showing us the disposition of atmospheric forces, is analogous to a map giving the exact position of enemy forces. The latter document would not perhaps reveal the true intentions of the enemy in a precise way; it would not always allow predicting with certainty the succession of events. But where is the general who, having it in his possession, would neglect to consult it?[16]

To defeat this inhuman foe — to predict its next move — required a continuing, military effort in synoptic meteorology. Alfred Angot's retirement in 1920 allowed the military to keep in peace what it had gained in war, albeit with a few administrative modifications. To the new Office national météorologique, directed by Delcambre and placed with the undersecretariat of the air in the ministry of the interior, went not only Angot's bureau but also the meteorological service of air navigation and the central services of military aviation.[17] Colonel Delcambre looked forward to controlling all facets of French weather. We shall see that he soon came to defer to two geophysicists — Charles Maurain and Henry Hubert — each of them heading an empire in exact sciences during the 1920s and 1930s.

14. Maurain, *Météorologie*, p. 18; *DBF*, s.v. "Delcambre"; "La carrière du commandant Rouch," Les Amis du Musée océanographique de Monaco, *Bulletin trimestriel*, no. 41, 1957, pp. 5–20. Rouch paints a vivid picture of his wartime meteorological exploits — where science was not much in evidence — in "L'offensive du 16 avril 1917," L'Amicale des écoles de perfectionnement de la marine, *Bulletin*, 30 June 1936.

15. Jules Rouch, "Historique du Service météorologique aux armées (mai 1915–février 1918)," *Revue maritime*, Feb. 1921, pp. 167–80; Philippe Schereschewsky, "La météorologie militaire pendant la guerre: Historique sommaire," *L'aéronautique* 1 (1920), 531–35; VAT, 5N 263, "Service météorologique," Bourgeois to Minister of War, 8 Jan. 1918.

16. Jules Rouch, "La prévision du temps," *Revue scientifique* 56 (1918), 391–400, on p. 400. In 1918–19 the Norwegians came up with a new cyclonic model featuring the "front," a term inspired by trench warfare (Robert Marc Friedman, *Appropriating the Weather: Vilhelm Bjerknes and the Construction of a Modern Meteorology* [Ithaca, 1989], pp. 122–37). The earliest use of the word *front* in meteorology, however, came from the geophysicist Ernst August Ansel in his doctoral dissertation of 1913 at the University of Göttingen (Aleksandr Khristoforovich Krgian, *Meteorology: A Historical Survey*, trans. Ron Hardin [Jerusalem, 1970], p. 212; *CIES*, pp. 86–87).

17. Jacques Dettwiller, *Chronologie de quelques évènements météorologiques, en France et ailleurs* (Paris, 1982) [Direction de la météorologie, Monographies, 1], pp. 41–42; Charles Maurain, "Répertoire de laboratoires français, suite (1)," Direction des recherches scientifiques et industrielles et des inventions, *Bulletin officiel* 2 (1921), 569–75, on pp. 574–75; Fierro, *Météorologie*, pp. 241–46.

Emile Delcambre (Courtesy of Météorologie nationale, Paris)

Dean and Seismologist

Charles Maurain, who rose to become France's dean of geophysics, was a normalien born into a family with high academic expectations. His brother Eugène attended the Ecole polytechnique and participated in the French expedition to remeasure the arc of longitude in Ecuador early in the twentieth century; a sister, Jeanne, graduated from the Ecole normale supérieure at Fontenay and went on to a career in secondary education. Maurain served as Eleuthère Mascart's *préparateur* at the Collège de France and turned in a doctoral dissertation, in 1898, consisting of separate theses on cathode-ray screens and hysteresis. After teaching for more than a decade in the provincial secondary schools and faculties, Maurain suddenly changed the direction of his research to take advantage of interest in the new field of aeronautics. The opportunism resulted in a call to teach the subject at Paris and direct the Sorbonne's new aeronautical institute at St-Cyr, endowed by Henry Deutsch de la Meurthe. The First World War brought prominence to the theoretician of aviation, and the scientific independence deriving from favorable publicity allowed him to move into geophysics,

Charles Maurain (Courtesy of Académie des sciences, Paris)

a field of his old professor Mascart's. In 1921 the fifty-year-old Maurain became director of a new Institut de physique du globe at Paris, the central ornament in a reorganization of France's string of geophysical observatories and stations. Maurain then won his way to a university chair in 1922 and became dean of the Sorbonne's Faculty of Sciences, a key position he held until 1941.[18]

Initially Maurain assumed responsibility for a central service in Paris, three outlying stations in France, and an office for terrestrial magnetism encompassing both the metropolis and the colonies; to this end, he was confided the parts of Angot's bureau that had dealt with magnetical questions.[19] By virtue of his location in Paris, Maurain became director of the Central Bureau for Terrestrial Magnetism and Atmospheric Electricity (replacing the American geophysicist Louis Bauer) and president of the International Commission of Terrestrial Magnetism (replacing the British magnetician Charles Chree). His operation, however, had nothing of the resources available to astronomers, who ran observatories that were, administratively speaking, on an equal footing with his. He complained as late as 1932 that in contrast to the director of an astronomical observatory, the Paris geophysical chief enjoyed no

18. Emile Thellier, *Notice sur la vie et l'oeuvre de Charles Maurain (1871–1967), membre de la section d'astronomie* (Paris, 1972); Charles Maurain and Anatole Pacaud, *La Faculté des sciences de l'Université de Paris de 1906 à 1940* (Paris, 1940), pp. xi – xii, 30.

19. Maurain, "Introduction," *AIPG* 1 (1922), 7–8; "Les observations magnétiques en France jusqu'en 1921," ibid., 9–19.

free lodging and received only a professorial salary; his central office for terrestrial magnetism was funded entirely through the geophysical institute, a situation without parallel among his astronomical colleagues.[20]

In the reorganization of geophysics that took place after the First World War, the ministry of public instruction set up a geophysical institute at Strasbourg. The impulse to endow Strasbourg with impressive facilities stemmed from a general policy of reintegrating Alsace into the French Republic. The decision to focus on seismology related to the location at Strasbourg, under the German regime, of the central office of the International Seismological Association. That office had shared facilities with the principal imperial German seismological station, both captained by Oskar Hecker. In Hecker's place came Edmond-Ernest-Antoine Rothé.[21]

Rothé, forty-eight years old in 1921, had taught physics at Grenoble and Nancy (where he succeeded René-Prosper Blondlot in the chair of general physics) before landing a directorship in his Alsatian homeland when Strasbourg passed into French hands. Like Maurain, Rothé had capitalized on aeronautics, teaching atmospheric physics at Nancy's aerodynamical institute; like Maurain, Rothé had become involved with wireless telegraphy, publishing a treatise on the subject; like Maurain, Rothé rose to become dean of his science faculty; like Maurain, Rothé was given responsibility for organizing observations in an entire field of study—in his case, seismology—in French territory and internationally; and like Maurain, by the time he rose to head a major institute, Rothé's professional allegiance had passed from research to administration.[22] Rothé's destiny was in fact linked to Maurain's, since their two institutes were overseen by a special interuniversity committee, consisting of a meteorologist, representatives from the Academy of Sciences

20. PIPG, Maurain to Director of Higher Education, 4 Apr. 1932.

21. Georges Perrier, "La coopération internationale en géodésie et en géophysique," Bureau des longitudes, *Annuaire pour l'an 1928* (Paris, [1928]), pp. C.1–C.120, pp. C.22, C.34–C.37 for the complicated absorption of the prewar international organization by the affiliates of the International Research Council and for Hecker's ceding to Rothé. On the history of seismology early in the twentieth century, see *CIES*, pp. 35–51; a recent survey of seismology by Benjamin F. Howell, Jr., *An Introduction to Seismological Research: History and Development* (Cambridge, 1990), may be consulted for concise information about many of the seismographs mentioned in the following pages.

22. "Edmond Rothé," *Bulletin géodésique*, nos. 69–76 (1941–42), 64–66; John E. Craig, *Scholarship and Nation Building: The Universities of Strasbourg and Alsatian Society, 1870–1939* (Chicago, 1984), p. 221, for Rothé as an Alsatian *revenant;* Georges Dubois, *La vie et l'oeuvre d'Edmond Rothé, géophysicien*, and Georges Cerf, *Edmond Rothé, citoyen, separata* in the Bibliothèque nationale et universitaire, Strasbourg. In his earliest publications, Rothé investigated Blondlot's notorious N-rays.

and the Bureau of Longitudes, the deans of the faculties at Paris and Strasbourg, and the two geophysicist directors.[23]

In one respect Rothé started out better than Maurain. Because of his operation's prewar heritage, Rothé commanded both a teaching staff, which was attached to his institute, and a research staff, attached to his various geophysical charges (including the meteorological service of Alsace-Lorraine). Rothé located all his enterprises under one roof (purchased with Fr 150,000 from the marquise Arconati Visconti), and in this way he had a larger central staff than Maurain's.[24] Rothé employed scientific workers, but he lacked a budget for his national and international mandates. Issuing his institute's various bulletins and exchanging seismograms with foreign observatories took a great deal of creative financing. And in this matter he was at a distinct disadvantage: he was not a member of the Paris Academy of Sciences, and he did not sit on all the interlocking committees where Parisian mandarins sliced up the pie of French science. Rothé in fact depended on Maurain for everything, from regulating his pension to receiving grants. He wrote to Maurain in 1932 to express his pain at remarks by Maurain and Jean Perrin to the effect that his research had been recommended to the Conseil supérieur de la recherche scientifique "by mistake."[25]

One of Rothé's projects to enlarge his geophysical niche involved inventing a new diploma, the *ingénieur-géophysicien*, of which he was the only grantor. For justifying the presence of his teaching staff the diploma was a godsend. "I am *very* worried about the future of seismology in France," he wrote to Maurain in 1931. "If I had not created instruction in applied geophysics at Strasbourg, I would have no more pupils."[26] The diploma, however, found a mixed reception. Charles Poisson, director of the Ambohidempona Observatory on Madagascar, noted in 1930 that "it does not seem to have a transcendent utility, nor does it involve anything of real difficulty for someone who would have completed the *certificat* in general physics."[27] The diploma drew hostile fire from Charles Maurain. In 1933 he criticized Rothé for offering the

23. PIPG, Maurain to Directeur de l'Enseignement supérieure (Coville), 8 Nov. 1921, accompanied by a draft letter for the ministry concerning the composition of the committee.

24. Ibid., Rothé to Maurain, 24 Mar. 1924, enclosing a copy of a letter from his university rector to his dean outlining the dual nature of his staff; PO, carton 3, II-B–1 c, Comité français de géodésie et de géophysique, minutes of general meeting of 13 Apr. 1921, report of the seismology section, where the grant is noted.

25. PIPG, Rothé to Maurain, 3 July 1935 about the pension, 24 May 1932 for the mistaken appropriation.

26. Ibid., Rothé to Maurain, 26 Aug. 1931.

27. Toul., FMa 956, Poisson to Provincial Superior, 8 Jan. 1930.

diploma to "young people who, not looking forward to [a career in] instruction and not being bacheliers, are not studying for a licence." Under these conditions, students working for Rothé's diploma depended on only Rothé for their notions of science; in contrast, Maurain's own students who had completed the three certificates required of a licence, showed that they had succeeded in difficult examinations of general knowledge. Rothé had contrived to allow his ingénieurs-géophysiciens into the ranks of the new colonial meteorological service (to be discussed shortly), while Maurain's geophysicists had to show evidence of two complementary certificates before being allowed entry. If Rothé's diploma really conferred privileged access to a respected career, Maurain concluded, then Maurain himself would have instituted it at Paris — a course that he had declined to take.[28]

The vacancy of two of Rothé's posts, late in 1937, led him to think about adding middle-level research staff to his institute. In a long letter to Maurain, he set out a plan of action, prefaced by a retrospective look at geophysics in France:

> The organization of the geophysical institute at Strasbourg, the first to be created, is unique among geophysical institutes. This derives from the date and the manner of its creation, just after the armistice. It was placed entirely under the administration of the faculties of sciences. It was set up differently from the geophysical institute of Paris, which inherited personnel from the Central Meteorological Bureau who belonged to special services. In this way, the director at Paris did not have from the beginning a *maître de conférences* or an *assistant* in the faculty, while at Strasbourg there were only maîtres de conférences and assistants, in addition to an assistant at the observatory.

Maurain, with energy and resources, had changed that.

> You have rectified this state of affairs by establishing a real teaching corps (*professeur, maître de conférences*, faculty *assistant*), the remaining staff occupying posts of *physicien-adjoint, aide-physicien, calculateur*, etc.

Rothé wanted to remove some of his vacancies from faculty control, in line with Maurain's organization. Maurain immediately circulated Rothé's proposal to his staff with an acerbic note to the effect that Rothé's history was not quite right. There can be no mystery about his feelings. Maurain had constructed his empire around a research staff. As late as 1928 he still lacked anyone to help him with university lecturing, his central staff consisting of one senior *physicien* who was a normalien and docteur ès sciences, three physiciens-adjoints who each held

28. PIPG, Maurain to Henry Hubert [?], 18 Dec. 1933.

a doctorat ès sciences, two aides-physiciens, two calculators, a mechanic, and a janitor.[29]

What exactly was the specialty that Rothé and Maurain governed? In his 1928 survey of physics in Paris, Charles Fabry identified geophysics as a "pure" specialty, enjoying the same status as physical chemistry, radiation physics, and mineralogy.[30] Geophysics, Charles Maurain wrote in the same year, was a special discipline requiring particular talents. It depended on data amassed at diverse, distant locations over long periods of time. The data were processed by different numerical and graphical techniques, and their synthesis depended on the results of laboratory experiments on the behavior of matter under various temperatures and pressures. Instruction in geophysics had to take these peculiar circumstances into account. With regard to teaching at the Sorbonne, geophysics enjoyed a close relationship with physical geography, applied geology, and meteorology—all four disciplines sharing rooms in a common building.[31] As dean of the Faculty of Sciences, Maurain used the condominium to advance his chosen field.

In the academical progress of Louis Cattala, a Jesuit missionary who spent three years at the Ambohidempona Observatory on Madagascar in the late 1920s, we find a record of what constituted training in geophysics for someone preparing to serve in the French colonial empire. Cattala went to Madagascar at age twenty-one, following two years as a novice at Mons in Gers and one year as a juvenate in the Jesuit seminary at Florennes in Belgium. He returned to study philosophy for three years at Vals. He went to Lille, where he completed the licence ès sciences. His trajectory then took him to Enghien in Belgium, where he devoted a number of years to theology in preparation for ordination as a priest in 1937.[32]

Senior Jesuit scientists were carefully groomed for their professional responsibilities. Charles Poisson, the Jesuit observatory director on

29. Ibid., Rothé to Maurain, 2 Dec. 1937; Maurain to staff, 4 Dec. 1937; Maurain to Minister of Public Instruction, replying to note of 13 Mar. 1928: Raphaël Roger Dongier the sixty-two-year-old senior man who had been a climatologist in the old Central Meteorological Bureau; Charles-Emile Pierre Jules Brazier, Marie Louis Eblé, and Ernest Labrouste the associate physicists; Jules Itié and Miss G. Homery the junior physicists. Labrouste was *professeur sans chaire* at Strasbourg, seconded to Paris.

30. Charles Fabry, "La physique à Paris," *Annales de l'Université de Paris* 3 (1928), 551–75, on p. 553.

31. Maurain, "Les études de physique du globe," ibid., pp. 18–30. The geographical institute belonged to the Faculty of Letters; meteorology entered the Faculty of Sciences under the guise of the Meteorological Society of France, which held its meetings in Maurain's building.

32. Bernard Leurent, "Le père Louis Cattala, 1904–1963," *Maduré-Madagascar*, no. 109, Oct. 1963, pp. 359–62.

Madagascar, wrote to his Toulouse superiors in 1931 about Cattala's scientific training. In Poisson's opinion, the training had to be tailored to Cattala's temperament: neither too advanced nor too long. Poisson settled on the Jesuit establishment at Lille to prepare Cattala for certificates in calculus and rational mechanics at the faculty of sciences there. This would take no more than a year. Then Cattala would go to Paris for two years, studying for certificates in general physics and in geophysics — the latter "having a very light program." Both physics certificates were necessary if Cattala wanted to obtain a doctorate.[33]

In 1934, Cattala wrote to his superiors at Toulouse, who surveyed his spiritual and temporal existence, about his progress toward a scientific career. Cattala emphasized that he had studied physics and mathematics during his time at Vals. At Lille he had attended courses in rational mechanics, calculus, special topics in mathematics, general physics, and fluid mechanics. His pace was slow and his accomplishments far from brilliant, but he had obtained certificates in general mathematics and rational mechanics, and he looked forward to taking a certificate in general physics. Success in obtaining these certificates would be insufficient for a thorough grounding in physics. A student really needed to follow specialized courses not only in mathematics but also in physics and chemistry. "This, Cattala observed, "is what students in mathematics and physics do who come out of the lycées," men who were ten years his junior. Even such a course of study provided nothing but basic knowledge:

> The programs for licence certificates in classical mathematics and physics are not really on the level of higher education, properly speaking. They are comparable to classes in grammar. Mathematical and physical theories are accessible only after these studies in technique and are taught only in several advanced courses at provincial faculties and at the Collège de France. But to follow them, it is necessary to have at base the programs of the licence or even the *agrégation*.

Individual ideas and personal reflection counted for very little on the level of the licence ès sciences, a situation in marked contrast to what happened in letters. Cattala mused about what he should do if he failed the physics examination. He would sit for it a second time, but if success did not crown his remedial efforts, he planned to satisfy himself with what he had already obtained. If he succeeded in the physics examination, his three certificates would give him a licence, but not the so-called teaching licence. For the latter he would have to sit for a fourth certificate, in calculus.

33. Toul., FMa 956, Poisson to Father Confessor, 27 June 1931.

Cattala had been approached by Joseph Kampé de Fériet, the professor of fluid mechanics at Lille, with a program of study in meteorology, but he was uncertain whether he should specialize so soon: "Is it necessary for me to begin to study the scientific branches that are the object of research at the Observatory of Tananarive before my departure from Europe, which will delay my departure, or is it good enough for me to specialize once I arrive?" If he had to write a thesis, Poisson might want him to work on it at Ambohidempona, drawing on local material, and Cattala thought that he would find in Poisson "for meteorology all the necessary competence" to direct him. Cattala did not have a high opinion of French meteorologists: "The official French meteorologists are generally inferior to the English, the Americans, the Norwegians, and even the Belgians, whether one is talking about the National Meteorological Office or the courses in geophysics at Paris or Strasbourg." Cattala believed that Kampé de Fériet was different. Well versed in the work of foreigners, he wanted to turn his institute at Lille into "the center of a school of meteorology, like the celebrated school of Bjerknes in Norway." Prestige would accrue to Tananarive through Cattala's having studied with Kampé.[34]

By 1938 Cattala was in his "third year" of theology at Amiens, making final preparations for joining Poisson at Ambohidempona. He had spent the summer of that year at the Brussels Observatory in Uccle, and Poisson insisted that he devote up to a year for additional training in geophysics before leaving Europe. With regard to magnetism, he was to study under Rothé at Strasbourg, learning how to prospect for minerals with a Schmidt magnetical balance. He was to study the magnetical properties of rocks with Emile Thellier at the Paris geophysical institute, and at either Strasbourg or Paris he would learn about making absolute magnetical measurements. Cattala would assimilate the general principles of gravimetry with Maurain and then receive instruction about the Holweck-Lejay gravimeter from Fernand Holweck himself. He would learn about solar-radiation measurement from Charles Brazier at the St-Maur Observatory near Paris.[35] So much for Cattala's required work. It was further suggested that he attend courses given by mineralogist Jacques de Lapparent at Strasbourg, paying special attention to microscopic examination of crystals; learn about analyzing seis-

34. Ibid., FMa 1001, Louis Cattala to Father Confessor, 7 May 1934.

35. The observatory in the Parc St-Maur, one of Charles Maurain's satellite institutions, began life in the middle of the nineteenth century as Emilien Renou's meteorological observatory; in 1900 electrical tramways forced evacuation of St-Maur's magnetical facilities to Val Joyeux in the commune of Villepreux (Fierro, *Météorologie*, pp. 115, 211). According to Alfred Angot, seismological recording began at St-Maur in 1908 (PMN, Angot to Robert M. Yerkes, 17 Mar. 1920).

mograms; and spend several days at the National Meteorological Office in Paris. Lower on the list of priorities was knowledge of atmospheric electricity, to be acquired in Paris. To his Jesuit superior in Toulouse, Cattala outlined a pilgrimage that would have him stop at each institution, beginning in the summer of 1939 and ending around April 1940. The authorities found his plan acceptable.[36]

Mapmaker and Signalman

Just as the navy provided careers for nineteenth-century French scientists, so the army provided special employment for scientifically inclined officers. The navy needed charts, the army needed maps. It assigned surveying to its Dépôt de guerre. Maps issued in a continuing stream across the first three quarters of the nineteenth century, both for overseas territory and for the hexagone, but the army surveyors enjoyed nothing approaching the cachet of their naval counterparts. François Perrier, a senior military surveyor, set out to rectify the situation. In 1881 he succeeded in assigning geographical work to a section under the army's Etat-major, which he captained. Six years later the Dépôt was abolished in favor of a geographical service directed by General Perrier. Under Perrier's guidance, the army began to undertake gravimetrical surveys; these continued with Perrier's successors, although the service focused increasingly on its North African possessions, linking them by topographical triangulation to French meridians.[37]

Wherever the French army camped, it sent a team of topographers to chart the surrounding terrain. Sometimes the military geographical services received orders from the council of a local regent (Tunisia), a governor general (Indochina and Madagascar), or an occupying army (Syria), but even in cases where civilians came to carry out the surveying (Morocco, Algeria, and Tunisia), military norms were omnipresent. Military geographers attempted to engage projects relating to pure science. The most ambitious of their projects – an early twentieth-century remeasuring of the eighteenth-century arc of longitude in Ecuador – did little more than advance the military career of men such as Robert Bourgeois and Georges Perrier, who slogged over South American hill and dale.[38] The Ecuadoran experience soured the geographical service on geophysics and astronomy. Bourgeois, who had commanded the ex-

36. Toul., FMa 1001, Louis Cattala to Father Confessor, 17 Dec. 1938, bearing a handwritten note dated 5 Jan. 1939 to the effect that the Father Confessor approved the course of study. The war prevented Cattala from reporting to Madagascar as planned.

37. *Armée*, pp. 35–62. For a survey of an earlier period, see Josef Konvitz, *Cartography in France, 1660–1848: Science, Engineering, and Statecraft* (Chicago, 1985).

38. *Armée*, pp. 63–120.

Georges Perrier (Courtesy of Observatoire de Paris)

pedition to Ecuador, rose to direct the service in 1911. He began the First World War with thirty geodesists in his service; by war's end, twelve had been killed, three had died through accident or disease, and a large number had retired or been transferred. Of the prewar staff, only Perrier remained.[39] During the interwar years, Perrier labored in vain to reconstitute a geophysical program for the military surveyors. Scientific progress eroded his claim to head the army's scientific elite.

Whereas in the nineteenth century military surveyors had pioneered in the construction of portable telescopes and the analysis of astronomical data, advances in technology lay at the root of the diminished scientific status of the geographical service. Two related developments reduced the geographers to little more than mapmakers: aviation and wireless. Aerial photography, which came to subtend mapping, depended on collaboration with the independent-minded (and from 1928 completely independent) air corps; furthermore, the corps's National Meteorological Office, which provided weather reports for airports, freely implicated itself in geophysical projects. Closely tied to aviation

39. Georges Perrier, "Les travaux actuels de la section de géodésie du Service géographique de l'armée française," in Prague, Masaryk Academy, Section of Natural History and Medicine, *Publication scientifique*, no. 47, 1929, pp. 35–46, on p. 37. Georges Perrier was the son of François Perrier.

was communications technology. Wireless, which provided the precise time signals that were essential for mapping, was the domain of the army's communications branch, whose resident expert was Gustave-Auguste Ferrié. He became the army's senior scientist.

After graduating from the Ecole polytechnique and the Ecole d'application at Fontainebleau, Ferrié served as a lieutenant at Grenoble and Besançon. Soon after he made captain in 1897, he was transferred to the central office of military telegraphy. In 1898 he began experimenting with Hertzian waves, and the following year he assisted Guglielmo Marconi in sending wireless signals between England and France. The ministry of war lost no time in having Ferrié organize a military radiotelegraphical service; over the next thirty years Ferrié directed the service's expanding fortunes. He wrote technical manuals, installed radios on airplanes and balloons, established radio communication with French colonies abroad, and lectured tirelessly to popularize his medium. At Ferrié's insistence, radio played a part in France's invasion of Morocco in 1907, and at his command the Eiffel Tower became a giant radio mast for broadcasting time signals. In close collaboration with astronomers at the Paris Observatory, he immediately made use of the time service to determine longitude differences in France and between Paris and Washington. Appointed lieutenant colonel in 1914, he set up a scientific research center that during the war employed a number of talented physicists, Henri Abraham, Eugène and Léon Bloch, Léon Brillouin, Marius Latour, Lucien Lévy, Edmond Rothé, and Ernest Labrouste among them. Colonel in 1915, brigadier in 1919, General Ferrié received the Osiris prize of the Academy of Sciences (Fr 100,000) and then took Alfred Grandidier's chair there in 1922. By this time he belonged to every major French committee exercising authority in the exact sciences.[40]

As Daniel Berthelot's successor in presiding over the section of terrestrial magnetism and atmospheric electricity in the Comité français de géodésie et de géophysique, Ferrié exercised discretionary power over funding. He corresponded with Charles Maurain about most geophysical mapping projects in the 1920s. When the *Annales de l'Institut de physique du globe et du Bureau central de magnétisme terrestre* strained Maurain's means (in 1927 the budget for his institute and its three dependent observatories at St-Maur, Val-Joyeux, and Nantes was fifty-two

40. See *La météorologie* 76 (1933): "Travaux et publications du Général Ferrié," pp. 355–62; P. Brenot, "Le général Ferrié," pp. 363–67; R. Bourgeois, "Le général Ferrié," pp. 373–76; Belhague, "Paroles prononcées aux obsèques du général Ferrié," pp. 376–80; Charles Fabry, "Gustave Ferrié," pp. 408–11; Maurice d'Ocagne, "Le général Ferrié," pp. 411–13; and R. Jouast, "Le général Ferrié," pp. 414–21.

Gustave Ferrié (Courtesy of Observatoire de Paris)

thousand francs), he enlisted Ferrié's support to ask the Caisse des recherches scientifiques for a special grant of twenty thousand francs to cover the cost of publication; the grounds for involving Ferrié concerned Maurain's project to compile and publish a new magnetical map of France. Maurain and Ferrié's schemes also implicated French physicists, such as Daniel Berthelot, Louis Dunoyer, and Aimé Cotton. Late in the 1920s, the Comité français de physique actually carried a budgetary attribution for magnetical surveying—the proper ownership of which occasioned complicated discussions between the geophysicists and the physicists.[41]

Maurain and Ferrié worked together for nearly twenty years on questions relating to aviation and radio, and from their mutual interests they formed common cause against outsiders like Delcambre, who might threaten their authority. Maurain had "excellent personal relations" with Delcambre, he wrote to Ferrié in 1930; the two families vacationed close by each other and went on walks together. But Mau-

41. PIPG, Maurian to Ferrié, 18 Feb. 1929, for example, where Maurain discusses the distribution of certain funds; 2 July 1927 and appended proposal of 1 July 1927; 8 Dec. 1927.

rain would not allow Delcambre to speak for French geophysical interests in general, and he was not above reprimanding Delcambre for conveying the wrong impressions at a meeting about French intentions toward the international geophysical station at Scoresby Sund in Greenland.[42] Maurain and Ferrié contested with Delcambre about the overseas projection of French geophysics, but by 1930 they all found themselves confronting a new and potentially menacing rival in the ministry of colonies, Marie-Henry Hubert.

Geophysicist Conqueror

In the period before the 1920s, geophysics in tropical Africa was carried out by military surveyors, by medical men hired to keep the surveyors healthy, and by Catholic missionaries who worked closely with France's professional killers.[43] The army's geographical service roamed the interior, generally concerned with establishing boundaries for political jurisdictions, and the navy sounded the coastal waters.[44] With the twentieth century, military geophysical services went under civilian direction. The demilitarization of colonial science proceeded haphazardly until Henry Hubert arrived in the ministry of colonies.

Hubert is the central figure in a reorganization of colonial geophysics between the two world wars. He was a physical geographer initially trained at the Ecole coloniale who had explored French West Africa and had come to inventory geophysical and meteorological phenomena there. In the early 1920s he leapfrogged over the traditional weathermen in his part of the world to head the colony's new meteorological service at Dakar. Hubert used his position to scatter reports in the local house organ, the *Bulletin* of the Comité d'études historiques et scienti-

42. Ibid., Maurain to Ferrié, 9 Sept. 1930.

43. VM, s.v. "Borius" and "Morio"; Brother Constantin (physical science teacher and head of the observatory in the secondary school at St-Louis du Sénégal), "Observations météorologiques: Moyennes conclues de 23 années d'observations," *BC* 13 (1930), 437–74; M. Pierre, "Nécrologie: Mugnier-Serand," ibid. 16 (1933), 329–31, for meteorological work by the lighthouse keeper in Tamara, at the harbor of Conakry in French Guinea.

44. Edouard de Martonne, "Le Service géographique de l'Afrique orientale française: Son fonctionnement, ses resultats (1922–1927)," ibid. 11 (1928), 321–51. Typical is the report of Bouvet, a captain in the colonial artillery: "Note sur des observations astronomiques faites à la Côte d'Ivoire," in PSG, colis 25, Bouvet [1906]. Among later military surveys, see Commandant Carrier, "Mission Ouadaï-Darfour, dirigée par M. le colonel Grossard; observations magnétiques," *AIPG* 3 (1925), 156–60; Capt. R. Rauchon (Colonial Infantry), "Première contribution à l'étude du magnétisme au Soudan français," ibid. 4 (1926), 119–32; Capt. J. Nevière (Colonial Artillery), "Première contribution à l'étude du magnétisme en Haute-Volta," ibid. 6 (1928), 139–41.

fiques of French West Africa, and to pepper the Paris Academy's proceedings with observations on air currents, earthquakes, and rainfall.[45]

Around 1926 Hubert returned to the hexagone with the rank of *administrateur en chef des colonies*, and he directed his attention toward persuading his ministry to support colonial geophysics. In 1927 he intrigued with Charles Maurain about setting up a station for terrestrial magnetism at Dakar. At Hubert's urging, Maurain circulated a proposal to Ferrié about the station. Maurain urged Ferrié to write to letter to Alfred Lacroix asking him to contact the minister of colonies. The following year, Maurain supported Ferrié's request that the ministry of colonies renew a grant of two thousand francs to finance the hourly wireless time signal sent out from Paris, a signal that was "from all accounts very important for the entire body of French possessions."[46]

This indirect politicking, which after two years netted magnetical equipment for Dakar, was cumbersome and uncertain, especially as the National Meteorological Office, under the new ministry of the air, sought exclusive domain over meteorology. We shall see that in the late 1920s Delcambre made his move to annex overseas meteorology. Hubert, as the most visible physicist functionary with overseas experience in the ministry of colonies, moved with dispatch to assert his own authority over colonial geophysics. He went to Delcambre's National Meteorological Office with the mission of constructing a new, autonomous colonial meteorological network.[47]

Hubert had before him a scheme for organizing such a service that had been drawn up in the early 1920s by Jules Rouch, the sometime head of military meteorology during the First World War. Rouch, who before the war had served the ministry of colonies, ended the war as general secretary of the Comité d'action des services météorologiques français, an umbrella committee uniting all military and civilian meteorological operations in France. The committee expired in 1920, before it could turn its attention to the colonies. Rouch, however, saw fit to propose a colonial meteorological service in 1921. He indicated the future distribution of meteorologists among the various colonies, and he specified their tasks; an office in Paris would assume responsibility for

45. Hubert, *Contribution à l'étude de la géographie physique du Dahomey* (diss., University of Paris, 1908); L. Welter, "Bibliographie météorologique de l'A.O.F.," *BC* 13 (1930), 475–82, where thirty-five of Hubert's publications are listed; *DBF*, s.v. "Hubert."

46. PIPG, Maurain to Ferrié, 22 Nov. 1927; Maurain, "Rapport à la commission des subventions aux établissements scientifiques" [1928].

47. Ibid., Maurain to Ferrié, 24 May 1929, for the equipment, credited on the budget of French West Africa; "Organisation de la météorologie aux colonies," *ACRACS* 4 (1928), 39–40, for Hubert's tour in the National Meteorological Office, authorized by Minister Léon Perrier.

Jules Rouch (Courtesy of Marine nationale, Etat-major, Service historique, Vincennes)

certifying them. Rouch's proposal gave no word to finances, and for good reason: the ministry of colonies would not add the salaries of hundreds of colonial civil servants to the budget of the central bureaucracy.[48] Hubert solved the conundrum that Rouch had implicitly posed a decade previously.

Hubert had the ministry of colonies work around the extraordinarily independent nature of colonial governments, where the governor general enjoyed almost unlimited latitude within the confines of his budget. Accordingly, the ministry concerned itself only with the creation of an intercolonial civil service that would be filled on the basis of merit and competitive examinations and would be overseen by a committee of metropolitan geophysicists and meteorologists. To join the corps, a candidate had to be an ingénieur-géophysicien or possess three certificates (effectively, the licence) awarded by a French faculty. Alternatively, one could be a lauréat of the Academy of Sciences for research in geophysics or a graduate of one of the grandes écoles.[49]

Two restrictions determined that the colonial meteorologists would remain second-class citizens in the world of science. First, the ministry left to the individual colonies the prerogative of erecting observatories and supplying local staff. A meteorologist sent to a minor posting—the Somali coast, for example—could expect to wait years for the construc-

48. Rouch, "Essai d'organisation d'un service météorologique colonial," *Colonies et marine*, Feb. 1921, pp. 120–24.

49. *Journal officiel de la République française*, 2 May 1929, pp. 5110–11; 12 May 1929, pp. 5402–4, under Minister André Maginot.

tion of his observatory; if the governor was not a scientific enthusiast, the meteorologist's budget required drawing local staff from the low end of the labor pool. Second, the colonial meteorological network could not operate in the environments most favorable for scientific research; excluded were Algeria (technically no longer a colony), Morocco and Tunisia (protectorates), Lebanon and Syria (mandates), and Shanghai (a concession). Algeria belonged to the ministry of the interior and received the usual benefits of the ministries of public instruction, agriculture, war, and the air; the rest came under the eye of the ministry of foreign affairs. Colonial functionaries were left to circulate around Africa south of the Sahara, Madagascar, Oceania, the Antilles, and Indochina. Hubert tried to centralize data emanating from *all* overseas locations, but his *Annales de physique du globe de la France d'Outre-Mer* remained something of a parochial house organ. The most distinguished overseas geophysicists – Pierre Lejay at Zikawei and Jean Coulomb at Algiers – studiously avoided placing their papers in it.

Following Hubert's elevation to the head of a colonial civil service, Charles Maurain included him on the central decision-making body of French geophysics, the Comité français de géodésie et de géophysique, in the section for terrestrial magnetism and atmospheric electricity. Hubert returned the consideration by providing Maurain's students with preferred positions in the corps. He informed Maurain in 1930 that he would be glad to appoint one man as an *ingénieur-météorologiste-adjoint stagiaire*, the lowest rank in the corps, rather than as an ingénieur-météorologiste-adjoint, the second lowest. Hubert emphasized that the aspiring colonial meteorologist would be able to attain the higher rank quickly, due to his military service. He urged Maurain to have his former student decide quickly whether he would accept the appointment, since it was being offered before the service opened up to formal competitions.[50]

A commission charged with setting the standards for ingénieur-météorologiste met in June 1933, read the candidates' written examinations, and questioned them viva voce. Hubert had indicated to Maurain the kind of subject for the written essays:

Part 1

a. Physical geography of Madagascar;
b. Political and administrative geography of French West Africa;
c. Economic geography of Indochina.

50. PIPG, Maurain to Ferrié, 24 May 1929; Hubert to Maurain, 20 Nov. 1931.

Part 2

a. Electromagnetism: action of the magnetical field;
b. Electrical measurements: units, intensity measurements, differences in potential, resistance, and energy measurements;
c. Electrical applications: telephone, wireless.

Part 3

a. Air temperature;
b. Atmospheric pressure;
c. Movements of air in the atmosphere.

A candidate would choose one topic from each part.[51] Geography, electricity and magnetism, and atmospheric physics were the legs of a colonial meteorologist's tripod.

For placing Maurain's students in his colonial corps, Hubert exacted the price of direct access to the Paris geophysical institute. When Hubert obtained the services of an *ingénieur des ponts et chausées* to carry out measurements of air temperature in the colonies, he asked Maurain to apprise the man of the latest techniques. And as Hubert answered to neither the ministry of the air nor the ministry of national education, he could orchestrate an interministerial commission on artificial precipitation on which both Delcambre's successor Philippe Wehrlé and Maurain sat. The commission had funds to distribute for worthy projects. Maurain, for his part, required evidence of serious scientific arguments before he would support a proposal.[52]

Maurain and Hubert's sweetheart arrangement went farther than funding and entry-level posts. It became a system of academic apprenticeship wherein talented meteorologists accepted a colonial post under Hubert's regime and then used their overseas resources to complete a doctorate with Maurain. Jean Pruvost, for example, ingénieur-météorologiste-adjoint at Yaoundé in Cameroun, wrote to Maurain in 1938 to discuss his plans for a doctorate of engineering. Pruvost had graduated from the Ecole supérieure d'électricité and had completed certificates in mechanics and mathematics at the Sorbonne, although his appointment in Africa – granted by Hubert following Maurain's support – had prevented him from finishing a geophysical certificate under Maurain's supervision. His thesis topic, which he had previously discussed with Maurain, concerned dynamic meteorology in Cameroun, particularly

51. Ibid., Hubert to Maurain, 2 June 1933 (with appendix) and 21 June 1933.

52. Ibid., Hubert to Maurain, 17 Apr. 1935; 3 May 1935; 24 June 1935, on which Maurain notes that he declined to support a proposal by Bernard Dubos because he had not seen "any serious argument" in Dubos's publications regarding artificial precipitation.

storms. Pruvost emphasized that he had "worked for three years on these questions."[53] Maurain cautioned:

> In principle, it may be said that any subject serves for a thesis, and the one that you indicate may well serve. But the subject is not the important thing; it is how the author does something original with the subject. A thesis must necessarily contain original points, either regarding methods and apparatus or regarding results and their interpretation. I want to be clear about this point, so that there is no misunderstanding. We cannot accept observations or measurements as a thesis — however well done and interesting these may be — if the work has no original parts.[54]

Here is the ethic of the German doctoral dissertation, firmly rooted at last on French soil and transposed to the colonies.

Inflexible routine characterized the French civil service. A colonial scientific functionary, however, was encouraged to use his resources to advance pure learning by completing a doctorate. Louis Lapayre, for example, a career meteorologist in central Africa, surveyed earth and sky for more than five years before defending a doctorate of engineering at the Sorbonne, with Maurain presiding, on the climate of Djibouti. During this time, Lapayre had organized a geophysical service and a central observatory for the Somali coast. In the opening line of his thesis, Lapayre indicated that his research had been undertaken while he was a colonial meteorologist.[55] This opportunity made the colonial scientific service an attractive career choice.

The word went out that someone marooned *in partes* could turn bitter isolation to geophysical virtue. René Champion, *surveillant général* at the Lycée Faidherbe in St-Louis-du-Sénégal, asked Maurain in 1931 whether there was a system of examinations in place for prospective

53. Ibid., Pruvost to Maurain, 24 June 1938; *APGFOM*, 1935, p. 96, for Pruvost's appointment in Cameroun; *ACRACS* 8 (1932), 58, for the transformation of Cameroun's meteorological service (until then under health services) into a geophysical service headed by an ingénieur-météorologiste; PIPG, Hubert to Maurain, 1 Mar. 1935, where Hubert notes: "Mr. Pruvost will surely go soon to Cameroun, which is what he wants."

54. PIPG, Maurain to Pruvost, 2 July 1938.

55. Louis Lapayre, *Le climat de Djibouti* (ing. diss., University of Paris, 1939), p. 10, for the creation in 1934 of a meteorological service on the Somali coast (from 1900 to 1933 the weather had been surveyed by health services); *APGFOM*, 1934, p. 63, for Lapayre's presence in Bangui during the International Polar Year of 1932–33. Djibouti's central observatory was completed in 1938, although Lapayre had previously taken measurements from temporary quarters (ibid., 1938, p. 96). See also Lapayre, "Secousses séismiques à la Côte française des Somalis," ibid., 1936, pp. 30–31, where Lapayre guesses at the severity of earthquakes, his promised seismometer not yet in service; and PIPG, Hubert to Maurain, 8 May 1935, where Hubert indicates Lapayre's desire to acquire instruments for measuring solar radiation.

geophysicists deprived of attending courses in Paris. Marcel Berthier, a science teacher at the Lycée Sadiki in Tunis, wrote to Maurain in 1932, also about beginning a thesis. Berthier had briefly visited Maurain's institute and had submitted a proposal to the Caisse des recherches scientifiques for money to buy matériel. A. André, a local expatriate employed in the meteorological service at Port Etienne in French West Africa asked in 1936 for information that would help in applying for a post of ingénieur-météorologiste-adjoint stagiaire.[56]

Edmond Since, a thirty-four-year-old physics teacher at the Lycée Schoelcher in Fort-de-France, Martinique, inquired of Maurain whether there was something he could study so that, several years in the future, he would be able to spend a year in Paris and obtain a *diplôme d'études* and a certificate at Maurain's institute. Since mailed his letter from France, just before steaming back to his Caribbean post. He was unable to meet Maurain in Paris, but he was glad to send his "young brother (*agrégé* in physics at the lycée in Saigon) to visit you. He wants to meet you to choose a thesis before his return to Indochina. You would be able to relay to him orally any explanations to orient each of us about analogous work." Maurain did not hesitate to reply in order to reach Since before his ship departed.[57] Here was a way to recruit geophysical observers in the Antilles and in Indochina, and at no cost.

There were limits, however, to an ecumenical outlook with regard to research students. Also addressing Maurain in the early 1930s was Georges Bonjean, a physics teacher at the Collège Moulay-Youssef in Rabat, Morocco. Bonjean had obtained a licence ès sciences from Bordeaux and in 1930 became agrégé in physics. Bonjean wanted to begin a thesis in Morocco. He envisaged work in terrestrial magnetism or gravimetry, but he could not afford to purchase the required equipment. The real problem, however, concerned local talent. To carry out magnetical or gravimetrical work "would perhaps require me to work with local societies there, while I would be very happy to remain in direct contact with metropolitan specialists." Atmospheric or telluric electricity seemed to be the ticket. He wondered whether a thesis could be produced from carrying out measurements and describing how a recorder functioned. Maurain, who insisted on relating data to theory and who held mathematics in high regard, did not approve; Bonjean received no doctorate under him. "The desirable prerequisites" for geo-

56. PIPG, Champion to Maurain, 27 Sept. 1931; Berthier to Maurain, 2 Oct. 1932; André to Maurain, 10 Nov. 1936.

57. Ibid., Edmond Since to Maurain, 18 Mar. 1935, on which Maurain commented: "Received letter Monday First April, wrote immediately."

physical work, Maurain wrote to the American Jesuit geophysicist James Bernard Macelwane in 1939, "are a thorough training in mathematics, physics and geology."[58]

Entrepreneurial exiles like the suppliants just mentioned would never succeed in scaling more than a few rungs up the ladder of colonial promotion, for they had not received Maurain's blessing. Furthermore, an ambitious geophysicist, much like an ambitious naval officer, would seek tours of duty at a number of foreign stations, a circumstance that discouraged colonial scientists from sinking their roots at any particular location. To imagine receiving direction of a major meteorological and geophysical service, say that of Indochina, a young man had to anticipate starting at the very bottom. The story of Jacques Ravet is typical.

In 1930 Ravet, a student at Maurain's geophysical institute, decided to present himself to Albert Lasserre, director of the Algiers meteorological service, for a posting to Lasserre's new branch observatory deep in the Sahara, at Tamanrasset. He met Lasserre in Paris and made a favorable impression. After learning more about the primitive conditions at Tamanrasset, however, Ravet declined to pursue the matter. The salary of *aide-météorologiste* was simply too low for him to entertain "such isolation," Ravet wrote to Lasserre, and he did not think that his wife "was suited to such an existence." Ravet had better prospects, for he was shortly appointed *chef de service* of meteorology on Tahiti.[59]

In 1931 Ravet arrived at Papeete, where he learned to live with inaction and inefficiency: "The bad will of the local higher administration, the principal obstacle, has meant the denial of funds for setting up the apparatus and for building an observatory. After ten months of useless discussions, I have set up a station in my personal quarters." Meteorology could be handled without assistance, but Ravet was unable to install the Mainka seismometer. Ravet had time for scientific research, he wrote to Maurain, and he asked to be assigned a topic that might eventually lead to a doctoral thesis; he also had ten thousand francs that he could spend immediately to acquire instruments related to the new "personal" research topic.[60]

That Ravet could express himself to Maurain so freely indicates that by this time it was quite normal for an overseas physicist to subordinate his administrative charge to the ends of a scientific career. Within a year Ravet had succeeded in obtaining money from the governor for study-

58. Ibid., Bonjean to Maurain, 1931; Maurain notes that he replied on 7 August; Maurain to Macelwane, 11 May 1939.

59. Ibid., Lasserre to Maurain, 22 Sept. 1930 (Lasserre and Tamanrasset are discussed in part 2 below); 15 Nov. 1930, citing a recent letter of Ravet's; Hubert to Maurain, 5 Dec. 1930, with word of Ravet's selection for Oceania.

60. Ibid., Ravet to Maurain, 22 May 1932.

ing high-altitude winds; in this way, Ravet could begin to understand atmospheric circulation generally. After five years Ravet took a leave in Paris, and he enlisted Maurain to obtain funding from the Caisse nationale de la recherche scientifique. Maurain's support garnered Ravet fifty thousand francs for research on terrestrial magnetism and atmospheric electricity.[61] By 1937 Ravet was taking measurements from two Mainka seismographs and sending meteorological notes to Hubert's *Annales*. He spent several years more on Tahiti and then moved up the ladder to Madagascar.[62]

After some time, Ravet turned over Tahitian research to Joseph L. Giovannelli, who had completed a licence ès sciences with Maurain in 1934 and who in 1936 was named ingénieur-météorologiste-adjoint stagiaire at Papeete. Giovannelli set to work producing a stream of meteorological notes for Hubert's journal and then, in August 1939, succeeded Ravet at the head of the service. Giovannelli, like Ravet, used his position to work on a doctorate. He wrote to Maurain in 1938 that he wanted to submit a thesis for a *doctorat d'université* on atmospheric pressure and winds in the South Pacific. By comparing atmospheric soundings in Papeete and Apia, he detected a simple relation between pressure and the rotation of winds in the upper atmosphere. Subjecting atmospheric pressure to a harmonic analysis, he identified a series of fluctuations that characterized the ocean climate. In 1939 Giovannelli sent Maurain a summary of four years' data, and he described his plan to compare observations from Apia with those recorded by the U.S. Weather Bureau in Hawaii.[63] Wartime isolation in the Pacific no doubt prevented Giovannelli from defending a dissertation.[64]

As geophysics organized itself on an imperial scale during the years between the world wars, Charles Maurain became increasingly involved

61. Ibid., Ravet to Maurain, [1932]; Maurain to presidents of the sections of meteorology and of terrestrial magnetism and atmospheric electricity of the Comité français de géodésie et de géophysique, [12?] May 1937; Ravet to Maurain, 14 Nov. 1937.

62. Jacques Ravet, "Remarques sur quelques enregistrements d'ondes à très courte période au cours de tremblements de terre éloignés à l'Observatoire du Faiere, Papeete," in Pacific Science Congress, Sixth, Berkeley, *Proceedings*, 6 vols. (Berkeley, 1940–43), 1:128–30; Tahiti, Service météorologique, *Résumé des observations, 1948* [Papeete, 1948], p. ii, for a list of publications by the observatory's staff since 1934; Ravet, *Atlas climatologique de Madagascar* (Tananarive, 1948).

63. *APGFOM*, 1936, p. 94, for Giovannelli's appointment (his name is misspelled in the *APGFOM* and elsewhere); Tahiti, Service météorologique, *Résumé des observations faites en 1939* [Papeete, 1939], p. i; PIPG, Giovannelli to Maurain, [1938], 10 July 1938, and 10 May 1939. A portion of Giovannelli's manuscript had appeared earlier: "Quatre années de sondages aérologiques en Océanie," *APGFOM*, 1938, pp. 102–5, 109–13. On geophysics at Apia (Western Samoa) between the world wars, see *CIES*, pp. 101–38.

64. Tahiti, Service météorologique, *Résumé des observations, 1948*, p. ii.

in colonial affairs. The ministry of colonies had financed a maître de conférences in colonial botany at the Sorbonne beginning in 1903, a commitment that Maurain, as dean, supervised.[65] Then, in 1928 the ministry of colonies created a special commission to decide how to support metropolitan scientific institutions. Maurain promptly advised the ministry about enterprises in the exact sciences, for example, the training school in astronomy at Montsouris and the Institut international du froid.[66] The ministry of colonies asked Maurain's opinion about using geophysical techniques to predict volcanic eruptions. By 1938 the ministry's commission was soliciting Maurain's opinion about awarding a grant to the Bureau of Longitudes for a time service at Bordeaux. Throughout the 1930s Maurain's institute supplied the ministry of colonies with examiners to evaluate candidates for the post of colonial meteorological engineer. To this explicitly colonial involvement must be added Maurain's commitment to apprise naval hydrographers of measurements in terrestrial magnetism, seismology, actinometry, and elementary meteorology at Maurain's two suburban observatories of Val-Joyeux and St-Maur (the army's geodesical service standardized its magnetical instruments at Val-Joyeux before sending them to the colonies).[67]

Momentum toward an imperial organization of scientific activity culminated in 1937 with the creation by Jean Perrin, then undersecretary of state for scientific research, of an overseas committee in the Conseil supérieur de la recherche scientifique. With this act, the ministry of national education (the new name of the ministry of public instruction) formalized the advisory role it had played in overseas territories. The first of the committee's seven sections handled geography, geodesy, astronomy, geophysics, and protection of sites (presumably

65. PIPG, Maurain, "Rapport à la Commission des subventions aux établissements scientifiques, cours de botanique coloniale (Sorbonne)" [1928], for a history of the chair. The grant from the ministry of colonies in 1903 was six thousand francs annually, of which thirty-five hundred francs went on the ledger of French West Africa and twenty-five hundred on the ledger of Indochina.

66. Ibid., Maurain, "Rapport à la Commission des subventions aux établissements scientifiques" [1928]; idem, "Rapport à la Commission de répartition des subventions à allouer à des organismes scientifiques" [1928]. The cryogenic institute was slated to receive nine thousand francs: four thousand from Indochina, three thousand from French West Africa, and two thousand from Indochina; Maurain indicates that Indochina wanted to increase its contribution to eight thousand francs.

67. Ibid., Service de préparation de la défense nationale, Ministry of Colonies to Maurain, 25 Oct. 1929, and Maurain's reply, 31 Oct. 1929; Director of the Institut national d'agronomie de la France d'Outre-Mer (Pernot) to Maurain, 3 July 1937; Director of Accounting, Ministry of Colonies, to Maurain, 11 July 1938; Maurain to Head of Hydrographical Service (Lazare-Eugène Fichot), 6 Jan. 1931, and Louis Eblé, note [1932].

those of archaeological significance). Membership in each section of the committee was allocated to the Conseil supérieur de la recherche scientifique (six out of a total of eleven seats), with other seats going to the various learned academies and special-interest groups. The ministries of colonies, foreign affairs, and the interior had veto rights over the Conseil's appointments and could also name a number of members *hors section.* The minister of colonies presided over the committee, with his vice-presidents coming from foreign affairs, the interior, and education's Caisse nationale de la recherche scientifique; real power resided with the *section permanente*, in which (according to a complicated formula) education and colonies had nearly equal representation, but in which foreign affairs and the interior held the balance of power. The section permanente received proposals from each of the seven disciplinary sections. It was empowered to spend money in the Caisse nationale de la recherche scientifique which had been placed there by the ministries of the interior, foreign affairs, and colonies, by the government of Algeria, by the colonies, by protectorates, and by mandated territories. On its own account, the ministry of colonies created a service for organizing scientific research (without, however, giving it a permanent budget or bureaucracy), and Henry Hubert pressed for an interministerial committee to coordinate *all* scientific activity overseas.[68]

If Maurain had wanted to become overseas science czar, few would have objected. But he kept colonial affairs at arm's length. He abstained from joining the organizing committee for the 1931 Congress for Scientific Research in the Overseas Territories, which was presided over by Alfred Lacroix, permanent secretary of the Academy of Sciences; at the 1937 congress he somewhat reluctantly chaired a section.[69] Maurain did allow himself to be named to the new overseas committee of the

68. "Le Comité de la France d'Outre-Mer auprès du Conseil supérieur de la recherche scientifique," *ACRACS* 14 (1938), 62–65, reproducing the foundation document published in the *Journal officiel* (2 Oct. 1937); Henry Hubert, "Rapport général présenté au Congrès de la recherche scientifique dans les territoires d'outre-mer," *ACRACS* 14 (1938), 26–32, on pp. 28–30 for Perrin's role, the independent service in the ministry of colonies, and the call for a grand interministerial committee to supervise overseas science. These developments took place at a time when the administration of research in France underwent changes leading to the creation of the Centre national de la recherche scientifique (Frédéric Blancpain, "La création du CNRS: Histoire d'une décision, 1901–1939," *Bulletin de l'Institut international d'administration publique* 32 [1974], 93–143; Jean-François Picard, *La République des savants: La recherche française et le CNRS* [Paris, 1990], esp. pp. 33–60. Picard paints a somber picture of science in France during the 1920s and 1930s, and he argues that various centralizing schemes were intended to catapult France to the forefront of research).

69. PIPG, Maurice Martelli-Chautard to Maurain, 17 Mar. 1931 (with an appended list of the organizing committee) and 8 May 1937; *Congrès*, pp. 148–53, for Maurain's sectional report.

Conseil supérieur de la recherche scientifique, where he sat with Emmanuel de Martonne of the Paris geographical institute and Jean Coulomb of the Algiers geophysical institute, among others having an interest in colonial affairs.[70] The dean of the Sorbonne's science faculty participated in imperial schemes to centralize the exact sciences, but he had little to gain from lending his name to showy displays of colonial fervor.

By the late 1930s, colonial enthusiasm had produced vigorous research programs at a number of scientific institutions. We shall see that Algiers and greater Shanghai hosted astronomical and geophysical complexes without peer beyond Paris. The Sherifian Institute of Morocco, an umbrella organization for one of France's North African protectorates, had a research budget that ranked in the top dozen French organizations. Laboratory research in nuclear physics went on at the Université St-Joseph in Beirut.

These institutions and programs enjoyed direct access to patrons in the hexagone, who participated in an imperial orchestration of science. Maurain's interest is directly revealed by the fact that throughout the 1930s the number of places at the Sorbonne for all kinds of physicist remained stationary—around 175 in its various institutes and laboratories; perhaps 15 percent of this number were unremunerated students or foreign guests. The colonial meteorological corps was Maurain's salvation. When Henry Hubert retired in 1938, his service had a cadre of forty-seven, arrayed in ranks ranging from two inspecteurs généraux down to five ingénieurs-adjoints stagiaires. The corps spread across the entire colonial realm.[71] Given the possibilities of offering positions to their students during a time of retrenchment, metropolitan scientists such as Maurain could not afford to abstain from associating themselves with the centralizing schemes of the ministry of colonies.

70. *APGFOM*, 1939, pp. 29–30.

71. Dominique Pestre, *Physique et physiciens en France, 1918–1940* (Paris, 1984), pp. 68, 92, for the Sorbonne census; beyond Paris, the number of faculty positions for physicists at universities (Algiers included; Hanoi, Zikawei, and Beirut excluded) rose from forty-five in 1913 to fifty-five in 1929 and then to sixty-four in 1938 (p. 14). See *APGFOM*, 1939, pp. 30–31, for the corps as of May 1938; only the two inspecteurs généraux did not have an overseas posting.

11 Wind, Rain, and Stars in the Major Colonies

Over the generation before the First World War, France became mistress of Indochina and the western half of North Africa. Ancient civilizations with learned institutions and extraordinary monuments fell to military rule and French administrative norms. Savants played useful roles in the conquest. They charted the new land and provided counsel about disease; they served as interpreters and ethnographers. In addition to these practical tasks, men and women at colonial institutions in the exact sciences undertook research of a disinterested nature.

Pure scientific research served two different strategies of cultural imperialism. On the one hand, physics and astronomy carried out in the French language radiated *French* cultural values. Autochthonous laboratory assistants and casual bystanders were invited to see the natural world through the lens of French convention. On the other hand, pure research extended the claim of advancing the principal mode of *Western* cognition. This part of the civilizing mission, providing results valued by colleagues in rival empires, was intended for establishing international hegemony or moral authority over the colony.

Material exploitation surrounded the two strategies of cultural imperialism. Researchers in pure science had to provide results of practical utility. They observed the stars and measured the movement of the earth—activities without lucrative import—while at the same time they inventoried rainfall and tracked storms. In Indochina and Tunisia, the dual mandate worked to the disadvantage of pure research; in Algeria and Morocco, it secured funds and facilities for undertakings of a general nature.

One might think that as France tightened its hold on the colonies, pure scientific research would be used less as a territorial marker than as an acculturating device. In fact both strategies of cultural imperialism operated undiminished across the first forty years of the twentieth century. Neither strategy produced the desired effect. Annamite and Arab assimilated the principles of rational mechanics and electromagnetism, but the knowledge did not lead them to view France with sympathy. Significant geophysical and astronomical research, presented at local and international conferences, did not forestall Japan's occupation of Indochina or Germany's invasion of North Africa. Although cultural prestige does not always change the course of a revolution or war, its hold persists when military garrisons have been dismantled. The following pages indicate why, today, French scientific norms play a part in the intellectual life of Southeast Asia and North Africa.

3

The Indochinese Gesture

Higher Learning for Lesser Peoples

Higher education in twentieth-century French Indochina provides a graphic illustration of the imperial policy of associationism, a notion identified by Joseph Buttinger as "a generous form of cultural imperialism." Uncivilized races (and the French continued to the last to govern Indochina as an amalgam of indigenous peoples with differing intellectual capacities and moral proclivities) were to become brothers of the French, *Français de couleur.* These races, inhabiting la France d'Outre-Mer, were not yet ready for the privileges of French citizenship. They were to be educated in colonial schools and trained for subordinate administrative positions; any native considered fit for advanced instruction would be sent to repeat his schooling in the metropolis. It was the French system of apartheid, but with a perverse variation. Existence in the colonies served the greater glory of France; in the colonies, one was a mere associate of French liberty and French learning. In Indochina, the French set up a primitive network of inferior schools in the vernacular and in the language of the conqueror. Where literacy had been widespread and valued, both in Vietnamese and in Chinese, and where morality had been clearly set by Confucian ethics, French policy produced a nation of illiterates inclined to rebel against authority.[1]

Associationism had no clear end. It was not evident how to graduate from conquered status. One could only follow one's destiny as a subject race, eventually to produce a culture that would respond gratefully to the inventories of metropolitan administrators — to invent an overseas

1. Joseph Buttinger, *Vietnam: A Dragon Embattled,* 2 vols. (New York, 1967), 1:75–76, 486–87. The triumph of associationist ideology, which replaced the notion of assimilationism, is persuasively recounted in Raymond F. Betts, *Assimilation and Association in French Colonial Theory, 1890–1914* (New York, 1961).

Alsace or a non-Latin Corsica. Knowledge and learning, then, had to be epiphenomenal at colonial universities. The main idea was to produce servility toward and reverence for the metropolis. As the ideals of the Third Republic were those of the First, their diffusion simply produced resentment and dissatisfaction among the few men selected to be future *collaborateurs*. The result, on the part of French educational administrators, was vacillation between merchandising the ideals of French culture and denying the existence of these ideals. Colonial education followed the whim of the governor general, and it rose or fell depending on the man who happened to be in power.

L'Université indochinoise at Hanoi was the invention of Henri-Emile Gourdon, appointed as Indochina's first director general of public instruction by Governor General Jean-Baptiste Paul Beau in 1905.[2] The proclamation of the new university in May 1906 indicates that it emerged as the pendulum of the administration swung toward a liberal maximum. The university was designed "to promote in the Far East, above all through the intermediary of the French language, knowledge of European sciences and methods." The university's governing council, presided over by Gourdon, included the directors of its various constituent schools and of the other scientific institutions of the colony, as well as professors designated by the governor general to represent various services and interests. University affiliation for existing institutions would be negotiated by the council. The grand vision included schools of law and administration, sciences, medicine, civil engineering, and letters to be located in Hanoi or Saigon. The university was for the indigenous peoples of Indochina, who had to have completed secondary school, as well as for European colonials, who enjoyed special standards of admission. The most apt Indochinese students were eligible for scholarships to continue their studies in France.[3]

The Hanoi branch of the university opened in October 1907 with sections in sciences, French literature, and French law. To judge from the courses offered, being perforce introductory ones, the level of instruction was barely that of a European secondary school. Mathematics and mechanics, taught by a graduate of the Ecole centrale des arts et manufactures, concluded with trigonometry and certain "industrial applications." Physics, remanded to Georges-Gilbert Audhuy, a military physician who had been an assistant in the subject at the University of Bordeaux, considered electricity, heat, optics, and acoustics from the point of view of "applications." A pharmacist with the colonial army who held a doctorate in pharmacy and had been an assistant in chemis-

2. Edmond Chassigneux, *L'Indochine* (Paris, 1932) [*L'histoire des colonies françaises*, ed. Gabriel Hanotaux and Alfred Martineau, 5], p. 513; *DBF*, s.v. "Gourdon."

3. Aix, Indochine 48042, Governor General, *arrêté*, 16 May 1906, no. 1514a.

try at the University of Nancy, taught the elements of chemistry. Another graduate of the Ecole central des arts et manufactures taught industrial chemistry. The chief medical doctor of the colonial army lectured on physiology. The only professor apprised of scientific research was Philippe Eberhardt, holder of a doctorat ès sciences naturelles, who taught botany and zoology.[4]

Governor General Beau returned to France in February 1908. As wheels turned slowly to provide Indochina with a new head man, Beau's conception of the Indochinese University sustained intense criticism. Henri Gourdon tried to deflect slander directed against the principal jewel in his educational diadem. In a long defense of his general educational policy, presented to interim governor Louis Alphonse Bonhoure in August 1908, Gourdon emphasized that the word *university* had given rise to much misunderstanding: "Words have their destiny, especially in France where one is attached, by national deficiency, more to words than to things." The university had "sustained, before the thing itself was in the process of being organized, the most acerbic criticisms." Beau had continually tried to phrase the new institution in terms of local needs and the ends of "foreign policy," but "the name of the establishment earned him the worst enemies." Had the course of study gone by another name, "no one would have thought to criticize it, and the ridiculous legends circulating about the university would never have seen the light of day." Gourdon listed and disposed of the most extreme slanders. He emphasized that the university cost virtually nothing, that its staff required no large salaries, that the ninety-four original students were drawn not far and wide from the working classes but from the pool of *compradores* in the neighborhood of Hanoi, that of those forty-six students who survived the first semester forty-one continued for the second, and that, finally, primary education had hardly been "sacrificed for extravagant, restive, and dangerous instruction."[5] The new governor general, Antony Wladislaw Klobukowski, arrived the next month and promptly shut down the university.

It has generally been argued that Klobukowski closed the Indochinese University because he saw it as a hotbed of sedition, fueling the uprising of 1908.[6] Klobukowski, however, was hostile in principle to advanced learning. He acted quickly to abolish the Permanent Scien-

4. Ibid., Henri Gourdon, *Université indochinoise, Section de Hanoi, règlement intérieur* (Hanoi, 1907); Audhuy, *Radiothérapie coloniale* (med. diss., University of Bordeaux, 1906). The Hanoi medical school, founded in 1903 and integrated into the university, had its own physics *cabinet*, valued at four thousand francs (Henri Galliard, "Le docteur Yersin et la fondation de l'Ecole de médecine de Hanoi," Hanoi, Faculté mixte de médecine et de pharmacie de l'Indochine, *Annales* 9 [1944], 1–24, on p. 23).

5. Aix, Indochine 7707, Henri Gourdon to Governor General, 14 Aug. 1908.

6. Buttinger, *Dragon*, 1:66.

tific Mission to Indochina, established in 1902 to inventory the natural wealth of Southeast Asia.[7]

The mission, following in the tradition of nineteenth-century missions in the Mediterranean and Mexico, comprised several biologists selected by and responsible to the Paris Academy of Sciences. Beginning in 1905 the mission cost the Indochinese colony some twenty thousand piastres for five European scientists and native staff. Between 1904 and 1908, European personnel and matériel alone came to over eighty-six thousand piastres.[8] The mission had a mandate to undertake pure research. Edouard Alfred Marie Broni, acting governor general in Beau's absence, wrote in 1902 that the geological and paleontological section of the mission was to conduct "exploration, from the point of view of pure science," that "its role is that of an organ of pure investigation, in the order of natural sciences, completing the program of explorations to be carried out by the botanists, the zoologists, and the anthropologists." This stood in contrast to the mining service, soon to be reconstituted, which sought "above all immediate, direct, and utilitarian application of knowledge" acquired about the colony's mineral wealth.[9]

Eight days after arriving in Indochina, Klobukowski met with Louis Boutan, the director of the scientific mission, who brazenly asked to double his budget of twenty-nine thousand piastres. When Klobukowski queried Boutan about the "practical character" of the mission, Boutan replied that utility would result by application of the scientific method of observation. Klobukowski then traveled to Saigon to meet with the other resident member of the mission, Armand Krempf, who asserted: "I do nothing practical, . . . I do pure science." Both Boutan and Krempf emphasized that they were responsible to the Natural History Museum in Paris, not to the colonial regime, and that the results of their research would appear in metropolitan journals rather than in colonial memoranda. Klobukowski concluded that Boutan was "consolidating his position in the academic world of the metropolis" and Krempf was holding down a lucrative post while preparing a doctoral thesis.[10] Only the third scientist of the mission, Philippe Eberhardt, escaped Klobukowski's wrath, for Eberhardt had had the good sense, as his teaching post at the Indochinese University evaporated, to obtain

7. The point is made in Jean-Baptiste Saumont, *L'oeuvre de Klobukowski en Indochine* (Hanoi, 1910), pp. 50–57. Saumont, a municipal councillor at Hanoi, was an intense critic of Klobukowski's policies.

8. PAN, F17/17261, Evaluation of the mission, 6 July 1908, "Dépenses de la Mission scientifique permanente de 1904 à 1908."

9. SOM, Indochine X 50 (1), Broni to Minister of Colonies, 2 Oct. 1902.

10. PAN, F17/17261, Klobukowski to Minister of Colonies, 10 Jan. 1909.

leave from the mission to become preceptor to the king of Annam.

The perpetual secretary of the Academy of Sciences, mathematician Gaston Darboux, wrote to the minister of public instruction emphasizing that England organized expensive scientific expeditions in its colonies, "an example to be imitated from all points of view."[11] The ministry of public instruction in France naturally protested Klobukowski's action – the mission, after all, figured on the colonial budget, not the educational one. The minister, Gaston Doumergue, repeated Darboux's argument to his colleague in colonial affairs.[12] Klobukowski, however, who decided the fate of the mission, would not be swayed. He wrote to the minister of colonies:

> Whatever may be the value of the arguments developed by the professors in the Faculty of Sciences at Paris, I persist in believing that in the present state of finances of the colony, it is not given me to allow the continuance of an institution (the scientific interest of which I do not fail to perceive) that, despite the allocation of an annual budget of nearly Fr 100,000, has not produced any practical result.[13]

The permanent scientific mission ended just as the Indochinese University had ended.

Boutan wanted practical results no less than Klobukowski, but they chose separate paths to practicality. Whereas Klobukowski insisted on recipes and useful gadgets, Boutan emphasized the moral authority of disinterested scientific research. By 1911 the former head of the scientific mission had changed his tune, emphasizing that science would do well to follow commercial and industrial demands. Boutan, by then *professeur titulaire* at the University of Bordeaux, wrote to the minister of colonies:

> Many enlightened Frenchmen believe that the sciences and their practical applications are separated by a large gap and that it is all to the advantage of scientists to carry out their work in isolation and quiet, sheltered from outside agitation.
>
> This aristocratic conception of the role of science and of the specialists charged with its advancement must seem to you, as it does to me, quite old-fashioned.

The German economic miracle and American agriculture showed how laboratory work lay at the base of national strength. France had to bring

11. Ibid., Darboux to Minister of Public Instruction, 17 Oct. 1908.

12. Doumergue, cited in Indochina, Gouvernement général, *Fondation de l'Institut scientifique de l'Indochine* (Saigon, 1919), pp. 5–6, in a section written by Auguste Chevalier.

13. PAN, F17/17261, Klobukowski to Minister of Colonies, 16 Apr. 1909.

science to its Asian colony as the Americans had done in the Philippines and the Dutch had done on Java.[14]

Klobukowski's successor, Albert Sarraut, slowly reversed the precipitous decline of science and higher education. He looked into those among the university's schools that had survived in other forms. In 1912 he received a report on the Ecole indigène des travaux pratiques, which functioned as the civil-engineering school, suggesting that "far from being abolished, this school should be reorganized on a more serious basis, in such a way as to form a nursery for technical personnel" who could replace French foremen, clerks, and technicians. Five years later Sarraut was once more governor general of Indochina. He reopened the university at Hanoi and revived the permanent scientific mission as the Scientific Institute of Indochina.[15]

The man charged with the renascence of scientific research in Indochina was Auguste Chevalier. In 1911 Chevalier, who had obtained a doctorat ès sciences from Paris in 1901 for a botanical thesis, was placed at the head of a permanent mission for colonial agriculture. In 1913 Chevalier toured the agricultural institutions of Indochina, Singapore, Kuala Lumpur, Java, and Ceylon, and he assembled material on British India. Beginning late in 1916, Sarraut engaged Chevalier to inventory the scientific wealth of the colony and to coordinate local agricultural efforts. The new scientific czar sought to produce a French-language version of the Dutch agricultural program at Buitenzorg on Java; his mandate was in fact limited to agricultural botany and commercial fisheries, and he received a tract of land in Saigon as the site for the institute's labor.[16]

Georges Maspero succeeded Sarraut as governor general in 1919, and a cool wind blew through higher learning. Chevalier quickly returned to France. He was succeeded at the Institut scientifique in rapid succession by the toxicologist Léon Garnier, who had taught in the medical faculty at Nancy, and then an understudy, Charles Lemarié, who had been trained at the Institut agronomique in the metropolis and

14. SOM, Indochine X 50 (1), Boutan to Minister of Colonies, 22 May 1911.

15. Aix, Indochine 43682, Head of Political and Native Affairs to Governor General, 19 July 1912; see Chassigneux, *Indochine*, pp. 542–44, for the reopening of the Indochinese University.

16. Chevalier describes his career in colonial agriculture in Indochina, Gouvernement général, *Fondation de l'Institut scientifique de l'Indochine*, pp. 7–10 and 17–18, the mission of 1911 on pp. 35–36, the aims and mandate of the institute on pp. 13, 39; see also Chevalier, *Monographie des myricacées: Anatomie et histologie, organographie, classification et description des espèces, distribution géographique* (diss., University of Paris, 1902). His survey of British India is in "Les institutions scientifiques, agricoles et forestières de l'Inde anglaise," *BEI* 20 (1917), 1–50, which, however, deals only with agricultural institutions.

who had formerly headed the agricultural service of Indochina.[17] The Institut scientifique was suppressed in 1925.[18]

Sarraut's university fared no better than Chevalier's institute, and advanced training played only a small part in its mandate. The director of public instruction for Indochina emphasized in 1924 to Governor General Martial Henry Merlin, himself no fervent believer in higher education, that he had reduced teaching personnel from 142 to 72, saving forty-one thousand piastres. "Some will accuse me — they have already done it — of megalomania, of dreaming to make a university after the model of the metropolis," the director wrote to Merlin. "You have already sensed, I am sure, the absurdity of these allegations. My wish is that a university be made here which responds to the needs of the country."[19]

In 1925 François-Amadée Thalamas, a historian, geographer, and former rector at Dijon, arrived to take charge of public instruction in Indochina. He reorganized the University of Hanoi into a collection of trade schools where future *collaborateurs indigènes* would acquire skills and discipline. Pure research figured nowhere in his plans.[20] His reorganization found supporters among old Indochina hands. One observer noted that higher education had until then "missed the point" by offering Indochinese students the rich complexity of French culture. "The indigene, a good memorizer, recites and imitates texts without really understanding them; he learns the sciences by heart without letting the scientific spirit sink in." The result had been a superficial and illusory grasp of French ideas.[21] By 1925, higher learning had no place in higher education. Etienne Patte, a former geologist in Indochina, emphasized in 1928:

17. See Saigon, Institut scientifique de l'Indochine, *Catalogue de la bibliothèque* (Hanoi, 1922), p. viii, for the succession (introduction by Paul Carton); and Indochina, Gouvernement général, *Fondation de l'Institut scientifique*, p. 46, for Lemarié.

18. Etienne Patte, *Le mouvement scientifique en Indo-Chine: Discours prononcé à la rentrée solennelle des facultés de l'Université de Poitiers, le 3 novembre 1928* (Poitiers, 1929), p. 8.

19. Aix, Indochine 51061, Director General of Public Instruction to Governor General, 2 Sept. 1924. On Merlin's opposition to advanced education see William J. Druiker, *The Rise of Nationalism in Vietnam, 1900–1941* (Ithaca, 1976), p. 132.

20. *Annales de l'Université de Hanoi* 1 (1933), 12; Chassigneux, *Indochine*, p. 544; John Sydenham Furnivall, *Educational Progress in Southeast Asia* (New York, 1943), p. 87; Thalamas, *La géographie d'Eratosthène* (Versailles, 1921). In 1905 Thalamas had incurred right-wing wrath by advocating atheism in his courses at the Lycée Condorcet in Paris (William R. Keylor, *Academy and Community: The Foundation of the French Historical Profession* [Cambridge, Mass., 1975], p. 154).

21. H. Délétie, "Le problème universitaire indochinoise," Académie des sciences coloniales, *Comptes-Rendus des séances, communications* 8 (1926–27), 247–62, on p. 256.

> Concerned with its civilizing role, France developed schools of all levels in Indochina, from small local schools to the university. The latter, addressing itself to a population only recently open to science and modern intellectual civilization, would not have the same ambitions as our universities in France. It trains doctors and veterinarians, prepares drivers for public works as well as schoolteachers, and teaches law to future functionaries.[22]

In view of this approach to university education, it comes as no surprise that exact sciences received little attention. Physics on the level of the P.C.N. (at the time the equivalent of an American undergraduate science education) was taught to students of medicine, pharmacy, and veterinary science along with science students. In 1928 the physics professor Bernard, who had been certified in France as an agrégé in science, counted ninety-five auditors in an amphitheater designed for two-thirds that number. His instruction was the equal of that in France, he crowed to the director of the medical school, and he was certain that many of his pupils "would succeed" in the examination of the P.C.N. certificate. Bernard disagreed with critics who claimed "that our advanced instruction is advanced in name only. . . . The legend of a 'feeble' university giving 'instruction on the cheap' should go join that of the failure of our school system."[23]

Like other imperialist physical scientists in Asia during the 1920s, Bernard was much concerned with the broader meaning of Western knowledge.[24] He believed that Indochina would benefit from a strong dose of the history of science, with special emphasis on the exact sciences. He deplored the most recent educational guidelines, which devalued historical treatment in physics and chemistry. Six elements were essential in any history of science, Bernard believed. First were biographies of scientists. The lives of great researchers—who lived only for "the disinterested progress of science"—provided moral instruction. Second were the great theories of science, constructed with courage and

22. Patte, *Mouvement scientifique*, p. 14.

23. Aix, Indochine 51063, Bernard to Director of Medical School, cited in Director General of Public Instruction to President of the Chamber of Commerce of Saigon, 16 Sept. 1928. Bernard is identified as the professor of physics in Aix, Indochine 51061, course assignments for 1924–25. Bernard is given as *agrégé des sciences* in Aix, Indochine 51136, where he is indicated as sitting on the P.C.N. examining jury for physics and chemistry in 1930. Instituted at the French science faculties in 1893, the P.C.N. was a short course of study in physics, chemistry, and natural sciences intended as general scientific preparation for medical students. By the eve of the First World War, one in four students at the science faculties in France was registered for the P.C.N. (Harry Paul, *From Knowledge to Power: The Rise of the Science Empire in France, 1860–1939* [Cambridge, 1985], pp. 147, 164).

24. *CIES*, pp. 287–89, for Joseph Drexler at the Tung-Chi University in greater Shanghai; *ER*, pp. 139–45, for Jacob Clay at the Bandung Institute of Technology on Java.

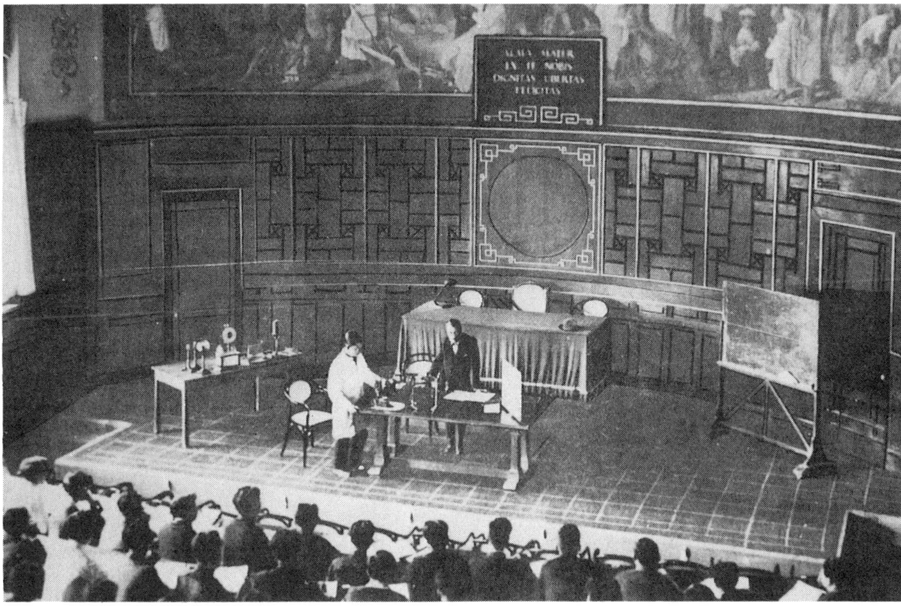

Neoclassical temple for natural law: physics lecture at the University of Hanoi (From *Annales de l'Université de Hanoi* 1 [1933], 16)

care. Third were scientific controversies, which provided evidence of extraordinary ingenuity and passion. Fourth were "scientific revolutions," which showed the "relativity of the way we understand the world, which seems to us so well established," and which demonstrated "the danger of believing that we have made better choices than our predecessors in all things." Fifth were the applications of science. These derived from gains of pure science as well as from mission-oriented research, although Bernard thought that the Germans had made great headway in applied science by directing their focus *away from* immediate applications. Finally, scientists illustrated the highest virtues of civilization by proposing "ideas of beauty and harmony that raise mankind well above immediate, practical interests and contribute to moral progress." In view of all these points, Bernard believed that "especially here, in Indochina, our instruction should give a large place to history of science." The students would quickly realize that "our country has always held the first place in the evolution of science and that our scientists have always been the first, and also the best." Here was precisely the lesson that needed emphasizing: "Is this not the most effective French propaganda?"[25]

25. C. Bernard, "La valeur éducative de l'histoire des sciences," Indochina, Direction

Did Indochina's advocate for the history of science have reason to be proud of his scientific operation? By 1928 Bernard had two agrégés in physical sciences – Eugène Delauney and A. Perucca – to help him run his laboratories. Delauney had obtained a doctorate in analytical chemistry under Georges Urbain; Perucca apparently had no research credentials.[26] The three men made little progress in research, although it may be recalled that many scientists avoided research at provincial universities in France during this time. Professorial salaries at the Indochinese University were around 10 percent lower than in Paris, just a bit under what a professor in a provincial university might have earned.[27] With the exception of the school of natural sciences, which had no quotas, places in university programs were strictly controlled according to estimates of positions to be filled by graduates; in 1929 the number to be admitted, exclusive of aspirants to the P.C.N., was 185.[28] Henri Gourdon thought that it would be good to have the science school offer serious preparation for the "grandes écoles techniques" of France and especially to begin "the initiation of the natives to scientific research."[29] But an Indochinese graduate of a grande école

de l'instruction publique, *Bulletin général de l'instruction publique* (Indochina) 8, no. 6 (1929), 85–88. The Indochinese Bernard, consistently given the first initial *G* in *Min.* and in the *Annales* of the medical school, may however be the medical doctor Claude Côme Luc Bernard, author of *Contribution à l'emploi des rayons ultra-violets, en oto-rhino-laryngologie* (med. diss., University of Paris, 1927). The first initial *C* is given in the byline to the 1929 article.

26. International Colonial Exhibition, Paris 1931, French Indochina, Science Section, *Le Conseil des recherches scientifiques de l'Indochine* (Hanoi, 1931) (hereafter cited as *Le Conseil des recherches*), p. 16, for the men and their diplomas. The council's correct title is Conseil *de* recherches scientifiques de l'Indochine. I infer that the Indochinese Delauney is Eugène Delauney (see his *Nouvelles méthodes d'analyse quantitative physico-chimique* [diss., University of Paris, 1926] and his related publications: "Sur une nouvelle méthode d'analyse quantitative par rayons X," *CRAS* 180 [1925], 1658–61; "Sur le caractère atomique de quelques propriétés des rayons X," ibid. 185 [1927], 193–95; "Sur une nouvelle méthode d'analyse quantitative applicable à un mélange de terres rares," ibid. 186 [1928], 850, all presented by Urbain). Perucca remained in the medical faculty at Hanoi until at least the late 1930s (*Min.*, 1936, s.v. "Hanoi").

27. Aix, Indochine 51057, Director General of Public Instruction to Governor General, 5 Feb. 1929, where salary scales are compared.

28. Ibid., Indochine 51069, Government decree of 5 Aug. 1929 specifying the number to be admitted. In "Université indochinoise: Nombre maximum d'élèves à admettre dans les Ecoles d'enseignement supérieure à la rentré [*sic*] d'octobre 1928," the chart labeled "Prévisions relatives au nombre des emplois qui peuvent être attribués par les divers services aux élèves diplômés de l'université en août, septembre, octobre, de l'une des années suivantes: 1929, 1930, 1931, 1932," dated 6 July 1928. The point is made in Chassigneux, *Indochine*, p. 544.

29. Aix, Indochine 51540, Reform proposals of Henri Gourdon, [1928?].

had to be made of strong stuff. His career choices were extremely limited.[30]

In general, university diplomas from Hanoi were not recognized in France and were thought less well of in the colony than foreign diplomas, for example, those from the Jesuits' Aurora University in Shanghai.[31] Beginning with the first graduating class, anciens élèves of Hanoi had to petition for the privileges of graduates from analogous schools in the hexagone.[32] A handful went on to further study during the 1920s. One graduated from the Ecole des mines, and three took a licence ès sciences. Two Vietnamese who finished a licence ès sciences physiques, Do Tri Le and Hoang Co Nghi, returned to become teachers at Vietnamese institutions. The number of these highly trained men during the interwar period seems to have been about the same as in the years before 1914.[33]

By 1930 the only serious part of the University of Hanoi was its school of medicine and pharmacy. It had been created in 1902 by Alexandre Yersin, then director of the Pasteur Institute at Nha-trang.[34] The faculty, the various Pasteur institutes in Indochina, and several zoologi-

30. Jean Chesneaux, *Contribution à l'histoire de la nation vietnamienne* (Paris, 1955), p. 198: "Upon returning the engineer trained at the Ecole polytechnique (there have been several) confronts the possibility of a monthly salary much lower, according to a frequently cited example, than that of the French janitor at the University of Hanoi." Virginia Thompson identified two Indochinese agrégés, Pham Duy Khiem (Ecole normale supérieure, 1934) and Hoang Xuan Han (Ecole polytechnique, 1930); and two docteurs ès sciences, Miss Hoang Thi Nga and Doan Khac Thinh (Thompson, *French Indo-China* [1937; reprint, New York, 1968], pp. 298–99). To judge from authors' names in the *Catalogue des thèses*, published periodically since the end of the nineteenth century, many Indochinese completed science doctorates in France.

31. Diep-van-Ky, "L'enseignement des indigènes en Indochine," Académie des sciences coloniales, *Comptes-Rendus des séances, communications* 5 (1924–25), 377–87, comparison on p. 383.

32. Aix, Indochine 51523, Note for the Secretary General, 15 June 1921, where the policy is defined.

33. Ibid., Indochine 51523, 51532, and 51537, for Indochinese students in France, a movement stretching back to before the First World War. According to documents in Aix, Indochine 51537, Pham-dinh-Viêm attended the Ecole nationale des arts et métiers at Aix in 1909 and then the Institut électrotechnique at Grenoble. In 1910 Bui vân Tiêm attended the Ecole d'électricité industrielle at Marseilles, Nguyên vân Cân *dit* Nglûa the Marseilles science faculty and also the Marseilles engineering school, and Nguyên vân Duong the Ecole d'arts et métiers of Angers.

34. Francisque Vial, *Le problème humaine de l'Indochine* (Paris, 1939), p. 91; Galliard, "Yersin"; Noël Bernard, *Yersin: Pionnier-savant-explorateur (1863–1943)* (Paris, 1955), pp. 120–21. It is well to remember that in the humanities, the language and archaeology school was justly renowned (Louis Malleret, *Aperçu d'un démi-siècle de travaux scientifiques à l'Ecole française d'Extrême-Orient* [Saigon, 1956], reprinted from *France-Asie*, Oct. – Dec. 1956).

cal laboratories cooperated closely over the first half of the twentieth century, and the medical researchers lent their substantial prestige to higher education.[35] After obtaining the right to confer P.C.N. certificates in the early 1920s, the medical school began awarding doctorates of medicine in the middle 1930s.[36]

Medicine threatened to dominate all features of scientific life in Indochina. Knowledge about modern physics was correspondingly restricted. In 1924 the director of the electrical workshop of the water and electricity monopoly of Indochina wrote to the dean of the Sorbonne's science faculty about the essence of physical reality. The electrical engineer wanted to know about the origin of magnetical energy as it was found in magnetized minerals; he asked why magnets had poles and how energy could be transmitted from natural magnets to magnetized bars; he wondered whether the table of natural forces in the universe, provided in the *Annuaire* for 1903 of the Bureau of Longitudes, had been revised or extended.[37]

Whether or not colonial engineers understood magnetism (the hapless man just cited certainly did not), they proved adept at taming a technical marvel that was based on electromagnetical fields—radio transmission. Indochina received a radiotelegraphical service in 1909. In 1911 Governor General Albert Sarraut decided to construct a broadcasting unit capable of raising France. The equipment, waiting for shipment on the docks at Marseilles when the First World War began, was rerouted by the army and installed at Lyons. After the war the colonial authorities, vexed at having to communicate with their superiors by way of cables controlled by foreign companies and concerned about powerful transmitters on Java and in Australia, pressed for a grand Indochinese antenna; geophysicists added their voice by emphasizing the utility of such a station for seismological reporting. The 500-kw station began life at Saigon in 1924 when Sarraut, promoted to minister of colonies, sent and received the first wireless messages. Shortwave re-

35. H. Marneffe, "L'oeuvre de l'Institut Pasteur en Indochine," *France-Asie* 13 (1956), 423–40; P-Noël Bernard, ed., *Les instituts Pasteur d'Indochine* (Saigon, 1922), p. 14 for Yersin and the medical school, p. 16 for interrelations among colonial biologists and healers. On the well-endowed oceanographical institute, founded in 1922 and confided to Armand Krempf, see International Colonial Exhibition, Paris 1931, French Indochina, Science Section, *L'Institut océanographique de l'Indochine* (Hanoi, 1931). According to Bernard, *Instituts Pasteur*, p. 20, Krempf worked at the zoological section of the Institut Pasteur in Nha-trang from the dissolution of the Permanent Scientific Commission in 1909 until 1918.

36. *Annales de l'Université de Hanoi* 1 (1933), 8; Pierre Gourou, *Le Tonkin* (Paris, 1931), p. 342. On the first doctoral theses in medicine, see Ecole supérieure de médecine et de pharmacie de l'Indochine, *Annales* 1 (1935–37), 19–21.

37. PIPG, Boucier to Charles Maurain, 3 Nov. 1924.

Prototype radio for frontier stations, Indochina. Instruments of imperial control, radio networks provided increasingly accurate weather forecasting in Southeast Asia. (From L. Gallin, *Le Service radiotélégraphique de l'Indochine de sa création [1909] à la fin de 1930* [Hanoi, 1931])

ception came in 1926, transmission in 1928. By the late 1920s Saigon was in radio communication with most parts of France's colonial empire and with stations around the Pacific basin. Supervising the operation of more than fifty European staff was a military engineer, Commandant L. Gallin.[38]

Colonial radiotelegraphy, basking in its high profile, provided the venue for original research in connection with the solar eclipse of 1929. In honor of the celestial event, General Ferrié sent two engineers to accompany France's astronomical expedition to the island of Poulo Condore, off the coast of Vietnam. Their task was to study the earth's electromagnetical field by shortwave, carry out radiogoniometry in col-

38. Lucien Gallois, "Inauguration d'une station de T.S.F. à grande puissance à Saigon," *Annales de géographie* 33 (1924), 188–90; L. Gallin, "Renseignements statistiques sur le développement des communications radiotélégraphiques en Indochine," *BEI* 32 (1929), 369–99; idem, *Le Service radiotélégraphique de l'Indochine de sa création (1909) à la fin de 1930* (Hanoi, 1931); Conseil de recherches scientifiques de l'Indochine, *Les recherches scientifiques en Indochine* (Hanoi, 1929) (hereafter cited by title only), p. 7. By 1921 Edmond Rothé was urging that a seismological observatory be set up in Saigon near the new radio station (PO, carton 3, II-B–1, Comité français de géodésie et de géophysique, minutes of general meeting of 13 Apr. 1921). For the staff, see *Le Conseil des recherches*, p. 26.

laboration with one of Gallin's stations, and measure the changing height of the Heaviside layer. They followed Edward Victor Appleton in varying a broadcast signal to detect the interference between the emitting signal and the one reflected off an atmospheric layer. Appleton had measured the Heaviside layer during the total solar eclipse of 1927, and Gallin wanted to compare the values in Indochina with those obtained by Appleton in England. Gallin also wanted to follow up on measurements of radio-wave echoes — observed by Jörgen Hals, Carl Störmer, and Balthazar van der Pol — which had been attributed to the existence of interplanetary matter. Gallin envisaged his service at the center of the technological and scientific life of Indochina.[39]

Gallin was not the only scientific entrepreneur eager to exploit technologies deriving from modern physics. In 1924 a radium institute came to Indochina. Its proprietor and guiding light was surgeon and longtime professor at the Hanoi medical school, Adrien-Charles Le Roy des Barres; its aim was to combat the peril of cancer, which according to Le Roy menaced Indochina just as it threatened France.[40] At a time when other Asian hospitals created radiological services with particular emphasis on diagnosis and prevention of local scourges such as tuberculosis, French Indochina focused on an affliction that had seized the imagination of metropolitan medical men.[41] Le Roy found shelter for his private institution in the Hôpital indigène at Hanoi while he waited for a building of his own. He generated X-rays by a Coolidge tube and a Gaiffe commutator, but the crown jewels consisted of 437 mg of radium, valued at some forty thousand piastres and divided into more than a hundred needles and tubes. By 1928 Le Roy had become head of a government-funded institute for cancer treatment.[42]

39. L. Gallin, "Les expériences de radioélectricité prévues à l'occasion de l'éclipse solaire du 9 mai 1929," *Bulletin général de l'instruction publique* (Indochina), 8 (1929), 136–42; Carl Störmer, "Sur un echo d'ondes électromagnétiques courtes arrivant plusieurs secondes après le signal émis; et son explication d'après la théorie des aurores boréales," *CRAS* 187 (1928), 811–12, where Störmer explains his radio echoes with the hypothesis of a toroid of charged particles confined by the earth's magnetical field. Henri Deslandres comments in the following page on the "exceptional importance" of Störmer's result.

40. Le Roy des Barres, *Etudes de pathologie chirurgicale exotique* (Paris, 1912).

41. On radiology in the Dutch East Indies during the early twentieth century, see *ER*, pp. 166–71.

42. The various "state" governments of Indochina and the general colonial secretariat provided a large part of personnel and administrative costs, and they gave land for a permanent institute building, well under way by 1927 ("L'Institut du radium à Hanoï," *L'éveil économique de l'Indochine*, 10 July 1927, pp. 10–13. See also "L'Institut Curie de l'Indochine," *L'Asie française*, no. 230, 1925, pp. 118–19; *Le Conseil des recherches*, p. 37; and *Recherches scientifiques en Indochine*, pp. 12–13). See Paul Rønne and Arnold B. W. Nielsen, *Development of the Ion X-Ray Tube* (Copenhagen, 1986) [Copenhagen, University Library, Acta Historica Scientiarum Naturalium et Medicinalium, 35], pp. 260–61, for a

The radium institute's man with practical experience in radiology, Paul-Charles-Mathias Heymann, had chosen a career in the naval health service. At age twenty-four, he completed a medical doctorate in thermotherapy at Bordeaux with Jean Alban Bergonié, one of many radiological pioneers who died from radiation-induced cancer.[43] Heymann's naval superiors sent him to Indochina, where he served as a radiologist at Hanoi before the war, doubling as a lecturer in medicine and medical physics at the Hanoi medical school. For more than a decade Heymann used radiation therapy on tropical illnesses, including liver abscesses and malaria. Le Roy promised Heymann better research facilities.[44]

According to the annual shareholders' meeting of the private clinic, Le Roy's building came on line in 1928. In 1927–28 the clinic treated 8 Europeans and 223 indigenes for various cancers; 7 of the European cases went into remission, and 40 cures were effected among the indigenes — the difference being ascribed to the unwillingness of Indochinese to put up with a complete radiological regimen. The cure was, if not worse than the affliction, dangerous to the health of the medical

---

Chabaud-Villard valve tube offered for sale by the firm of Ladislas Adolphe Gaiffe of Paris in 1909. The well-known Coolidge tubes had an incandescent cathode in a hard vacuum.

43. Heymann, *Contribution pratique et clinique aux applications de la thermothérapie localisée* (med. diss., University of Bordeaux, 1908), biographical information from the title page, dedications, and preface; Heymann served as Bergonié's préparateur. Heymann had attended the school of the naval health service in Bordeaux, which was affiliated with the university medical school (Philippe Masson, "L'Ecole de santé navale et coloniale de Bordeaux," in *Histoire des médecins et pharmaciens de marine et des colonies*, ed. Pierre Pluchon [Toulouse, 1985], pp. 163–84; Hermann Holthusen, Hans Meyer, and Werner Molineus, eds., *Ehrenbuch der Röntgenologen und Radiologen aller Nationen* [Munich, 1959], pp. 78–81, on Bergonié; and *Min.*, 1928, s.v. "Hanoi," for Heymann, erroneously listed with the first initial *W* [Heymann consistently gave his first initial as *P* in his articles dated from Hanoi]). Radiation therapy, including that deriving from infrared and ultraviolet light, enjoyed a vogue at the beginning of the twentieth century; Niels Ryberg Finsen won a Nobel prize in 1903 for phototherapy.

44. Heymann, "De l'examen radiologique du foie: Sa valeur dans les cas d'abcès," *Bulletin de la Société médico-chirurgicale de l'Indochine*, no. 8, Oct. 1913, abstracted in *Journal de radiologie et d'électrologie* 1 (1914–15), 99; "Traitement par les rayons X des ulcères phagédéniques tropicaux," *Annales d'hygiène et de médecine coloniales*, no. 1, Jan. — Mar. [no year given], pp. 87–96, abstracted in ibid., p. 358; "Du traitement par les rayons X des ulcères phagédéniques tropicaux," *Archives d'électricité médicale et de physiothérapie* 25 (1917), 375–82; "Abcès du foie et pneumopéritoine," ibid. 29 (1921), 196–200; "De la valeur du pneumopéritoine dans le diagnostic des abcès du foie," ibid. 31 (1923), 358–75; "Productions ostéophytiques de la colonne vertébrale," *Bulletins et mémoires de la Société de radiologie médicale de France* 13 (1925), 142–43; "Du traitement des rates paludéennes par la radiothérapie," *Journal de radiologie et d'électrologie* 9 (1925), 480 (abstract), appearing *in extenso* in ibid. 10 (1926), 481–92.

staff: one male nurse died of radiation sickness, and another was hospitalized because of it. It is not unreasonable to blame the occupational hazard on accelerating use of X-rays: in 1928–29, 157 indigenes were treated by radium and 69 by X-rays; in 1929–30, X-rays were trained on about half of the 192 indigenes who received treatment. Le Roy's operation thrived. By 1929 the shareholders' report indicated that the clinic had received a grant of forty thousand piastres from the Indochinese government and obtained permission for a fundraising lottery; in 1930 the government guaranteed forty thousand piastres annually for five years. About that year, Le Roy set up a branch operation in Saigon under Heymann, who had gone south to direct the Cholon Hospital.[45] Le Roy encouraged his radiologists to join his longstanding investigation of diseases of the sexual organs, especially the penis and scrotum. The group broadcast their techniques both locally and internationally. Le Roy's research catapulted him to the directorship of the Hanoi medical school.[46]

Benevolent nods by an engineer and a surgeon in the direction of scientific research called for direct supervision by the colonial bureaucracy, especially in the wake of the hugely successful Pan-Pacific Science Congress, held at Tokyo in 1926. To provide a sense of proportion and control, in 1928 the colonial government instituted a Scientific Research Council for Indochina. It was an omnibus gathering, informally overseen by the Paris academies of sciences and colonial sciences, with a representative from every learned institution in the colony; permanent members were encouraged to consider themselves as belonging to a learned academy. The council would provide advice to the colonial government and represent the colony in the world of science.[47]

The Scientific Research Council also provided entree for Vietnam-

45. Hanoi, Institut du radium de l'Indochine, *Procès-Verbal de l'assemblée générale ordinaire du 27 octobre 1928; Procès-Verbal . . . du 26 octobre 1929; Procès-Verbal . . . 1930*, all presumably printed in Hanoi and all available in PBN.

46. Much of Le Roy's textbook, *Pathologie chirurgicale*, is devoted to the male sexual organ. See also Le Roy des Barres, Heymann, and Bablet, "Le cancer de la verge au Tonkin," *Journal de radiologie et d'électrologie* 11 (1927), 89–100; idem, "Radiumpuncture et appareils externes dans le cancer de la verge," ibid. 12 (1928), 364–71; Le Roy des Barres and Heymann, "Le traitement des cancers de la verge," Far Eastern Association of Tropical Medicine, *Transactions of Congresses: Calcutta 1927* 1 (1928), 251–66; Le Roy des Barres, Heymann, and Bablet, "Radium Puncture vs. External Appliances in Cancer of the Penis," *American Journal of Physical Therapy* 5 (1929), 447–50; Le Roy des Barres, *Le cancer de la verge* (Hanoi, 1931), published for the International Colonial Exhibition of that year, providing a bibliography with other publications by Le Roy's group on radium therapy applied to cancer of the penis. The symbolic overtones here are too obvious to merit discussion. For Le Roy as director of the medical school, see *Annales de l'Université de Hanoi* 1 (1933), 75.

47. *Le Conseil des recherches.*

ese who wanted to carry out research in physics. The first elected Vietnamese member of the council, in fact, traveled to France late in 1937 to attend two scientific meetings. The aspiring scientist, Nguyên-công Tiêu, observed France's geophysicists assembled for the scientific congress of that year devoted to overseas research. Tiêu wrote to physicist Aimé Auguste Cotton, in Cotton's capacity as a member of the Academy of Sciences, that he very much needed to consult with a senior man about a radiometer that he had designed to measure solar radiation. "I love scientific research," Tiêu emphasized. "I am in a country where I have no guide. I would be discouraged if, in France, all the scientists' doors, in front of which I wait impatiently after having made a sea voyage of thirty-five days, were hermetically closed." He thanked Cotton for one audience and asked for another. He was just "a humble Indochinese, avid for French science," who had "great need of your enlightenment to illuminate my path in my present and future research." Tiêu, a graduate of the Ecole supérieure d'agriculture et de sylviculture, worked in the agricultural service at Hanoi.[48] He returned to Indochina with an improved instrument, courtesy of Charles Maurain's intervention.[49]

Geophysics in Southeast Asia

During and after nineteenth-century wars of conquest in Southeast Asia, the French navy sent astronomically adept hydrographers to chart the new colony and determine its place on the globe. In this way, the sky and the coast of Indochina received regular attention from the invaders. Naval hydrographer Philippe Eugène Hatt, who served in Saigon during the late 1860s, observed the solar eclipse of 1867 in the Straits of Malacca; he later rose to the heights of the Academy of Sciences. In his wake in 1870 came the naval physician Alfred-Emile Borius, fresh from a year on Madagascar as a medical climatologist. Borius spent two years at Saigon before being transferred to Dakar and St-Louis in Senegal. With climatological monographs under his belt, he joined the French expeditionary corps to Tonkin in 1883. His task, late in December 1884, was to set up a meteorological service for Indochina; Borius met death at Haiphong in 1885. Not to be outdone by the navy, the occupying army created a topographical bureau in 1886. The bureau made itself the beneficiary of maps drawn by various exploring

48. *Congrès*, p. 157, where the man's name is given as Tien; PIPG, Tiêu to Cotton, 16 Sept. 1937; *Le Conseil des recherches*, p. 17.

49. Cotton transmitted Tiêu's query to Maurain, who commented on it (PIPG, Cotton to Maurain, 18 Sept. 1937, and subsequent postscript).

Structure erected at Saigon for observing the passage of Venus, 1874 (From G. Héraud, *Rapport sur l'observation du passage de Vénus à Saïgon* [Paris, Académie des sciences, *Recueil de mémoires, rapports et documents relatifs à l'observation du passage de Vénus sur le soleil du 9 décembre 1874* 2, pt. 2, no. 5])

and boundary expeditions, such as Auguste-Jean-Marie Pavie's from 1888 to 1895.[50]

Permanent institutions in the exact sciences appeared in Indochina at the turn of the century. In 1897 the Marseilles professor of geology and mineralogy Gaston Vasseur set up a colonial geological service.[51] A geographical service, independent of the army's surveying corps, began in 1898–99 under Lt. Col. Jules Clément Ladislas Lubanski. Deciding that a program of both astronomical and geodesical measurement would be a waste of time, Lubanski concentrated simply on a network of meridian and parallel links determined geodesically.[52] Over the next

50. VM, s.v. "Hatt" and "Borius"; Georges Maspero, ed., *L'Indochine: Un empire colonial français*, 2 vols. (Paris, 1929–30), 2:98–99, for the late nineteenth-century military mapmakers; International Colonial Exhibition, Paris 1931, Gouvernement général de l'Indochine, *Service géographique de l'Indochine: Son organisation, ses méthodes, ses travaux* (Hanoi, 1931) (hereafter cited by title only), pp. 8–9.

51. Maspero, *Indochine*, 2:124.

52. *Service géographique de l'Indochine*, p. 10, for the restriction. But see Aix, Indochine 17538, note on the creation of the geographical service, 5 July 1899, where geodesy and

forty years the service produced maps for the colony and little else.[53] General Georges Perrier, one of the army's most distinguished geodesists, wrote in 1938:

> What we have done in Indochina, where no scientific research has been carried out by the geographical service of the colony, contrasts singularly with scientific activity carried out by the "Survey of India," which for at least a century and a half has carried out quite lovely work; the latter's collection of publications shames the almost nonexistent publications of the geographical service of Indochina, which restricts itself to a small, summary, annual report.[54]

With ambitions greater than those of the geologists or the geographers, colonial physicists transformed a primitive meteorological network of six main stations and twelve secondary stations into a central meteorological service.[55] The organization of meteorology in Indochina followed the instructions of Eleuthère Mascart, head of the Central Meteorological Bureau in Paris, and Louis Froc, director of the Zikawei Observatory in Shanghai. After surveying the Indochinese coast, Froc identified the hill of Phu-liên, some nine kilometers from Haiphong, as the appropriate site for an observatory.[56]

The French administration in Indochina wanted a meteorological observatory to serve the practical needs of the colony, but Governor General Paul Doumer realized that metropolitan financing could be arranged for a more ambitious undertaking that included geophysics and astronomy. To this end he approached the principal astronomical maecenas of late nineteenth-century France, Raphaël Bischoffsheim, who had handsomely endowed the Nice Observatory and who was providing substantial support to French astronomical enterprises from China to Ecuador. Doumer wrote to Bischoffsheim in 1899 that funding had been secured for the observatory building. Bischoffsheim promised to raise the money for the observatory's instruments. After

---

astronomy constituted one of the service's three founding sections. After he was no longer in charge of the Indochinese geographical service, Lubanski published a manual of practical astronomy for surveyors: *Instruction pratique d'astronomie de campagne* (Paris, 1903). In his sketches about life in Indochina, *Tonkinades* (Paris, 1902), written under the pseudonym Jean Star, Lubanski is silent about his geographical work.

53. PSG, colis 41/4125, Lt. Col. Solichon, "Les travaux du Service géographique de l'Indochine depuis la guerre de 1914–1918."

54. Perrier, "Congrès de la recherche scientifique dans les territoires d'outre-mer, première section, rapport définitif," *ACRACS* 14 (1938), 53–58, on p. 55.

55. VM, s.v. "Fitte," Minister of Colonies to Minister of the Navy, 13 Apr. 1899, for the meteorological network; Maspero, *Indochine*, pp. 99–100, 123–25.

56. Maspero, *Indochine*, pp. 123–24.

five months Doumer asked Bischoffsheim for ten thousand francs to cover meteorological and magnetical instruments only. The governor general sought to replicate in Indochina what the Jesuits had done at Zikawei, and following this vision he had Froc help design the observatory buildings. The cost of construction came to around forty-seven thousand piastres.[57]

As Indochina was the creation (although not the domain) of the navy, the colonial regime first thought of naming a naval officer to direct the new observatory. Jean Alexandre Justin Paul Eugène Fitte, a thirty-five-year-old lieutenant with fifteen years' service in Indochinese waters, fit the bill. But the navy would not allow Fitte to become director while on active service – it would have meant paying for something "not having an essentially maritime character" – and Fitte did not yet want to sacrifice a naval career by going over to the colonial ledger.[58] Doumer settled on C. J. B. Ferra to direct the observatory, which was placed under the colonial civil affairs portfolio.[59]

The new man held a licence ès sciences and had been employed in the Paris Observatory and the Faculty of Sciences. He traveled to Zikawei for instructions from Froc and then returned to set his house in order. According to the budget of 1903, Ferra's European staff cost Indochina some thirty-seven thousand francs. This sum went to support himself, a meteorologist, and an assistant. He wanted to add an officer in the Messageries maritimes to take charge of the time service and a licencié ès sciences from Poitiers to supervise the magnetical measurements. His meteorological service was unable to send reports to Zikawei for the Jesuits' daily weather bulletin. Ferra had his observatory, but little to put in it. He steamed off to France to press Bischoffsheim for the astronomical hardware that had been promised.[60]

57. Aix, Indochine 4472, Doumer to Bischoffsheim, 2 Apr. 1899; Doumer to Bischoffsheim, 4 Sept. 1899 (telegram); Alfred Angot to Doumer, 6 Sept. 1899; Doumer to Mascart, 20 May 1900. Ibid., Indochine 4469, *Cahier des charges pour la construction d'un observatoire à Phu-Lien* ([Hanoi?] 1900).

58. VM, s.v. "Fitte"; ibid., Minister of Colonies to Minister of the Navy, 13 Apr. 1899, and Fitte to Director of Underwater Defenses at Rochefort, 16 May 1899. Fitte retired from the navy in 1905 as a capitaine de frégate and accepted a post in the Indochinese civil service.

59. Aix, Indochine 4470, "Dépenses de l'observatoire météorologique en 1902," for the dependency. The archival documents substantially modify retrospective accounts of the observatory's origin, e.g., in Indochina, Section des services d'intérêt social, *Le Service météorologique et l'Observatoire central de l'Indochine* (Hanoi, 1931), p. 7, where the colonial director of agriculture and commerce is given credit for it.

60. Aix, Indochine 4502, Governor General to Minister of Colonies, 16 Sept. 1909, for Ferra's scientific experience; Indochine 4493, Ferra to Secretary General of Indochina, 9 Feb. 1903; Indochine 4473, Report by Ferra, 15 Aug. 1903; Indochine 4476, Governor

Ferra directed the institution with an eye on his approaching retirement. Early in 1905 he signaled his intention to step down when he had given twenty years' service to the colony, and to this end he asked the governor general to call in a first-class meteorologist who could be groomed as a successor. The man he supported was thirty-nine-year-old Georges Le Cadet, then *astronome-adjoint* at the Lyons astronomical observatory in St-Genis-Laval. Ferra believed that Le Cadet had the moral authority in the world of research to bring Phu-liên to world attention.[61]

A beneficiary of Bischoffsheim's generosity, the Lyons Observatory was a reservoir of talent for overseas positions. (We shall see that it provided astronomers for Algiers, Santiago de Chile, and Quito.) Georges Henri Pierre *dit* Le Cadet was one of the Lyons alumni who represented France in distant corners of the world. His father was a servant, his mother a chambermaid. He attended the Ecole de la Martinière, a technical school in Lyons. At the age of fifteen Le Cadet was introduced to the director of the Lyons Observatory, Charles André, who employed him as a préparateur. Over the next twenty years Le Cadet worked his way up the academic hierarchy at Lyons, from the baccalauréat to (in 1898) the doctorate for a thesis on atmospheric electricity. He grew passionately attached to balloon ascents, which provided him with scientific data. By the time he presented himself as Ferra's understudy, Le Cadet claimed to know Mandarin Chinese.[62] Le Cadet's appointment as a colonial meteorologist, which followed directly from his meeting Ferra at Chamonix, came in 1906.[63]

According to a pattern typical in colonial institutions, where men at the top had the opportunity of extended leave in the metropolis, Director Ferra invested Le Cadet as interim chief and then departed for France. For three years Le Cadet directed the observatory, waiting for Ferra's retirement or transfer. Le Cadet did what was required, observing the solar eclipse of 1907 with his 13-cm Dallmeyer equatorial tele-

---

General Beau to Director General of Post and Telegraph, 23 Oct. 1903, citing extracts of a letter by Froc; Indochine 4472, Ferra to Governor General, 3 Feb. 1903, where Ferra asks for leave to canvass Bischoffsheim in France.

61. Ibid., Indochine 4797, Ferra to Governor General, 16 Mar. 1905.

62. *Astron.*, s.v. "Le Cadet"; Joanny Philippe Lagrula, "Georges Le Cadet," *L'astronomie* 49 (1935), 385–89; Aix, Indochine 4497, Le Cadet to Governor General, 11 May 1905, where Le Cadet describes his career; Le Cadet, *Etude du champ électrique de l'atmosphère* (diss., University of Lyons, 1898); André, "L'observatoire universitaire de Lyon," Lyons, Académie, *Mémoires de la classe des sciences* 23 (1878–79), 53–71, p. 59, for Bischoffsheim's gift of Lyons's meridian circle.

63. Aix, Indochine 4497, Governor General, *arrêté*, 4 Apr. 1906; Lagrula, "Le Cadet," for Le Cadet and Ferra at Chamonix.

scope.[64] Ferra returned to Phu-liên, investigated the possibility of changing places with Fitte (who had retired from the navy to become the administrative head of the province of Quang-Yôn), and then ceded the observatory to Le Cadet, whose salary increased substantially.[65]

In 1908, when Ferra returned to Hanoi, there were six Europeans on the observatory's payroll. Ranking below Ferra and Le Cadet was Guillaume Le Lay, a meteorologist second-class, who had spent thirty years in the navy, the customs, and the colonial alcohol commission before joining the agricultural service as a meteorologist in 1898; his annual salary of Fr 9,750 would be raised to Fr 15,000 as a gesture anticipating his imminent retirement. Then came aide-météorologiste Paul Durand, who received Fr 8,000. Antonin Beljonne, a wayward clerk, garnered Fr 5,700, and a clerk-accountant received Fr 2,500.[66]

Le Cadet became permanent director about 1909 and continued in this capacity until 1927.[67] During his tenure he focused on rainfall and basic climatological data, establishing a network of recording stations throughout Indochina, and in 1924 he installed two horizontal Mainka seismometers.[68] But Le Cadet was neither a climatologist nor a geophysicist at heart, and he showed little interest in advising plantation owners and developers. His observations on tropical weather were derivative and qualitative. His most significant insight occurred in the years before the war, when he tried to explain why typhoons generally exhibited no electrical discharges; in his view, lightning derived from the motion of a vortex that was horizontal to the earth's surface, and typhoons generally presented vertical vortices.[69] Le Cadet failed to act

64. Aix, Indochine 4502, Alfred Angot to Minister of Colonies, 16 Sept. 1909, reporting on Le Cadet's effective directorship of the observatory since 1906, when Ferra went on leave while retaining the title of director; Le Cadet, *Observation de l'éclipse de soleil du 14 janvier 1907* (Hanoi, 1907), where Le Cadet gives his title as interim director.

65. *Min.*, 1909–10, gives Ferra as director of Phu-liên. See also Aix, Indochine 4500, Resident Superior at Tonkin to Governor General, 22 Sept. 1908, for the proposal to exchange Ferra for Fitte, the administrator of Quang-Yôn; and ibid., Indochine 4496, observatory budget for 1908, according to which Ferra earned twenty thousand francs as director, while Le Cadet earned fifteen thousand francs as a meteorologist first-class.

66. Aix, Indochine 4496, Ferra to Governor General, 4 Apr. 1908; "Relève des services de M. Le Lay"; budget for 1908. Ibid., Indochine 4492, J. Beljonne to Secretary General of Indochina, 6 Feb. 1903, for Ferra's sanctions against Antonin Beljonne for poor performance.

67. According to ibid., Indochine 51136, in 1927 a "Cadet, agrégé des sciences," is listed as the physics examiner for the P.C.N. at the medical school of the University of Hanoi.

68. Le Cadet, "Régime pluviométrique de l'Indochine," *BEI* 20 (1917), 1–50; *Le Conseil des recherches*, p. 23, for the seismological service.

69. Le Cadet, "Le climat du Delta du Tonkin," *BEI*, no. 92, 1911, pp. 757–76. See also his *Etude du champs électrique de l'atmosphère* (Paris, 1908) and "Sur l'origine des manifes-

The Phu-liên Observatory, near Haiphong. The extravagant tower, of little scientific utility, served as an indicator of French authority. (From Indochine française, Section des services d'intérêt social, *Le Service météorologique et l'Observatoire central de l'Indochine* [Hanoi, 1931])

on a commission from the Parisian physicist Jules Louis Gabriel Violle to report on atmospheric electricity.[70] With his effective retirement in 1925, the Phu-liên Observatory was reorganized under the longtime observer Paul Durand, who served as interim director. A separate section for climatology and meteorological botany came into being under thirty-five-year-old Paul Carton, an agronomist who had previously worked for the agricultural service and who had served as librarian of the short-lived Institut scientifique.[71]

---

tations électriques des orages, à l'occasion de l'observation des cyclones en mer de Chine," *CRAS* 153 (1911), 985–86.

70. PO, carton 3, II-B–1, Comité français de géodésie et de géophysique, minutes of general meeting of 13 Apr. 1921. Violle also notes that Le Cadet did not report on research for which he had been awarded a grant by the Academy of Sciences.

71. Institut scientifique de l'Indochine, *Catalogue de la bibliothèque* (Hanoi, 1922), pp. v – vi, for Carton's pedigree; *Le Conseil des recherches*, pp. 22–23, for Durand's interim direction and the reorganization. One assumes that meteorologist Paul Durand was the interim chief, although it is possible that chief naval clerk J. Durand, who handled the administration of the oceanographical institute at Nha-trang beginning in 1928, is the man in question. See *L'Institut océanographique de l'Indochine*, pp. 89, 117, for J. Durand.

Le Cadet's permanent replacement, Del Marie Etienne Bruzon, brought new spirit, enterprise, and funding to the observatory. Bruzon, only thirty-two years old in 1926, arrived in Indochina eager to place Phu-liên on the scientific map by issuing regular publications.[72] His arrival coincided with the creation of an Indochinese Committee for Geodesy, Geophysics, and Astronomy, an ad hoc structure called forth by the 1926 Pan-Pacific Science Congress that prepared the way for creating the Scientific Research Council of Indochina two years later. The French National Committee for Geodesy and Geophysics extended warm greetings to the new colonial cousin, urging it to join the International Geodesical and Geophysical Union; since votes in the international union were apportioned on the basis of the populations of countries served by the member committees, Indochina would contribute greatly to French international influence. The way would be paved, furthermore, for having all the Indochinese nations under French protection—Annam, Cambodia, and Tonkin—join the International Research Council and its unions. This expansion did not take place, but anticipation of it brought metropolitan geophysicists to look with benevolence on France's Asian colony. Bruzon, as a member of the colonial geophysical committee, received favorable exposure.[73]

In the late 1920s Bruzon doubled the number of Indochina's climatological recording stations. He encouraged Carton and his météorologistes-adjoints, Goux and Bernard Guilmet, to publish regularly and rapidly.[74] Bruzon issued a monthly bulletin and yearly *Annales*. Carton published a series of descriptive climatological studies and tried to relate his data to Władysław Gorczyński's broad generalizations about "continentalism."[75] Bruzon assisted the Jesuit observatories at Zikawei

72. Bruzon is cited as director at Phu-liên in the observatory's *Bulletin pluviométrique, année 1925* (1926).

73. PO, carton 3, II-B–1, Comité national de géodésie et géophysique, section de géodésie, draft of minutes for meetings of 21 Apr. 1926 and 8 June 1926.

74. For Bruzon's aspirations and his staff, see *Annales du Service météorologique de l'Indochine, année 1928* (1930), masthead and preface.

75. Carton, *La météorologie et ses applications dans les pays tropicaux* (Hanoi, 1930); Bruzon and Carton, *Le climat de l'Indochine et les typhons de la mer de Chine* (Hanoi, 1930); Henry G. S. Morin and Carton, "De l'influence des facteurs climatiques sur la répartition de l'endémie palustre en Indochine," Far Eastern Association of Tropical Medicine, Nanking, Ninth Congress (1934), *Transactions*, 2:145–58, pp. 149–50 for the appeal to Gorczyński; Carton, *Le climat de l'Indochine* ([Hanoi?] 1934); idem, *Nouvelle classification des climats; applications à l'Indochine* (Hanoi, 1934); P. Chevey and Carton, "Les courants de la mer de Chine méridionale et leurs rapports avec le climat de l'Indochine," *Notes et mémoires de l'Institut océanographique de l'Indochine*, no. 26, 1926; Carton, "Le degré de continentalisme de Gorczyński et son application dans la climatologie de l'Indochine," reprint from *Comptes-Rendus des travaux de l'Institut des recherches agronomiques en 1932–33* (Hanoi, 1934); idem, *Considérations sur l'emploi de formules utilisant les valeurs des extrêmes*

and Tananarive in the French project to remeasure global longitudes in 1933, but he knew that advancement in the new colonial meteorological service, to which he belonged, had nothing to do with astronomy, Jesuitical or otherwise; when France sent out a team to observe the 1929 solar eclipse on the island of Poulo Condore, Bruzon elected not to join them.[76] Bruzon assiduously measured solar radiation following the standards and instructions of Charles Maurain and his associates.[77] He cultivated the patronage of Henry Hubert, who used data from Indochina to advance general propositions about tropical weather.[78] Bruzon published three articles in the *Annales de physique du globe de la France d'Outre-Mer*, a creation of Hubert's, received the Tchihatchef prize from the Academy of Sciences in 1937, and then, in 1938, succeeded to Hubert's position as head of colonial meteorology.[79]

When Bruzon left for Paris, his observatory was a complex union of civilians and military men with broad responsibility for weather forecasting and climatology.[80] In practice, however, the observatory had become "almost completely absorbed by the protection of air navigation and sea navigation," with the largest part given over to the hourly bulletins needed by military and civilian air transport.[81] The 1930s saw a major turn away from agriculture – the focus of the reorganized obser-

---

pour obtenir les moyennes approximatives de température et d'humidité relative (Hanoi, 1934); idem, *Le climat de l'homme* (Hanoi, 1935).

76. F. Brachet, "L'éclipse de soleil du 9 mai 1929," *Bulletin général de l'Instruction publique* (Indochina) 8 (1929), 77–84. Bruzon resented the Jesuit monopoly on weather forecasting for East Asia. He disputed with Ernesto Gherzi, chief of forecasting at Zikawei, about standards for transmitting data by radio (SMB, Bruzon to Gherzi, 17 May 1928 and 3 Aug. 1928).

77. PIPG, Maurain to unknown, 1930: "I am returning the documents that you so kindly sent me. They show that the meteorological service of Indochina brings the greatest care to its actinometrical measurements, which are in fact rather delicate if some precision is desired." Bruzon had asked for instructions from Charles-Emile Brazier (director of the geophysical observatory in the Parc St-Maur) and F. Volochine (the Czech director of the International Actinometrical Laboratory at Trappes), which Maurain found entirely correct, and "there was nothing" in Bruzon's request that "might shock" anyone. On Volochine's operation, funded by Czechoslovakia, see SMB, Emile Delcambre, "Note sur l'installation et le programme du laboratoire actinométrique de Trappes (S & O)," 1929.

78. See, e.g., Hubert, "Sur la circulation générale de l'air atmosphérique au-dessus de l'Indochine," *CRAS* 198 (1934), 1055–57.

79. See *APGFOM*, 1939, p. 30, for Bruzon's promotion; on his modest independent scientific work, see "Note sur le Service météorologique colonial français dans le Pacifique," in Pacific Science Congress, Sixth, Berkeley, *Proceedings*, 6 vols. (Berkeley, 1940–43), 3:751–54, on p. 753.

80. SMB, Ernesto Gherzi to Pierre Lejay, 4 Jan. 1937, where Gherzi comments on the Phu-liên Observatory.

81. "Note sur le Service météorologique colonial," p. 751.

vatory in the 1920s — as prices and production declined during the economic depression. Bruzon's assistant Guilmet, whose professional training had been in agriculture, accordingly returned to France for more meteorological training under Charles Maurain.[82] The meteorologists, in fact, looked to agronomy for future staff. An agronomist, Charles Cloix, set off to pass a year at Phu-liên in 1935 before going to Saigon as meteorological liaison to agronomists and "above all to organize a service for meteorological forecasting required by the development of aviation in the Far East." Cloix anticipated setting up a geophysical station in Saigon. To this end he planned to spend his three-year colonial tour of duty studying for a second licence in mathematics, physics, and geophysics.[83] Arriving at Phu-liên, Cloix reported to Maurain on the realities of colonial geophysics:

> The government of Indochina wants to support a scientific establishment only to the extent that this establishment can provide practical services, and immediately so. The crisis, with the budgetary restrictions that it imposes, has further accentuated this tendency. The meteorological service conforms to this law, the same as the other technical services. Its activity is almost entirely absorbed by weather forecasting for aviation and the navy, current climatological observations, and operating a seismological station at the central observatory. The other branches of geophysics are almost completely abandoned, lacking personnel and funding.

Cloix regretted that under these circumstances he could not think of carrying out the observational work required for an undergraduate certificate in geophysics.[84] Such restrictions on the possibilities of carrying out independent geophysical research at the level of the licence explain why no physicist assigned to Indochina transformed his livelihood into a doctorate. Cloix, at least, remained associated with the Phu-liên meteorologists for several years.[85]

What Charles Cloix could not think of accomplishing in Indochina, an Indochinese student succeeded in finishing at Paris. Hoang ngoc Cang received a licence ès sciences from the Sorbonne in 1938 for certificates in physics and chemistry. Called home by the illness of his mother, Hoang applied for a meteorological post at Phu-liên. Follow-

82. D. G. E. Hall, ed., *A History of South-East Asia* (London, 1981), p. 829; PIPG, Guilmet to Maurain, 13 Nov. 1933, where Guilmet signs himself as an *ingénieur-agronome*.

83. PIPG, Cloix to Maurain, 29 Apr. 1935.

84. Ibid., Cloix to Maurain, 10 Sept. 1935.

85. See *Annales du Service météorologique de l'Indochine, année 1937* (1939), where Cloix is listed as a staff member.

ing a recommendation by Charles Maurain, he was confident of receiving the appointment.[86] What happened is unclear; the Indochinese aspirant does not appear on the observatory's masthead in the next yearly reports. He would, in any event, have found himself among other qualified, Indochinese candidates. By 1937 there were indeed two Indochinese météorologistes-adjoints at Phu-liên, Nguyen Xien and Dô-dinh Cuong.[87]

When Bruzon left the Phu-liên Observatory to direct the colonial meteorological service in Paris, Guilmet took charge until André Romer arrived as a permanent replacement in 1939. Like Bruzon before him, Romer was a career meteorologist in Hubert's colonial corps. He had parlayed direction of the Martinique geophysical observatory into the Phu-liên post; as lagniappe, he received the Gay prize of the Paris Academy of Sciences.[88]

The imperial seat cascaded rewards on loyal scientist administrators in Indochina. Metropolitan scientific centers were more circumspect. Research in colonial Asia to the north and to the south overshadowed the observations that French physicists managed to publish from Hanoi and Saigon. In the view of Alfred Lacroix, perpetual secretary of the Academy of Sciences, caprice of colonial resolve was responsible for the lack of solid accomplishments by colonial institutions. "The malaise that afflicts them, and that sometimes kills them, is the lack of continuity in the designs of higher-level administration." One governor general would formulate an ambitious policy and endow an institution, only to see his plans completely reversed by his successor. "All this is an almost inevitable result of the regime in our colonies, where the heads change too often" and where not all governors general had Paul Doumer's enlightened approach to science. The malaise could be cured by the tonic of metropolitan control. "The roots of these scientific institutions must be planted in stable soil, that is to say, in a permanent metropolitan form, sheltered from local reactions." This metropolitan patronage, already given to the Ecole française d'Extrême-Orient and to the Indochinese Pasteur institutes, would free scientists overseas from impecunious supervision by pettifogging colonial administrators. With security and stability, scientists would be able to devote all their

86. PIPG, Hoang ngoc Cang to Maurain, 14 Oct. 1938 and 16 Dec. 1938.

87. *Annales du Service météorologique de l'Indochine, année 1937* (1939), masthead. Cuong became an author: P. Cauvet-Duhamel and Dô-dinh Cuong, *Nouvelles tables psychométriques et de calcul de la tension de la vapeur d'eau* (Hanoi, 1941).

88. *APGFOM*, 1939, p. 30, for the prize, awarded "for the totality of his works regarding geophysics." But *CRAS* 207 (1938), 1285, has the prize awarded "for the organization and the starting up of the observatory on Martinique."

energies to their calling.[89] Some scientists may find comfort in receiving instructions formulated by senior colleagues ensconced in academical bowers half a world distant, but it is unclear whether the most direct path to new knowledge lies in such bondage.

Indochina, the most distant of France's true colonies, was a contrary stepdaughter in the world of exact sciences. Institutes emerged only to founder; courses of study materialized and then evaporated. The lack of a consistent colonial policy for higher learning certainly contributed to mediocre scientific performance, but we would be mistaken to identify either physical climate or administrative inconsistency as a first cause for scientific torpor. In later chapters we shall find French physicists situated in North Africa and China who undertook significant research projects. Before Henry Hubert's colonial meteorological service recruited physicists on the basis of merit, Indochina attracted adventurous men with encyclopedic, if superficial, interests. The ruthless pursuit of a clear research problem remained foreign to the physicists and astronomers who disembarked on the shores of the South China Sea.

89. Lacroix, "La recherche scientifique dans les territoires d'outre-mer," *ACRACS* 14 (1938), 1–13, on pp. 6–7.

4

Algeria: The Overseas Projection of Metropolitan Terrain

Practical Demands of Pure Learning

After France conquered the Algerian coast in 1830, occupying forces set out to explore the new land. Inventory belonged to the spirit of the post-Laplacian age. Enumeration and classification of a thing are not identical with possession of it, but knowing the particular form of an object is essential for turning it to practical use. The July Monarchy subtending a nation with considerable scientific prestige, the Algerian command came up with a scheme that would have the army share in a mission to civilize the new territory by taking its physical measure.

The governor Marshal Sylvain Charles Valée proposed in 1837 to create a commission "charged with scientific research in our possessions in Africa." Following instructions from the academies of sciences and of inscriptions and belles-lettres, its special mission would treat "topography, geography, botany, mineralogy, archaeology, and architecture." Because the commission had "as its aim mainly the progress of science," Valée expected that ministerial colleagues in public instruction and commerce would help with the costs.[1] The military authorities in Paris preferred to focus on useful results. The minister of war complimented his subordinate in Algiers on the proposal, adding the clear proviso that the work accomplished by such a commission be practical rather than artistic or literary.[2]

Behind the Algerian scientific commission lay the Egyptian Institute of Napoleon I. Charles Coulson Gillispie observes: "The Napoleonic occupation of Egypt may be considered the first instance of nineteenth-century imperialism in that it comprised a cultural component lacking in the mercantile colonialism that preceded it. What the French later

1. Aix, F80/1590, Director of Algerian Affairs to Minister [of War], 14 Aug. 1837.
2. Ibid., Minister of War (Simon Bernard) to Valée, 14 Aug. 1838.

came to call their 'civilizing mission' had its origins partly in the Enlightenment and partly in revolutionary ideology." The Egyptian interlude, Gillispie continues, "marks the beginning of the spread of European science and its appurtenances to African and Asian societies under the aegis of military conquest and political power." The weight of the exotic Egyptian drama, however, lies more with the overseas presence of the central actor Napoleon than with the value or utility of its published record. The latest chronicler of Napoleon's Egyptian scientists, in fact, is uncertain whether "the Expedition was reciprocally dignified by the incorporation of science." As we shall see, this incorporation is a central cog in the mechanism of French cultural imperialism. Egypt is precedence, not prescience.[3]

The Algerian commission's members had been recommended by ministers who were to share in the cost. Public instruction named Henri Milne-Edwards as the naturalist; commerce proposed Auguste Daubrée, then a pupil at the Ecole des mines, as the geologist. Jean-Baptiste-Georges-Marie Bory de Saint-Vincent, a veteran of scientific expeditions to Australia and Morea (Greece) and a member of the Paris Academy, signed on as the expedition's leader. And as the man in charge of "sedentary observations" — meteorology and the like — the Academy of Sciences called Georges Aimé, a normalien who was then a teacher of mathematics at the newly formed Collège d'Alger.[4]

Aimé's father worked at the artillery school in Metz and had been assigned to Napoleon's scientific commission in Egypt; his uncle had attended the Ecole polytechnique and then rose to become a lieutenant colonel in the artillery. After lycée and a poor performance at the Ecole normale supérieure, Aimé began his research career in 1834, at age twenty-one, with an article on fluorine in the *Annales de chimie et de physique*. Licencié in 1834, he worked toward a doctorat ès sciences, which he obtained from Paris in 1837. He issued a series of punctual investigations in mechanics, chemistry, electricity, and magnetism, appearing in the *Annales* and also in the Paris Academy's *Comptes-Rendus*.

3. Gillispie, "Scientific Aspects of the French Egyptian Expedition, 1798–1801," *Proceedings of the American Philosophical Society* 133 (1989), 447–74, on pp. 473–74; the "exotic episode" of Egypt in *Histoire de la France coloniale*, 2 vols. (Paris, 1990), vol. 1 (until 1914) by Jean Meyer, Jean Tarrade, Annie Rey-Goldzeiguer, and Jacques Thobie, p. 297. The New World and Asia, of course, saw earlier colonial scientific academies.

4. See Aix, F80/1590, for the various correspondence, including Minister of War to Secretary of the Academy of Sciences, 22 Aug. 1838, about Aimé; it is unclear whether Daubrée joined the commission. See Jean Mélia, *Histoire de l'Université d'Alger* (Algiers, 1950), pp. 18–26, on the creation (in 1835) and early years of the Collège d'Alger; in 1848 the collège acquired the rank of lycée.

He despaired of finding a post in Paris that would allow him to continue scientific research, and in 1838 he accepted an appointment in Algiers.[5] With his assignment to Algiers, Aimé progressively oriented his research to the physics of the local environment.[6] By 1839 he had set up a meteorological observatory and was regularly sending memoirs to Paris. At the same time, he studied the temperature, currents, tides, and waves of the nearby sea. This ingenuous reporting was intended to keep his name before powerful patrons in Paris and in this way gain him a berth on the Algerian scientific commission. The appointment won, Aimé took leave to consult with his mentors in Paris.[7]

Aimé indicated his aspirations in a letter to the minister of war:

> Since my installation in the Collège d'Alger as a teacher of mathematics, I have brought in at my own cost some physics instruments from Paris, and these have allowed me to begin, around three weeks ago, a physics course for the pupils in the collège. These instruments and some others that I put together in Algiers have also allowed the means for establishing an observatory in the collège building, where I take observations six times daily.[8]

The physical measurements dragged on for years. Aimé's torrent of notes to the Paris Academy slowed to a trickle, and these concerned routine meteorological or magnetical data along with occasional forays into the physical properties of Algiers's harbor and neighboring waters.[9]

5. For Aimé's publications in the *Annales* see *Pogg.*, vol. 1. Aimé's experiments from 1834 and 1836, on the electromagnetical properties of steel and on electrically induced motion in fluids, found an admirer in the American physicist Joseph Henry (*The Papers of Joseph Henry*, vols. 1–5 ed. Nathan Reingold [Washington, D.C., 1972–85], 2:146; 4:62). Aimé's early notes in the *Comptes-Rendus*, frequently less than a page long and not infrequently nothing more than the announcement of a memoir received, may be retrieved from the annual indexes of the journal. See J. Thouet, "Georges Aimé," Monaco, *Bulletin de l'Institut océanographique*, no. 897, 1946, pp. 3–25 (reprinting a publication of 1898), p. 6 for Aimé's poor performance as a student; pp. 34–35 for Aimé's diplomas and credentials; p. 39; pp. 40–41 for letters from Provensal to Thouet, 1 Feb. 1899, and Hanriot to Thouet, 1 July 1896, on Aimé's family and his time in Paris.

6. He continued to work on chemistry, however, and he considered all natural history to be his realm (see "Nouveau composé provenant de l'iode," *CRAS* 4 [1837], 181; "Note sur les cotons cultivés en 1837, à la ferme Rahraya [Algérie]," ibid. 6 [1838], 500; Aimé's shipment of a mineral of sulphurated lead found in the neighborhood of Bouzaréah, in ibid. 7 [1838], 246; and "Corail à l'état fossile conservant encore une teinte rougeâtre," ibid., 903).

7. For Aimé's meteorological observations at his Algiers collège, for all of 1838, received, see ibid. 9 (1839), 87; for his appointment, ibid., 735; for his request to go to Paris, 10 (1840), 11; and for the request granted by the minister of war, 11 (1841), 181.

8. Aix, F80/1593, Aimé to Minister of War, 17 Mar. 1838.

9. Aimé, "Sur le mouvement des vagues," *CRAS* 9 (1839), 218–19; "Mémoire sur le mouvement des vagues," ibid., 409–10; "Courants de la Méditérranée," ibid., 608–9;

In 1842 Bory de Saint-Vincent apprised the minister of war about progress in Algerian physics. Observations in the interior were charged to an army officer, while Aimé remained on the coast. The physicists had to observe at times prescribed by Dominique François Jean Arago. Their progress was limited by difficulties in transporting the instruments safely. In Bory's residence, Aimé constructed a cabin of glass and copper where he could take magnetical data without disturbance. The young physicist passed many nights there taking delicate readings. Bory asked that Aimé continue as a member of the commission until the following year. This would cost the government nothing more than lamp oil and candles, since Aimé worked without remuneration. Bory requested, furthermore, that the instruments be left in place at the end of Aimé's tenure. Bory emphasized that Aimé had undertaken many different kinds of measurement — the depth of the sea, water currents, tides, and atmospheric variations — for which expensive matériel had been provided; the purchases would be rendered useless if Aimé had to end his research before firm conclusions were reached.[10]

When he had assembled the data, Aimé obtained leave to spend three years in France to publish his findings in two volumes, one for magnetical observations and one for various memoirs on geophysics. With the volumes in print, Aimé returned to Algeria, where in 1846 he died, the victim of an accident in Blida.[11] Had he lived (he fell to gangrene some time after the accident), geophysics in Algeria might have taken a very different direction. His memoirs inventoried the physical climate with ingenious inventions; in his last researches he built an instrument for measuring the motion of clouds, later known as a mirror nephoscope.[12] Aimé's geophysical data constitute the starting point for the exact sciences in French North Africa.[13] All of them derive from the ambitions of the ministry of war.

"Mémoire sur les variations du niveau de la Méditérranée, dans le port d'Alger," ibid., 701. See also ibid. 10 (1840), 11, for Aimé's meteorological data received; "Mémoire sur le magnétisme terrestre," ibid. 17 (1843), 1031, with Aimé's data over nineteen consecutive months having been received; "Température moyen d'Alger déduite des températures maxima et minima observées journellement," ibid. 14 (1842), 72–73; "Note sur les températures observées en Algéria," ibid. 20 (1843), 1101–3; "Observations sur l'eau de mer puisée à différentes profondeurs," ibid. 16 (1843), 749, where Aimé measures the amount of air dissolved at depths up to 1,600 m; and "Mémoire sur la compression des liquides," ibid., 1165–69, Aimé's longest and most theoretical publication with the academy.

10. Aix, F80/1593, Bory de Saint-Vincent to Minister of War, 5 Mar. 1842.

11. *CRAS* 23 (1846), 612, for the announcement.

12. W. E. Knowles Middleton, *Invention of the Meteorological Instruments* (Baltimore, 1969), pp. 268–69.

13. See Jules Rouch, "Un océanographe algérois: Georges Aimé," France, Comité des

The magnetical part of Aimé's mandate related to problems in pure science, particularly to the global surveys of terrestrial magnetism that had seized the imagination of Carl Friedrich Gauss, Alexander von Humboldt, the British admiralty, and finally Arago. The rest of his mandate derived from greed, as knowledge of climate and topography is essential for exploiting a new land.[14] Yet cultivation of exact sciences in Algeria also stemmed from fear, in this case from the fear of contracting disease.

Medical men accompanied France's Algerian armies, erecting hospitals and sermonizing about sickness.[15] In addition to dispensing quinine and vaccinating against smallpox, the learned doctors sought to ward off diseases whose origins were vaguely ascribed to bad air or miasmas. They accordingly studied how running water might interact with a fetid miasma, how trees might keep it away, and how a tight roof might provide a barrier against it. They recorded winds, temperature, humidity, and atmospheric pressure as essential medical data.[16] Casimir Broussais (son of the influential Restoration physician François Joseph Victor Broussais), for example, published a long study of the relationship between climate and sickness in Algeria generally; Eugène Grellois, who became president of the Meteorological Society of France, did the same for a particular Algerian town.[17]

---

travaux historiques et scientifiques, *Bulletin de la section de géographie*, 1954, pp. 161–65, for a concise evaluation of Aimé's scientific work. Thouet, "Aimé," provides a detailed summary of Aimé's geophysical papers.

14. Aimé indicates that he is conforming to the guidelines of the Royal Society of London in magnetical matters in his "Mémoire sur le magnétisme terrestre," *CRAS* 17 (1843), 1031. On the British magnetical research program, see John Cawood, "The Magnetic Crusade: Science and Politics in Early Victorian Britain," *Isis* 70 (1979), 493–518. On the geological explorers during the nineteenth century, see G. Bétier, "La continuité de l'oeuvre scientifique française au Sahara," in *Regards sur l'Algérie* (Paris, 1956), pp. 187–90; and *GFOM*, pp. 11–15.

15. Yvonne Turin, *Affrontements culturels dans l'Algérie coloniale: Ecoles, médecines, religion, 1830–1880* (Paris, 1971), pp. 335–36; Paule Laberge, *Politiques scientifiques au Maghreb: L'implantation du système scientifique dans les sociétés maghrébines de 1830 à 1980* (diss., University of Montreal, 1987), pp. 135–38.

16. Malaria was the principal object of attention (see Turin, *Affrontements culturels*, pp. 400–405; J-B. Vincent, *Les médecins de l'Algérie au temps de la conquête* [diss., University of Algiers, 1914], esp. pp. 64–72, 85–90; and Jean Tremsal, *Un siècle de médecine coloniale française en Algérie* [Tunis, 1929], p. 16).

17. Broussais, "Notice sur le climat et les maladies de l'Algérie," *Recueil de mémoires de médecine, de chirurgie et de pharmacie militaires* 60 (1846), 1–134; Grellois, "Esquisse sur la topographie médicale d'Hammam-Meskhoutin," ibid., 347–61. On the Broussaises, see Jacques Léonard, *La médecine entre les savoirs et les pouvoirs: Histoire intellectuelle et politique de la médecine française au XIXe siècle* (Paris, 1981), p. 129; on Grellois, Jacques Damiens, "La météorologie au XIXe siècle," *La météorologie*, no. 17, 1987, pp. 18–27, on p. 25.

By the middle nineteenth century, French medical men had been won over to the power of statistics, precision, and experimentation. It followed naturally that eleven years after the creation, in 1857, of the School of Medicine and Pharmacy of Algiers, the physicians there asked for a school of sciences. In 1877 the head of education in Algeria proposed a literary and scientific institute that could serve as the nucleus of a university. It remained to Paul Bert, university physiologist and anticlerical minister of public instruction under Léon Gambetta, to table a proposal in the Chamber of Deputies for organizing higher education in Algeria. Bert's proposal had little to do with the advancement of learning. In his view, knowledge was to serve the practical ends of the colonizer, and he resisted creation of a true Algerian university. In 1879 the Senate approved new preparatory schools of law, sciences, and letters – all in the realm of higher education but none privileged to be part of a "university." [18]

The Algiers School of Sciences had as its mission, an early administrator wrote in 1880, the study and elucidation "of the riches hidden within Algeria, of which the surface has only as yet been grazed." [19] The idea was to produce technicians and practical men. As the head of the school wrote in 1900, the "disinterested" aim of the school had only the prosperity of Algeria in mind.[20] The reason for not creating an engineering school followed from the logic of higher education in the metropolis. State engineering schools destined graduates for state engineering corps – an implicit promise of employment. The state, having no desire to certify Algerian engineers in competition with metropolitan ones, preferred to generate local technical competence with a minimum of attractions or promises. To this end it gave courses leading to a licence, following a pattern common to minor, provincial universities in the hexagone. In 1886–87, for example, twenty-one students were registered in the school, and fifteen of these proposed to take a licence; the number of registered students rose to thirty in 1895.[21] In 1909 the various Algiers schools received the rank of university faculties without, however, enjoying any of the legal or financial independence of metropolitan faculties; the *university* alone was a legal corporation, and the university would apportion funds to the faculties as it saw fit.[22] Under these circumstances, the exact sciences received little encouragement from university authorities. Enterprising astronomers and physi-

18. Mélia, *Histoire*, pp. 39–57.
19. Ibid., p. 150.
20. Ibid., p. 151, paraphrasing Professor Louis Charles Thomas.
21. Ibid., pp. 63, 65, for statistics and comparison with metropolitan universities.
22. Ibid., p. 105.

cists nevertheless brought Algeria to the attention of the scientific world.

Charting Heaven's Vault

Astronomy was the principal ornament in the crown of higher education at Algiers. The observatory began as early as 1858 in the form of a time service for mariners, although its first director, Charles Bulard, having worked at the Paris Observatory, had higher ambitions. He wrote to Marseilles astronomer Benjamin Valz in 1862 that he was counting on the intervention of the emperor to speed construction of his observatory. Until then, he was working with a Foucault telescope of 50-cm aperture to photograph the sun and moon. "The Algiers Observatory is to be an observatory where physical astronomy, in particular, will be done. . . . The study of nebulas, double stars, the physical constitution of stars, etc., . . . asteroids, comets." Bulard was at the time concentrating on asteroids because of the popular interest evoked by such work.[23] Ten years later he summarized his research for Charles-Eugène Delaunay, the director of the Paris Observatory. Bulard sent photographs showing his Foucault telescope, which had been set up with the assistance of the municipal authorities in Algiers, and he promised soon to produce a string of "selenological, photographical, and meteorological" works. The point of the communication, however, was to speed Bulard's exit from "this brutish [*chamelante*] and equivocal post" in which he found himself. His telescopical mirrors had oxidized, and he wanted to travel to Paris to have them resilvered. He had been away from the seat of empire for five years: "I am like an exile here. I do not know if my parents are dead or alive."[24]

Delaunay's death some four months later cut short any metropolitan patronage that Bulard might have received, for the new director at Paris was none other than Delaunay's enemy and predecessor Urbain Jean Joseph Le Verrier. During his colonial career, Bulard published occasionally and managed, into the late 1870s, with only a janitor; his budget was on the order of Fr 10,000, and his salary oscillated from Fr 4,500 to Fr 6,924.[25] When the geographical service of the French army placed Algiers on the map in the middle 1870s, they avoided Bulard and set up their own observatory.[26] We are left to speculate about Bulard's finest

23. *Astron.*, s.v. "Bulard," for the extracts from the letter, dated 15 June 1862, which is located at the Marseilles Observatory.

24. PO, "Alger, 1872–1902," Bulard to Delaunay, 25 Apr. 1872.

25. PAN, F17/3753, Algiers Observatory budgets for 1874, 1879, and 1880.

26. *Armée*, p. 49.

North African cyclops: Charles Bulard mounting his 50-cm Foucault telescope at Algiers, 1872 (Courtesy of Observatoire de Paris)

moment: it is possible that he was responsible for designing an early astatic pendulum seismometer that recorded the shocks of an 1867 Algerian earthquake on smoked glass.[27]

At the time of the consolidation of higher education in Algiers, Admiral Mouchez, of the Paris Observatory, facilitated the decision to relocate the astronomers to land situated on a promontory called Bou-

27. M. Chesneau, "Note sur les tremblements de terre en Algérie," *Annales des mines, Mémoires* 1 (1892), 5–46, on p. 19. The pendulum consisted of a rigid rod supporting a copper ball. The ball "is pierced by a vertical hole," in which a steel needle moved to trace curves on smoked glass when the pendulum moved. If the needle was attached to the copper ball, and if the rod was erected vertically, the seismometer would have resembled one designed and constructed by James Forbes in 1844 (see James Dewey and Perry Byerly, "The Early History of Seismometry [to 1900]," *Bulletin of the Seismological Society of America* 59 [1969], 183–227, on pp. 187–89).

zaréah, some 342 meters above sea level. Instruments for the new site included an equatorial refractor of 31-cm aperture and a meridian circle of 19-cm aperture.[28] Bulard continued to command the skies until 1881, when Charles Trépied arrived, with his assistant Vincent Charlemagne Rambaud, to take charge.[29]

In 1875, at the age of thirty, Trépied had put aside a career in provincial lycée and university to work at the Montsouris Observatory in Paris. He then obtained a position as assistant in the Bureau of Longitudes, where he published on geodesy and photometry and earned a reputation as a mathematical adept. In 1877 he rose to membre-adjoint, and from then until 1888 he helped to edit the bureau's journal *Connaissance des temps*. In 1880 his Parisian masters tapped him to organize the Algerian observatory. He had temporary installations and limited assistance until the permanent opening of the Bouzaréah site in 1890. His specialty became the solar system – planets, comets, asteroids, and the sun; he took more than twenty thousand observations, and he made more than a thousand solar photographs. His great talent for photography led him to participate in the international conferences, beginning in 1887, to formulate policy for Mouchez's photographical survey of the skies.[30]

In 1883 Trépied informed the ministry of public instruction in Paris, which supervised the Algerian educational hierarchy, about astronomy at Algiers. He was proceeding with routine observations, but he sought to begin a study of the rotation and physical constitution of the sun. Undertaken lightly, such a study would be of little importance; undertaken with the right instruments, however, it would absorb the resources of his entire budget of Fr 12,900. Trépied protested against the Algerian military command's desire to have the new observatory buildings constructed closer to Algiers so that they could be defended more easily. In Trépied's view, an isolated observatory was more easily guarded and could in any event be surrounded by a wall.[31] The wall soon became "an absolute necessity" to prevent theft of the observatory's equipment.[32]

Trépied began a spectroscopical survey of solar hydrogen in 1885.

28. Mélia, *Histoire*, pp. 195–96.

29. See PO, "Trépied," for three summaries of Trépied's career, the first from 1877 and the latest (written no earlier than 1902) by Maurice Loewy. Ibid., "Alger 1872–1902," Trépied to Tisserand, 19 Apr. 1893, indicates that Rambaud had worked at Algiers for twelve years.

30. See *Astron.*, s.v. "Trépied" and "Rambaud," and PO, "Trépied," for three career summaries.

31. PAN, F17/3753, Trépied to Minister of Public Instruction, report for 1883.

32. Ibid., Algiers Observatory inspection, Mar. 1885.

By 1890 he was collaborating on spectroscopy with the Algiers professor of physics Louis Charles Thomas, and in 1893 he extended hospitality to astrophysicist Eugène Jean Hubert Spée, of the Brussels Observatory, who took spectroscopical measurements during a solar eclipse.[33] By this time Trépied's work on the sky chart commanded much of his attention. He installed objectives furnished by the brothers Paul and Prosper Henry, took delivery of the Gautier measuring machine (at Fr 11,900), and assigned his assistants Frédéric Sy and François Renaux to record the sky quadrant assigned to Algiers. Sy, who had apprenticed at the Paris Observatory, and Renaux, a normalien, spent decades on the task, finally publishing their catalogue of 9,997 stars in 1924.[34]

The sky chart gave Trépied a measure of independence from the Algiers educational authorities, his immediate superiors. Surveying the heavens provided the occasion for asking Mouchez's successor at Paris, Félix Tisserand, to support promotions for his three assistants. Renaux, Trépied informed Tisserand in 1893, had received a salary of only three thousand francs for the preceding five years; Sy had worked at Algiers for six years and Rambaud for twelve, each earning only twenty-five hundred francs. Tisserand intervened with Louis Liard, director of higher education in France, and obtained increases of five hundred francs for Rambaud and Sy and a promotion to astronome-adjoint with an increase of six hundred francs for Renaux.[35] In 1894 the ministry of public instruction provided Trépied with forty-five hundred francs for the sky-chart work, a third of his operating budget. By 1898 the budget had almost doubled, and more than a third of it, or nine thousand francs, came as a special payment for the sky chart. As Trépied hit his stride, spectroscopy dropped from sight. Responsibilities for the sky chart and the time service compromised observations of planets and comets.[36]

It is difficult to imagine how Trépied could have taken his job more seriously. He supervised the local baccalauréat examinations even

33. Ibid., 22 Mar. 1890; PO, "Alger 1872–1902," Trépied to Tisserand, 19 Apr. 1893, where Trépied sends notes about the eclipse from him and from Spée for insertion in the Paris Academy's *Comptes-Rendus.*

34. *Astron.*, s.v. "Sy" and "Renaux," On the Gautier, see PAN, F17/3753, Algiers Observatory inspection, 8 Apr. 1893. According to the consistent record in *Min.*, Renaux's first name was François, but *Astron.* gives it as Joseph.

35. PO, "Alger 1872–1902," Trépied to Tisserand, 19 Apr. 1893 for the request, 28 May 1893 for the promotions.

36. PAN, F17/3753, Algiers Observatory inspections, 26 Mar. 1894 and 11 Mar. 1898; Trépied to Minister of Public Instruction, 29 Jan. 1896, where, in a description of the work of the observatory, spectroscopy does not appear.

though his position freed him from teaching responsibilities.[37] By 1890 he had published fifty-six notes in the *Comptes-Rendus;* he attended meetings in Paris of the sky-chart committee, acting as the secretary for technical discussions.[38] In 1895 he provided the star cartographers with a reliable means of establishing photographical magnitude, and in 1900 he once more gave technical instruction to his colleagues on practical matters relating to the project.[39] Dedication and fealty brought him, in 1902, the Paris Academy's Lalande prize for the totality of his work.

Trépied's devotion to the pet project of Admiral Mouchez guaranteed that his observatory would face no serious competitors for funds devoted to overseas astronomy. But by the same token, metropolitan authorities would not allow him to enlarge his operation into an empire of his own; they opposed expanding anywhere in North Africa. The ministry of public instruction refused Trépied's request, shortly after he arrived at Algiers, to study the feasibility of establishing an observatory at Cairo.[40] Nor would a metropolitan entrepreneur be allowed to expand into Egypt. When in 1895 the independently wealthy astronomer Aymar, comte de La Baume Pluvinel, sought to enlist the ministry of foreign affairs to build an observatory on the Muqattam Hills outside of Cairo, the Parisian astronomers would have none of it. Félix Tisserand condemned La Baume Pluvinel's proposal to the minister of public instruction. La Baume Pluvinel, then associated with Jules Janssen's astrophysical observatory in Meudon, claimed that the Cairo facility was needed because solar observations there would be better than at Meudon; furthermore, conditions in Egypt would allow for discovering smaller asteroids than could be found from Tisserand's new branch observatory at Nice. Tisserand firmly noted that Meudon was not the only observatory in France engaged in solar astronomy and that the results

37. PO, "Alger 1872–1902," Trépied to Tisserand, 5 May 1893, where he mentions that the date of the upcoming sky-chart conference in Paris should be changed so as not to conflict with the baccalauréat examinations at Constantine.

38. PO, "Trépied," second career summary. One of the publications of his secretarial responsibility is Trépied, *Commentaires des décisions prises par les conférences internationales qui se réunirent à l'Observatoire de Paris en 1887, 1889 et 1891 pour l'exécution photographique d'une carte du ciel* (Paris, 1892).

39. Trépied, "Sur la nécessité de constituer des types photographiques conventionnels des étoiles des grandeurs 9, 11 et 14," *Bulletin du Comité international permanent pour l'exécution photographique de la carte du ciel* 1 (1892), 343–49; idem, "Détermination des grandeurs photographiques des étoiles du catalogue," ibid. 2 (1895), 383–428; idem, "Sur la construction des échelles de grandeurs pour les cartes à trois poses de trente minutes," ibid. 3 (1900), 82–105.

40. PAN, F17/3010, Note, Ministry of Public Instruction, 21 Oct. 1884. According to the proposal, if the Cairo observatory proved feasible, Trépied was to be placed at its head.

from Nice were not yet in.[41] Paris controlled Nice and Algiers, and it would not permit an enterprising philanthropist, much less a colonial functionary, to undercut its authority.[42]

When Trépied died suddenly in 1907, his position went to fifty-one-year-old François Gonnessiat, veteran of French observatories at home and overseas. Gonnessiat began his long career at the Lyons Observatory, where during the course of twenty-two years he rose from stagiaire to astronome-adjoint. He labored at the routine meridian observations required for the local time service, turning necessity to virtue by establishing expertise in positional astronomy. He worked up practical notes on the "personal equation" of observers and presented them as a doctorate in 1892. When, at the very beginning of the twentieth century, the French army decided to remeasure the arc of longitude in Ecuador, Gonnessiat followed in their train by occupying the vacant directorship of the Quito Observatory. There he seconded the military geodesists; after they left in 1905, so did he. Following a short stay at the Paris Observatory, Gonnessiat fell into the Algiers directorship.[43]

By this time the observatory budget had burgeoned to Fr 28,000 for all the salaries, Fr 8,200 for general operating expenses, and Fr 12,500 for the sky chart. Renaux was astronome-adjoint, Rambaud and Sy were the aides-astronomes, and Eugène Gaultier, Noël Villatte, and Léon Thivin worked as calculators.[44] Gonnessiat continued the programs that Trépied had undertaken: asteroids, the time service, and especially the sky chart. Rambaud moved up the ladder to astronome-adjoint, and Villatte to aide-astronome, following the French tradition according to which apprentice astronomers advanced at guild-regulated speed. Villatte traveled to the central Sahara in 1904 for astronomical observations relative to mapping the area around In-Saleh; in 1908 he took astronomical and magnetical determinations there.[45] Gonnessiat slowly expanded the observatory's mandate into seismology and atmospheric

41. Ibid., F17/3776, Tisserand to Minister of Public Instruction, 10 May 1895.

42. On the history of the Nice Observatory, founded by Raphaël Bischoffsheim and then presented to the Paris astronomers, see A. Clorennec, R. Feldman, and J-P. Zahn, eds., *Cent ans d'astronomie à l'Observatoire de Nice* (Nice, 1981).

43. Georges Perrier, "François Gonnessiat (1854–1934)," *Bulletin géodésique*, no. 44, 1934, pp. 298–311. Gonnessiat was named as Trépied's replacement on 26 Nov. 1907 (PAN, F17/13582, Algiers Observatory inspection, 10 Mar. 1908). Gonnessiat's activity in Ecuador is treated below, in chapter 11.

44. *Min.*, 17 (1908), s.v. "Algier." Here, however, I follow PAN, F17/17292, s.v. "Villatte," for the spelling.

45. PAN, F17/17292, s.v. "Villatte." The second expedition involved four men in addition to Villatte; it took a year to reach the Niger and return (Gonnessiat to Minister of Public Instruction, 1 July 1909).

electricity, taking charge of the colony's meteorological service upon the consolidation of Algiers's various higher faculties into a university. Meteorology, however, received little support from the university, its new patron. After four years Gonnessiat abandoned his meteorological charge without having been able to make much of it, although he retained responsibility for atmospheric electricity and seismology.[46]

A clever administrator will surround himself with bright young talent. Gonnessiat consistently recruited the best staff available. In 1912, with meteorology in the hands of a specialist colleague who ran an independent operation, Gonnessiat named Fernand Baldet as assistant and then, in 1914, aide-astronome. A watchmaker who in 1907 obtained a diploma in physics under Jules Violle at the Conservatoire des arts et métiers, Baldet worked with La Baume Pluvinel and undertook research projects at the observatories of Pic-du-Midi and Juvisy. Persistence brought him a licence ès sciences in 1910 and an appointment the following year as a stagiaire in the Paris Observatory. From 1912 to 1922, under Gonnessiat, Baldet divided his time between the meridian service of the Algiers Observatory, observations of terrestrial magnetism and atmospheric electricity, and spectroscopy. The astronomical part of his charge went to positional stellar astronomy and the local time service. His spectroscopical experiments, cut short by the war, continued after 1922, when he won his way to the Meudon Observatory. There he became one of the world's experts on astronomical, and especially cometary, spectroscopy.[47]

As Baldet's interrupted program in spectroscopy suggests, the progress of astronomy at Algiers, like the enterprises of metropolitan colleagues, suffered during the war and well after the peace. In 1922 Gonnessiat still had the same staff positions he began with in 1907: one astronomer and three aides and assistants; among the twelve major observatories in France and Algeria, only that of the Pic-du-Midi, a station dependent on Paris, had fewer senior people.[48] With the sky-chart work finished, however, Gonnessiat looked to wider hori-

46. Albert Lasserre, "Historique de l'Institut de météorologie et de physique du globe de l'Algérie," introduction to Paul Seltzer, *Le climat de l'Algérie* (Algiers, 1946), pp. vii–xxiii, on p. viii. Atmospheric electricity and seismology continued under the astronomers into the 1920s (PAN, F17/13582, Algiers Observatory report, 1921).

47. Charles Bertaud, "Fernand Baldet (1885–1964)," *L'astronomie* 79 (1965), 65–69; idem, "Fernand Baldet," Bureau des longitudes, *Annuaire pour 1968* (Paris, [1968]), pp. B.1–B.4.

48. Henri Deslandres, "Rapport sur l'inspection des observatoires français décidée par l'ordre ministériel du 29 avril 1922." A copy of this document was kindly lent to me by Philippe Véron, St-Michel-l'Observatoire. The observatory at Strasbourg had the same number of staff as Algiers, although its equipment was inferior.

The Algiers Observatory at Bouzaréah, postcard sent by François Gonnessiat to William Wallace Campbell, 22 January 1912 (Courtesy of Mary Lea Shane Archives of the Lick Observatory, Santa Cruz, California)

zons. He had four refractors with apertures of 25 cm or greater, and the 50-cm-aperture Foucault, which was so poorly figured that it remained out of service.[49] Fortune provided him with the means for expansion.

Two Russians came Gonnessiat's way, and he was quick to make use of their talents. He first picked up the services of Benjamin de Jekhowsky, who had studied at the universities of Warsaw and Moscow before traveling to Paris to work at the Bureau of Longitudes and the observatory. With a doctorat d'université from Paris on the celestial mechanics of small bodies in the solar system, Jekhowsky became an assistant at the Paris Observatory. He came to Gonnessiat in 1919 as an aide-astronome detached from Paris, a situation that was regularized only in 1926, one year before he defended a doctorat ès sciences at the Sorbonne for a thesis in pure mathematics.[50] The second Russian, Lionel Filippoff, had been educated as an astronomer at the University of Dorpat; when Gonnessiat found him, Filippoff was a naval-lieutenant refugee in the St-Jean camp at Bizerte, the internment center for a White Russian squadron that the French navy had assisted in evacuat-

49. Jean-Louis Lagrula, "Le télescope Foucault de l'Observatoire d'Alger," *Journal des observateurs* 15 (1932), 39–42.

50. *Astron.*, s.v. "Jekhowsky"; Jekhowsky, *Exposé succinct, suivi d'un exemple numérique d'une méthode nouvelle pour la détermination d'une orbite parabolique* (univ. diss., University of Paris, [1916]); idem, *Etude sur les transcendantes Fourier-Bessel à plusieurs variables* (diss., University of Paris, 1927).

ing from Crimea. Gonnessiat promptly put him on the rolls as a temporary aide-astronome. He graduated into the permanent (but lower) position of assistant in 1926.[51]

Gonnessiat was delighted to have this cheap foreign labor. In his annual report for 1924, he emphasized that Jekhowsky was a hard worker who knew how to use photographical telescopes for observing asteroids and noted that Filippoff was good on the meridian and equatorial telescopes and could calculate orbits. He contrasted the two Russians with his French staff. Renaux, who had served at Algiers for a generation, "has always believed himself misunderstood; he has always demonstrated a certain misplaced pride that has him attribute overly much importance to his work." Renaux's mathematical doctorate had hardly advanced in twenty years. Nor did Gonnessiat's new assistant director, Joanny Philippe Lagrula, measure up to the Russians. Lagrula had been Gonnessiat's student and then colleague at Lyons, submitting a doctorate there in 1901 on lunar occultations; he succeeded Gonnessiat for a year at Quito before a revolution persuaded him to return to France. He found a berth as astronome-adjoint at the Nice Observatory in 1910, and he remained there until Algiers called him in 1924. In Gonnessiat's opinion, Lagrula was rather limited in range: "Stepping ashore at Algiers to take up the post of assistant director, he found himself, from the practical point of view, in a state of inferiority with respect to the personnel whom he would watch over and guide." Lagrula retreated to his home and did not implicate himself in the life of the observatory.[52]

A combination of low salaries and low prestige conspired to turn French scientists away from Algiers. Even directorships often went to people who had already headed an observatory *in partes infidelium* — the case for Gonnessiat and Lagrula — or who had roots in the colony — the case of Albert Lasserre and Jean Coulomb, to be discussed shortly. When Jekhowsky left for Bordeaux in 1927, Gonnessiat could not attract a French replacement of comparable caliber. He petitioned the educational authorities in Algiers to allow him to hire, again as a temporary aide-astronome, Boris Veselovsky. The new man, who had been an artillery lieutenant in the White Russian army of Pyotr Nikolayevich, baron Wrangel, spent most of the 1920s in Prague, where he obtained a doctorate in 1928 for a thesis on celestial mechanics. Veselov-

51. See PAN, F17/13582, Gonnessiat to Rector of the Algiers Academy, 23 June 1921, for the provenance and position of Filippoff; see also *Astron.*, s.v. "Filippoff."

52. *Astron.*, s.v. "Lagrula"; PAN, F17/13582, Gonnessiat, "Observatoire d'Alger: Rapport, 1924, deuxième partie," for evaluation of Jekhowsky, Filippoff, Renaux, and Lagrula. Lagrula's thesis was *Etude sur les occultations d'amas d'étoiles par la lune, avec un catalogue normal des pléiades* (diss., University of Lyons, 1901).

sky received a permanent position of assistant in 1947, and he advanced to aide-astronome two years later only after the French educational authorities awarded him a licence ad eundem by virtue of his Prague doctorate.[53] Gonnessiat's Russian émigrés suffered from a complicated charade, elaborated over the century between the July Monarchy and the fall of the Third Republic, for disguising the lack of prestige and authority carried by French diplomas.

Low prestige guaranteed that no large government observatory would come to the mountains of Algeria, despite their favorable window on the sky. When in 1922 Henri Deslandres called for creation of a great new mountain observatory to rival those in California, he categorically rejected Algeria as a site for it, even though Algeria was climatologically much better suited than the hexagone for astronomy. Algeria, which resembled California in some respects, was "not civilized and populated like the United States; the station, chosen for its astronomical attributes, would be far from great intellectual centers; and providing matériel and personnel would certainly be difficult."[54]

The mountain observatory that did come resulted from the initiative of an amateur, René Jarry-Desloges. In the years before the First World War, Jarry-Desloges had equipped an observatory on Mt. Revard in Savoie; he directed his attention to observing planetary surfaces, especially (with the model of Percival Lowell in mind) Mars. He sought, however, a better location. His interest then turned to Algeria, where he investigated a number of sites. He settled on Sétif, located some one hundred kilometers west of Constantine, with an elevation of more than a kilometer. In 1911 he set up a Schaer refractor of 37-cm aperture there. In 1913–14 Emile Schaer, astronome-adjoint at the Geneva Observatory, traveled to Sétif to replace his 37-cm objective with a 50-cm one. A second station, two hundred and fifty kilometers due south of Algiers at Laghouat, received Jarry-Desloges's 29-cm Merz refractor. In 1922 he substantially expanded the Sétif operation by adding the Merz refractor. From the Sétif Observatory, made permanent in 1924, Jarry-Desloges and his assistants produced a continuing stream of publications about planetary surfaces. In the early decades their observations appeared in the form of hand-drawn sketches.[55] An abstention from appealing to the broader periodical literature and an apparent horror of

53. *Astron.*, s.v. "Veselovsky."

54. Deslandres, "Rapport," p. 46.

55. *Astron.*, s.v. "Jarry-Desloges." According to PSG, colis 2 *ter*, 1580, Jarry-Desloges first set up a temporary observatory at Sétif in 1911 and 1912. G. Rouvray, "L'oeuvre d'un grand amateur français d'astronomie, René Jarry-Desloges," *Astronomie* 61 (1947), 181–85; G. Fournier, "René Jarry Desloges et ses stations astronomiques," ibid. 65 (1951), 388–95; Observatoires Jarry-Desloges, *Observations des surfaces planétaires* 3 (1913, for

any kind of theory suggest why both Paris and Algiers viewed Jarry-Desloges with detachment, not to say studied neglect.

Gonnessiat's operation was just what metropolitan French astronomers required for their major project of the 1920s, the revision of world longitudes by means of radio and positional astronomy. It had a respectable set of telescopes, a long tradition of observation, and a complement of talented astronomers and calculators. Algiers had successfully acquitted itself of the sky chart, and its director, approaching retirement, had no research ambitions of his own. Algiers became one of three principal vertices chosen for the longitude project. The second vertex, located about 120° away, was the Jesuit observatory at Zikawei, and it shared each of Algiers's virtues. As no observatory in California was willing to devote substantial resources to the project, the French astronomers leaned on the United States Naval Observatory to send astronomers and equipment to the San Diego Naval Air Station, which *faute de mieux* constituted the third vertex of a spherical triangle. Gonnessiat supervised the observations on his large meridian circle.[56] The distance of his observatory from China was henceforth known with unprecedented precision, but Gonnessiat, held back from retirement by the project's calculations, found this fact devoid of interest.

When Gonnessiat finally stepped down in 1931, Lagrula succeeded him. By then sixty-one years old, Lagrula had ceased active research; in 1938 he retired completely. In his place as director came Georges Meyer, agrégé in mathematics, from Bordeaux. Lagrula's alter ego at Algiers was his son, *polytechnicien* Jean-Louis Lagrula. Born in Quito in 1906, Lagrula *fils* landed an appointment as aide-astronome in the same year that he took an engineering diploma from the Ecole supérieure d'optique and, coincidentally, the year that saw Lagrula *père* rise to the directorship. When Meyer arrived, Jean-Louis Lagrula rose to astronome-adjoint. The junior posts were occupied by Louis Boyer (who had been aide-astronome since 1928), Odette Bancilhon (at the observatory as an assistant since 1932), and the Russian émigrés. Jean-Louis Lagrula, it may be assumed, spent the first years of his time at Algiers working toward a Sorbonne doctorate, defended in 1937.[57]

1911 and 1912), 307–17 for temporary installation at Sétif, 4 (1921, for 1913 and 1914), 1–3, 279–301 on Sétif and Laghouat.

56. Gonnessiat, *La participation française à la révision des longitudes mondiales: II. Opérations de l'Observatoire d'Alger, grand cercle méridien* (Orléans, 1929) [Algiers Observatory, Publications, 9, fasc. 1]. The longitude project is considered in chapter 6 below.

57. *Astron.*, s.v. "Meyer," "Boyer," "Bancilhon"; PAN, F17/13591, Algiers Observatory inspection of 1938; Lagrula, *Contribution au perfectionnement des méthodes photographiques en photométrie astronomique différentielle* (diss., University of Paris, 1937).

Climates of Expansion and Oppression

Over the first twenty years of French occupation, meteorological observations by military engineers appeared in the annual report of the administration. The minister of war, the comte de Vaillant, issued instructions for systematic observations in 1858, but the earliest attempt to create a permanent record of the Algerian climate dates from 1864, when the governor general, Marshal Aimable-Jean-Jacques Pélissier, duc de Malakoff, confided a network of some thirty meteorological stations, established four years previously, to the astronomer Charles Bulard. At least for the decade 1871–81, Bulard published meteorological reports in local newspapers.[58] The assignment was unappreciated in the metropolis.

At the time that Charles Sainte-Claire Deville was struggling to separate French meteorology from the pernicious tutelage of the Paris Observatory, he constituted an Algerian meteorological network under polytechnicien Henri Pierre Jean-Baptiste Brocard, a captain in the engineering corps and a mathematician. Between 1873 and 1876 Deville undertook four tours of inspection in Algeria. By the last inspection, Algeria was producing a daily weather bulletin and monthly reports printed in six hundred copies. The early years of data were synthesized by Brocard and also by scientists in the metropolis. Then, the basic meteorological dimensions of the colony having been chronicled in print, interest waned. Deville died in 1876, and the Algerian engineers asked to be relieved of their responsibility for the weather. In 1884 the service was remanded to the Algiers School of Sciences and placed in the hands of Antoine Thévenet, the school's professor of mathematics. He installed himself on the terrace of the Algiers city hall. As he expanded his service from 42 to 150 recording stations, Thévenet published various climatological surveys.[59]

Thévenet had the services of a secretary, a lithographer, a clerk, and

58. A. Hée, "Note sur la fréquence des tremblements de terre ressentis en Algérie," *APGFOM*, 1936, pp. 23–26, on p. 23.

59. Lasserre, "Historique," pp. vii – viii; PAN, F17/13590, G. Boulogne, "Rapport au conseil du gouvernement," 1920, where a chronology of meteorological initiatives is provided. More reliable on Deville's Tunisian involvement is Alfred Lacroix, "Charles Sainte-Claire Deville (1814–1876)," in *Figures de savants*, 4 vols. (Paris, 1932–38), 1:45–57, on pp. 56–57. For an attempt at synthesis, see PMN, H. Brocard, "Etude de la période météorologique du 5 au 18 octobre 1875 en Algérie et en Europe." See also R. B., "Henri Brocard," *Nouvelles annales de mathématiques* 1 (1922), 357–58, which indicates his numerous publications in that journal and in the *Intermédiaire des mathématiciens.* Brocard's correspondence with historian of science Paul Tannery in Tannery, *Mémoires scientifiques*, 13, ed. A. Diès (Toulouse, 1934), 241–76.

an office boy.[60] He put up with competition, however, from officially sanctioned adventurers. On the local side, he faced a group of French intellectuals and medical men who had established a society to advance knowledge of the Algerian climate. Taking form in 1863, it was called, appropriately enough, the Algerian Society of Climatology. The society's general secretary, military physician Emile-Louis Bertherand, obtained prefectorial permission to hold meetings with more than 20 people, and by 1867 the society claimed 65 regular members and 160 corresponding ones. Between 1868 and 1888 the climatologists received annual grants from the ministry of public instruction of between three hundred and five hundred francs.[61]

From the metropolitan side, the central meteorological authorities sent out talented observers to gather basic information that they presumed would not be obtained by the men in Algiers. Léon Teisserenc de Bort, the head of general meteorology in the Central Meteorological Bureau at Paris, himself took magnetical readings across Algeria and Tunisia during the middle 1880s.[62] Direct intervention had the effect of undercutting colonial authority. When he needed intelligence about native Algerians, Teisserenc de Bort wrote for advice to E. J. Garnier, who lived in Touggourt; he relied on advice regarding expedition garb and geography from the Saharan veteran Henri Duveyrier; and when he required barometrical readings of the Sahara, Teisserenc de Bort called on explorer Fernand Foureau rather than on Thévenet and his local staff.[63]

In 1907, the year of Thévenet's retirement, the post of secretary transformed into that of aide-météorologiste at thirty-five hundred francs.[64] Then the service passed into the hands of the Algiers astronomers, under Gonnessiat. When Gonnessiat asked to be relieved of his role in meteorology, the colonial bureaucracy, from university to public works, formed a commission to reorganize things. The university pro-

60. Boulogne, "Rapport."

61. Ibid., F17/13491(2), Le Tourneux to Ministry of Public Instruction, 23 Sept. 1867, for history and state of the society; internal Public Instruction note of 21 Feb. 1872 for permission to hold meetings in 1864; Bertherand listed as secretary general on letterhead of 1872. E-L. Bertherand is not to be confused with his brother François-Alphonse Bertherand, médecin principal at Algiers from 1852 to 1887 and director of the Algiers medical school from 1859 (*DBF*, s.v. "Bertherand"; cf. Turin, *Affrontements culturels*, p. 336).

62. PAN, F17/3008, Teisserenc de Bort to Minister of Public Instruction, 6 Nov. 1883.

63. PMN, Garnier to Teisserenc de Bort, 19 June 1884; Duveyrier to Teisserenc de Bort, 6 Apr. 1887 and 7 Nov. 1887; Foureau to Teisserenc de Bort, 22 Feb. 1889 and 22 Jan. 1895.

64. Boulogne, "Rapport."

fessor of physics and meteorology Louis Charles Thomas served as an interim director until the commission persuaded Albert Lasserre, a physics teacher at the local lycée, to assume charge.[65] His meteorological empire, shorn of responsibility for the interesting parts of geophysics, was to serve agricultural needs. Furthermore, the meteorological director would not sit on the newly created university council along with faculty deans, as the director of the astronomical observatory did.[66]

Lasserre negotiated a personal salary increase to eight thousand francs.[67] He insisted that his service figure in the rolls of the ministry of public instruction as one of its observatories, just as the Algiers astronomical observatory was listed.[68] He rearranged the library, relocated the service in new university rooms, and constructed an observation station in the botanical garden. He pulled together a staff from local talent. He brought in a mechanic and an assistant, both from his lycée; he had Léon Thivin, formerly at the astronomical observatory, appointed as his principal calculator. To the meteorologist Georges Poulin, who had been appointed under Thévenet's directorship, Lasserre added Sirot, a physics teacher at the collège in Mostaganem, as aide-météorologiste.[69] He labored to standardize and supervise the network of weather stations that his predecessors had strung out across northern Algeria.

The war threw Lasserre's plans into disarray. He lost all his senior men, recovering only one position in Julien Loisel, who came as aide-météorologiste in 1916 from Camille Flammarion's private observatory at Juvisy-sur-Orge. He received help from astronomical colleagues at Algiers, notably Fernand Baldet and his wife, in an attempt to keep up with incoming data. Lasserre's staff worked overtime, and two assistants died at their post. Dedication earned support in the colonial bureaucracy. In 1918 his salary rose to around Fr 9,000, with corresponding increases for everyone. Lasserre's staff increased their take from Fr 43,900 in 1913 to Fr 61,300 in 1914, and to Fr 66,340 in 1919–20. In

65. Thomas gives his charge, including physics and meteorology, in *Adressbuch der lebenden Physiker, Mathematiker und Astronomen des In- und Auslandes und der technischen Hilfskräfte* (Leipzig, 1909), p. 3.

66. For the composition of the university council, see Mélia, *Histoire*, p. 104.

67. Boulogne, "Rapport."

68. Ibid., Rector of the Algiers Academy to Minister of Public Instruction, 27 Apr. 1915, citing Lasserre's letter of 23 Apr. 1915 objecting to his service's exclusion from the latest bulletin issued by the ministry.

69. Ibid. Rector of the Algiers Academy to Minister of Public Instruction, 27 Apr. 1915, where Sirot's appointment is given as 1913, although in "Historique" Lasserre writes that Sirot was appointed sometime after the initial staff reorganization. Perhaps the meteorologist is André Sirot, who in 1929–30 was chargé des travaux in applied chemistry at the Algiers science faculty.

1920 the director pulled in Fr 18,000.[70] In 1921 the salaries went up again.[71] Inflation aside, the principle behind these salary increases was parity between Algerian meteorologists and those working in the central bureau at Paris; but in accepting equality, the Parisian ministry questioned the 25 percent bonus generally received by Algerian functionaries for colonial service.[72]

Meteorology assumed a higher profile early in the 1920s as meteorologists became involved with both military and commercial aviation. Lasserre benefited from this heightened visibility by acquiring a second post of météorologiste-adjoint in 1921. The post went to Lucien Petitjean, a product of Edmond Rothé's geophysical institute at Strasbourg, who was trained in the latest meteorological theories of Vilhelm Bjerknes. Disaster struck just as the Algerian weather service regained momentum. Loisel's wife died in 1922; the loss quite devastated him, depriving Lasserre of his services.[73] A second catastrophe came in the shape of the crash of the airship *Dixmude*. The press blamed Lasserre's service for not having issued adequate weather reports.[74]

With the Algiers operation caught off guard, metropolitan meteorologists seized the opportunity to "annex" it. Colonel Emile Delcambre, head of the National Meteorological Office, tried to bring Lasserre to heel; Lasserre fought back and retained control of his service, but not without suffering losses.[75] In 1926 Delcambre assimilated Petitjean and placed him in charge of air navigation in Algeria. While remaining on Lasserre's staff, Petitjean took directions from Paris. After two years the ambiguous situation proved intolerable to Lasserre, who arranged for a divorce. Air navigation and its attendant posts went to Delcambre, Lasserre reserved the right to establish his own stations in the desert, and Petitjean was formally detached from Lasserre to become, for the next five years, Delcambre's regional inspector for Algeria and Tunisia.[76] Delcambre felt confident in extending his influence in part because in 1928 his office came under an independent ministry of the air.

70. Boulogne, "Rapport," where a history of the salary schedules is provided. By 1920 the franc had plummeted in value with respect to sterling and the dollar, and prices inflated dramatically (René Sédillot, *Histoire du franc* [Paris, 1979], pp. 119–23).

71. PAN, F17/13590, Governor General of Algeria to Minister of Public Instruction, 6 Jan. 1921.

72. Boulogne, "Rapport."

73. PIPG, Lasserre to Maurain, 19 June 1923.

74. See Lasserre, "Historique," p. xv, for repercussions of the *Dixmude* crash.

75. PIPG, Lasserre to Maurain, 7 Sept. 1925, where Lasserre writes about "the attempt to annex my service" by the French National Meteorological Office.

76. Lasserre, "Historique," pp. xv – xvi. It is possible that Petitjean was carried from 1924 to 1926 on the budget of the geographical service of French West Africa (Edouard

Before his call to serve Parisian interests, Petitjean seemed destined to become Algeria's first serious meteorological researcher. He published a great deal about Saharan weather, relating his work to the theory of fronts.[77] He began a doctoral dissertation with Charles Maurain about weather prediction.[78] But Petitjean's situation on the cutting edge of meteorological theory evoked hostility and jealousy from his colleagues, and especially from his theoretically unapprised superior Lasserre, for whom routine compilation was the greatest good. In 1929, at the moment of Petitjean's detachment, Lasserre observed that he was "detested" by the meteorological staff.[79] Petitjean, who had seen his engagement by the National Meteorological Office as the beginning of a brilliant administrative career, failed to carry out his mandate as the economic crisis deepened.

Late in 1934, Petitjean thought to return to Lasserre's institute at Algiers under the protection of Henry Hubert's colonial meteorological service. In his cause, Petitjean enlisted Philippe Wehrlé, who had just replaced Delcambre as director of the National Meteorological Office. Wehrlé explained to Maurain that Petitjean would gladly serve Hubert, provided that a "suitable administrative formula" allowed him to begin at a higher rank in the colonial meteorological corps.[80] But Hubert made no exceptions for Delcambre's former protégé, even if by making exceptions Hubert could establish a presence in North Africa. In 1935 Petitjean returned to Lasserre's service, "the causes of his disgrace" known to the entire staff.[81] He courageously resisted the efforts of Lasserre to impose disciplinary sanctions against him. He likewise declined a post in Martinique, offered him by Hubert.[82] His reintegration at Al-

de Martonne, "Le Service géographique de l'Afrique orientale française: Son fonctionnement, ses resultats [1922–1927]," *BC* 11 [1928], 321–51; on the table facing p. 324 there is a Sergeant L. Petitjean, engaged as a draftsman for this period).

77. Among Petitjean's many publications from this time are "Surfaces de discontinuité en Algérie et au Sahara," *Cahiers du Service météorologique d'Algérie* 1 (1923), 13–16; "Sur l'application de la frontologie aux dépressions sahariennes," *CRAS* 179 (1924), 64–65; "La discontinuité Nord-Africaine," *La météorologie* 68 (1925), 446–59; "Sur la répartition des forces au voisinage d'une discontinuité," *CRAS* 181 (1925), 429–31; "La méthode norvégienne en Afrique du Nord," *La météorologie* 69 (1926), 463–64; "L'air actif et l'air passif dans les discontinuités atmosphériques," ibid. 70 (1927), 145–57; "La dépression saharienne," ibid. 71 (1928), 145–54; "Dix années de sondages à Alger (1918–1928)," ibid. 72 (1929), 214–49.

78. PIPG, Petitjean to Maurain, 15 Nov. 1922.

79. Ibid., Lasserre to Maurain, 6 Oct. 1929.

80. Ibid., Wehrlé to Maurain, 28 Dec. 1934.

81. Ibid., Lasserre to Maurain, 25 May 1935. The staff protested against his reinstatement "with intense emotion" (Lasserre to Maurain, 17 Nov. 1935).

82. Ibid., 28 June 1935 and 10 Aug. 1935.

giers resulted in a good deal of creative financing on the part of the director of higher education in Paris to ensure that geophysics in Algeria continued with the personnel engaged during his five-year absence.[83] In 1936 Petitjean persuaded the Algiers rector that the problems plaguing Algerian meteorology derived from Lasserre's advanced age and lack of energy and that his reintegration would precipitate Lasserre's resignation in protest.[84] Lasserre in fact stayed in his post only through 1937.

In 1936 Petitjean sent a short note on synoptical meteorology to Maurain for insertion in the *Comptes-Rendus* of the Paris Academy. He added: "I hope that the future will give me the means of rapidly pursuing the work undertaken, to which I have dedicated myself passionately, a passion the disappointments of life have not succeeded in extinguishing."[85] By 1937 Petitjean was perceived as the institute's assistant director, to the anguish of Lasserre. Petitjean defended his prerogatives as head of the weather-prediction section, insisting on his rights to promotion and travel.[86] Honors usually result if one holds a job long enough. When the outbreak of war in September 1939 deprived Algiers of its geophysical director, the institute passed into the hands of the senior man left on staff—Petitjean. Julie Malbos, a longtime assistant, complained to Maurain that given "the demoralizing presence at the head of the service of the agent chased out by the National Meteorological Office, it may be asked whether failure is not better rewarded than effort."[87] Petitjean displayed remarkable equilibrium and steadfastness in view of the forces arrayed against him.

The signal characteristic of Petitjean's writings—slavish adumbration of metropolitan discourse in an attempt to attract notice by metropolitan savants—extended over a broad layer of scientific researchers in French North Africa. Exceptional thinkers, however, used creative leisure to reflect on questions of general importance. We shall shortly meet with one such thinker, the playwright Albert Camus. In the exact sciences, Camus's counterpart may be found in Auguste Souleyre. A high-ranking engineer in Algerian public works, Souleyre advocated various schemes to support the imperial grandeur of France (a trans-Saharan railway and a defensive canal from the Rhine through Belgium). Cosmology and cosmogeny were also on his agenda. In 1923,

83. Ibid., Jacques Cavalier to Maurain, 31 Jan. 1935, probably misdated for 1936.

84. Ibid., Lasserre to Maurain, 5 Feb. 1936.

85. Ibid., Petitjean to Maurain, 31 Dec. 1936.

86. Ibid., Lasserre to Maurain, 2 Apr. 1937, where Lasserre notes that Petitjean was listed as assistant director in the March issue of *Sciences*, published by the French Association for the Advancement of Science; Petitjean to Maurain, 30 May 1937.

87. PIPG, Julie Malbos to Maurain, 10 Nov. 1939.

near the end of his life, he summarized his ideas in a massive monograph that treated, among other problems, the relation between tides, climatical change, and solar cycles. In his erudite work, Souleyre came down firmly against Alfred Wegener's theory of continental drift; the geological evidence, he believed, simply did not support it. He also opposed Darwinian evolution, the essence of which he believed to be slow, gradual, and continual transformation in environment and species. He cited paleontological authorities – Louis Agassiz, Charles White, Oswald Heer, Jules Tissot, René Zeiller, and Oliver Lodge – to support his view that such transformations occurred for the most part as punctuated explosions in an otherwise equilibrated ecosystem. He concluded that life began on earth some hundreds of thousands of years ago.[88]

In the late 1920s, Lasserre obtained two younger meteorologists. One, Paul Queney, came as a temporary replacement during Petitjean's years with the ministry of the air. Queney worked hard, finally defending a doctorate at Paris in 1936.[89] The second, Paul Seltzer, was the son of the director of a technical school in Strasbourg. He studied geophysics under Rothé before joining Lasserre in 1928 as aide-météorologiste.[90] Then, from 1932 to 1934, Lasserre allowed him to return to Strasbourg on a scholarship to complete a doctorate.[91] Each young meteorologist had a particular style. Seltzer was a talented instrumentalist who focused on local thermal contributions to weather. Queney had a more mathematical bent, seeking in his dissertation research to predict coastal weather from first principles in fluid mechanics; when this proved too ambitious, he worked up an experimental supplement based on Algerian data that was weakly related to his initial calculations. Early in 1937 Seltzer became physicien-adjoint, the position vacated by Loisel's retirement, and Queney received Seltzer's post before returning to the hexagone with a rank equivalent to Seltzer's.[92]

88. Souleyre, *Les oscillations océaniques et les oscillations climatiques dans le passé et dans le présent, le rythme de contraction du soleil* (Bône, 1923), pp. 280–89 for evolution, pp. 337 and 374 for continental drift; C. Duprat, "Auguste Souleyre," *Bulletin de la Société astronomique de France* 40 (1926), 147.

89. Queney, *Recherches relatives à l'influence du relief sur les éléments météorologiques* (diss., University of Paris, 1936).

90. PIPG, Edmond Rothé to President of the Caisse nationale des recherches scientifiques (physique), 13 June 1932 (copy), where Rothé asks for a scholarship to allow Seltzer to return to Strasbourg for additional study.

91. Seltzer, *Etudes micrométéorologiques en Alsace* (diss., University of Strasbourg, 1935); Rothé proposed Seltzer for a grant that allowed him to complete his dissertation (p. 2).

92. PIPG, Queney to Maurain, 6 June 1935 and 9 Nov. 1935; Seltzer to Maurain, 8 Mar. 1937, where he applies for the post; Lasserre to Maurain, 2 Apr. 1937, for the pro-

Both Queney and Seltzer had competed unsuccessfully to follow Lasserre as director.[93]

Lasserre knew how to exploit researchers blown toward his shore. Petitjean's complicated situation was only the most troublesome of his strategic engagements. A foreign geophysical engineer from Strasbourg, Oussaroff, went to take charge of the regional station at Maison-Carrée although at a greatly reduced salary. And when he could not make appointments at bargain-basement rates, Lasserre assimilated observers from other services. By the late 1920s he had drafted Boris Veselovsky from the astronomical observatory to head up his seismological service.[94]

Fortune came begging in the form of Polish meteorologist Władysław Gorczyński, and Lasserre did not turn him away. In 1902 Gorczyński had taken a candidate's diploma at Warsaw and then in 1906 a doctorat ès sciences at Montpellier for a dissertation on solar radiation. From 1902 to 1928 he directed various meteorological bureaus in Warsaw, eventually rising to head the Central Meteorological Institute in independent Poland. During the middle 1920s, however, he effectively retired to the French Riviera, from which location he traveled around the world measuring solar radiation and classifying climates.[95] Among his first stops were points in North Africa.

Gorczyński presented himself to Charles Maurain in 1923. Maurain arranged for the loan of a Michelson actinometer and introduced the Polish scientist to Lasserre. Gorczyński ran down to Algiers for a few weeks, took measurements, and returned to Paris, where Lasserre had instructed him to purchase several instruments on behalf of the Algiers meteorological service. Lasserre then financed an expedition of Gorczyński's to the interior of Algeria; he provided free rail transport and a grant of fifteen hundred francs. Lasserre also promised extensive support for Gorczyński's plan to travel deep into the Hoggar, although he doubted that Gorczyński's prism spectroscope would survive six weeks on the back of a camel.[96]

---

motions. Queney became météorologiste-adjoint at the Puy-de-Dôme Observatory in Clermont-Ferrand.

93. Ibid., Queney to Maurain, 13 Feb. 1937.

94. Ibid., Lasserre to Maurain, 30 Nov. 1926; *ASPG*, 1934, p. vi, where "B. Vesselovsky" is noted.

95. Gorczyński, *Liste des travaux scientifiques publiés de 1898 à 1935* (Warsaw, 1935), which also provides a career summary; *Polsky Słownik Biograficzny*, vol. 8 (Wrocław, 1959), s.v. "Gorczyński."

96. PIPG, Société anonyme des établissements Jules Richard to Maurain, 20 Dec.

Lasserre felt deeply wounded by the appearance of Gorczyński's results, published in 1925. There Gorczyński indicated an early trip to Algiers in 1922, and he thanked Delcambre and Petitjean, while remaining silent about the Algiers meteorological service.[97] Lasserre reminded Maurain about all that had been done for the Polish visitor. "So General Delcambre and Mr. Petitjean receive all the honor for the help that we provided." Lasserre continued: "I am not at all surprised that Gen. Delcambre wants to eliminate me and have 'Meteorology' all to himself. But don't you think that it would be well to warn overconfident colleagues about this sort of undertaking? You, who sent me Mr. Gorczyński, you who are without question the authority, could you not, in short, clarify this point again? You would do everyone a favor, especially me."[98]

Lasserre's sour mood may have derived in part from Gorczyński's decision to base himself in neighboring Tunisia and to publish detailed climatological accounts in the house organ of the Tunisian Botanical Service.[99] Gorczyński knew the value of cultivating meteorologists in the ministries of the air and the colonies. By the middle 1930s, Henry Hubert decided to provide his entire colonial service with pyrheliometers and solarimeters of Gorczyński's design.[100]

Lasserre seized hold of opportunity, but at the very end of his tenure one man escaped his net. Georges Liandrat, a physicist with a doctorat ès sciences, was teaching—and unhappily so—at the Algiers lycée. In 1936 he declined to be appointed aide-physicien at the Puy-de-Dôme Observatory, holding out for a higher rank at the Pic-du-Midi or elsewhere; if he could flee Algiers for France, he emphasized to Charles Maurain, he would begin research into cosmic rays and atmospheric ozone. He presented himself unsuccessfully for the post of physicien-adjoint at Puy-de-Dôme, recently vacated by Jean Coulomb.[101] Then

1923; Maurain to Lasserre, 21 Dec. 1923; Lasserre to Maurain, 4 Jan. 1924 and 21 Mar. 1924.

97. Gorczyński, "Solarimètres à lecture directe et enregistreurs," *La météorologie* 70 (1927), 402–9. Gorczyński's fatuous acknowledgments on p. 403 are remarkable, even in the context of French literary convention; the omission of Lasserre is an intentional slight.

98. PIPG, Lasserre to Maurain, 25 Oct. 1927.

99. Tunisia, Direction des affaires économiques, *Annales du Service botanique et agronomique* 3 (1925), 5–87, with Gorczyński's data from Algeria and elsewhere, Gorczyński's analysis of Tunisian climatology according to its "continentalist" isobars, and the comments of Polish meteorologist Jean Paul Rychliński on Tunisian precipitation and Gorczyński's continentalism (compared with data from Australia, South America, and Poland).

100. PIPG, Hubert to Maurain, 8 May 1935.

101. Ibid., Liandrat to Maurain, 25 Oct. 1936, 10 Apr. 1937, and 3 Nov. 1937.

he received word from his patron Pierre Auger that a university-level position was opening up in physics at Kabul, Afghanistan. He thought to apply for it, provided that a three- to six-year stint in Central Asia would not lead back into the secondary schools of greater France. There was no other way that France could send first-rate people abroad, he noted. The lack of promises for advancement at home had led France to send second-rate people to Istanbul, for example, with the result that the university there was "completely Germanized." If Liandrat succeeded in going to Kabul, he would do whatever Maurain suggested, as "the only thing that a physicist may do in Central Asia is geophysics."[102] But the educational hierarchy in Paris would not be bound to return Liandrat to a university post after an Afghan tour of duty. Liandrat applied for a position as maître de conférences in geophysics and meteorology at Strasbourg, which Coulomb, by then directing geophysics in Algiers, had told him about. Geophysics, however, remained the province of geophysicists. Liandrat finally received a professorial appointment in physics at the University of Aix-en-Provence.[103]

The man who edged out Liandrat for a post at Puy-de-Dôme was Lasserre's loyal employee and Maurain's loyal student Paul Queney. In 1935 Queney, météorologiste-adjoint at Algiers in charge of weather forecasting, started writing his doctoral dissertation under Maurain's direction. He began to study the friction between the atmosphere and the earth's surface on the Algerian coast. Then he moved to consider the effect of coastal mountains on weather. Maurain queried Queney's mathematical presentation of the problem, but Queney insisted that it would all go toward more accurate prediction. Maurain circulated Queney's draft to the Sorbonne's professor of fluid mechanics, Henri Villat, who wrote directly to Queney. The doctoral student then realized "that the theoretical problem leads to considerable difficulties, and it is better for me to extend as far as possible over the experimental part." Queney set to work on a revision for Paris and asked for instructions regarding the subject of his second thesis. Within six weeks Queney had sent the new text to Maurain, Villat, and the secretary of the Paris Faculty of Sciences. Queney followed Maurain's advice in devoting the experi-

102. Ibid., 3 Nov. 1937. According to Liandrat, the Kabul post was vacated by Auger's student Gabriel E. Monod-Herzen (see Monod-Herzen, *Recherches sur les neutrons* [diss., University of Paris, 1935]). Liandrat's conditions could not be met, no doubt, because Kabul had no proper university; its faculty of sciences came only in 1941 (Vartan Gregorian, *The Emergence of Modern Afghanistan: Politics of Reform and Modernization, 1880–1946* [Stanford, 1969], p. 355).

103. PIPG, Liandrat to Maurain, 8 Jan. 1938 and 7 Feb. 1938; *Min.*, 1966, s.v. "Aix-en-Provence."

mental part of his work to Algerian rainfall. He began looking for a place to print up the result – whether in a French or German periodical or at his own expense. By December 1935 Queney had decided to issue the dissertation privately. All firms in Algiers would have to send the portions with equations to Paris. Queney thought that a Parisian firm might do the job for less than five hundred francs, the lowest Algerian quotation. He wrote to Gauthier-Villars, who wanted to see the manuscript; Queney asked whether Maurain could forward his copy to the publishers.[104] He defended the final result in the middle of 1936.[105]

The thesis contributed to Queney's promotion dossier at Algiers, and it fanned his desire to be considered as the next director of the geophysical institute there. He applied for both a promotion and the directorship, and he was twice disappointed. He did not mind having normalien Jean Coulomb come to direct the institute, but he did resent seeing a coworker, Paul Seltzer, ranked before him in that competition. Queney asked Maurain whether there were not a post for him somewhere else – in geophysics, fluid mechanics, or in the French weather service. Maurain having arranged for him to be entered on the *liste d'aptitudes* for university teaching, Queney applied to follow Georges Rempp at Strasbourg.[106] He wound up in Coulomb's post of météorologiste-adjoint at the Puy-de-Dôme Observatory.

The Institute and the Desert Wind

Delcambre's absorption of a part of Lasserre's operation provided impetus to a reorganization that propelled Algerian geophysics into the first ranks of the scientific world. Just as his metropolitan colleagues had responded to a similar situation in 1919, Lasserre reacted quickly to his newly found freedom from routine weather watching. The opening allowed him to learn about magnetical observations, and to this end he traveled to the hexagone for instruction from metropolitan masters.[107] He learned quickly, setting up magnetical stations across northern Algeria. When he sent a summary of five years' measurements to Maurain in 1928, he exclaimed: "I do not at all claim to be independent, quite the contrary! But I do not want to be dependent on General Delcambre. On Maurain, right away!"[108] The dependence on Maurain had already proved valuable, for in 1927 Lasserre had obtained a seat on

104. Ibid., Queney to Maurain, 28 Mar. 1935, 6 Apr. 1935, 6 June 1935 (quotation), 25 July 1935, 9 Nov. 1935, and 2 Dec. 1935.

105. Ibid., Lasserre to Maurain, 30 May 1936.

106. Ibid., Queney to Maurain, 26 Nov. 1936, 13 Feb. 1937, and 4 June 1937.

107. Ibid., Lasserre to Maurain, 7 Nov. 1925 and 29 June 1926.

108. Ibid., 16 June 1928.

the French national council of geophysical institutes and observatories.[109]

The ministry of the air in Paris, which supervised the National Meteorological Office, then sought to extend its authority to all overseas meteorology. It convened a conference in 1930 to settle the matter. The colonial meteorologists, who had just set up their own civil service under Henry Hubert, vociferously objected. Algeria, under the Bureau of Algerian Affairs in the ministry of the interior and already party to a meteorological entente, decided not to contest what Delcambre wanted, provided that the meteorological service be elevated to a geophysical institute. The change, voted in 1930, took place in 1931. It resulted, among other things, in a salary increase for Lasserre's staff. Lasserre could henceforth require greater qualifications when positions were filled: a doctorate or the agrégation for the post of physicien-adjoint or météorologiste-adjoint, a licence for the post of aide-météorologiste. In principle, geophysicists at Algiers had achieved parity with the astronomers, and with the situation in the metropolis. In practice, however, the minister of finances in Paris used every available dodge to delay granting the salary increases.[110]

Governor General Paul Bordes found it convenient to support Lasserre's struggle for independence, since Lasserre fell under his jurisdiction. Paris having taken responsibility for meteorological stations near the Mediterranean, the way was open to use geophysics as a means of asserting imperial authority in the southern Sahara. "The governor general has taken the initiative," Lasserre wrote Maurain in 1928, "of creating two observatories for 1930, one at El Arfiane, near Touggourt, and the other at Tamanrasset, in the Hoggar. The task of setting them up has fallen to me. It is necessary to do 'something good with little money.'" The observatory at Tamanrasset, furthermore, was "to have a world interest." Besides meteorology, Lasserre projected studies of solar radiation and terrestrial magnetism for Tamanrasset, some two thousand kilometers south of Algiers.[111]

The choice of Tamanrasset was symbolic. Its activation coincided with the centenary of the French invasion of Algeria. Tamanrasset was also the final missionary outpost of Charles Eugène, vicomte de Foucauld, a Jesuit-educated veteran of the Algerian wars and a former

109. Lasserre, "Historique," p. xvii.

110. Ibid., p. xviii; PAN, F17/13590, Council of geophysical institutes and observatories, session of 8 Mar. 1930; Rector of Algiers Academy to Minister of Public Instruction, 23 Nov. 1931; note to Minister of Budget, 9 Apr. 1930; PIPG, Lasserre to Maurain, 18 Feb. 1930, where Lasserre still waits for salary increases authorized by the ministry of public instruction in 1926 and 1928.

111. PIPG, Lasserre to Maurain, 1 July 1928.

Trappist monk who had published accounts of his travels across Morocco's Rif in the 1880s.[112] He attempted to accumulate meteorological data before he fell victim to a revolt of the native Tuaregs in 1916.[113] Governor General Bordes sent a mission south to survey a site for the observatory near Fort Laperrine. Lasserre found a candidate to take charge of the new operation in Jean Dubief, a twenty-six-year-old graduate of the Algiers agricultural institute and also a veteran officer in the Algerian army, who had been working as an Algerian meteorologist for several years.[114] Lasserre, Lasserre's trusted assistant Miss Julie Malbos, the dean of the Algiers science faculty Léon Rouyier, and a medical doctor set off in two trucks to inspect the construction.[115]

The Tamanrasset expedition took thirty days, of which only four were spent at the destination. The meteorological and residential pavilions received their first occupants, aide-météorologiste Dubief and calculator Lauriol, in 1931. Lasserre led a second expedition there early in 1932; this time he brought five tons of material and equipment. He found Dubief and Lauriol camped in half-finished buildings, lacking plumbing and cooking facilities, their only link with Algiers being the nearby military radiotelegraph. Lasserre furnished radio equipment and two radio technicians to end their isolation; he enabled Tamanrasset to begin magnetical observations and hoped soon to provide a shed for work in atmospheric electricity.[116] Late in 1932, Lasserre traveled south for a third time to inspect the observatory, which had been damaged by torrential rains.

Dubief remained the only observer who could live with the altitude (1,350 m) and the isolation. A *mutilé de guerre* lasted seven weeks before retiring; an astronomer licencié from Paris and Besançon left after a year.[117] Queney, temporarily engaged as météorologiste-adjoint in 1931 to replace the detached Petitjean, took his turn at the southern out-

112. See André Boucher, *La vie héroïque de Charles de Foucauld* (Paris, 1931), p. 35, for de Foucauld's affinity with Marshal Louis-Hubert-Gonzalve Lyautey.

113. Jean Coulomb, "Note sur les observations météorologiques du père de Foucauld (avec six lettres inédites)," *Bulletin de l'enseignement public du Maroc* 28 (1941), 112–22.

114. PIPG, Lasserre to Maurain, 27 Mar. 1929, with Dubief's qualities; PMN, Dubief to Emile Delcambre, 9 Nov. 1926, where Dubief asks to join the Meteorological Society of France and provides Petitjean as a nominator.

115. Lasserre, "Historique," p. xix. PIPG, Julie Malbos to Director of the Caisse nationale de la recherche scientifique, 11 Nov. 1939, where Malbos indicates that she began serving the state in 1912; it is unclear whether the years of service were all in meteorology.

116. PIPG, Lasserre to Maurain, 23 Jan. 1930 and 25 Apr. 1932.

117. Lasserre, "Historique," p. xx. PAN, F17/13590, René Emile Elie Malfroy to Minister of Public Instruction, 23 Nov. 1933, for his application to go to Tamanrasset; Malfroy to Maurain, 11 June 1935, for his request to leave the southern post.

post.[118] Dubief brought the new station, dignified with the name Observatoire Jules-Carde, to world attention. His observations on physiological acclimatization brought interested inquiry from at least one English reader.[119]

As Tamanrasset was a symbol of French presence, it figured on international agendas. A suitable occasion came during the International Polar Year of 1932–33, whose organizing committee encouraged geophysical measurements from all points of the globe. The National Meteorological Office seized the occasion as an excuse to expand its minor station at Tamanrasset, near Lasserre's geophysical observatory. The Paris meteorologists sent out an expedition, funded by more than one hundred thousand francs, to scan the upper atmosphere using balloons equipped with radio emitters. The men ordered into the desert were Paul Bruyère, a graduate of the Ecole centrale and a meteorologist in Paris, and Roger Faillettaz, a Swiss student of radioelectricity who was in his sixth year of an appointment at the Paris meteorological office. They packed five tons of material in two convoys and by November 1932 found themselves in the Hotel Catelan at Tamanrasset, the command center booked in advance by Petitjean. They set up the usual meteorological recording devices in temporary outdoor installations and made their radio surveys for upper-atmospheric temperatures twice weekly. The data were compared against simultaneous soundings at other French locations – Paris and Rabat.[120] Several publications – notably Faillettaz's doctoral dissertation – resulted from the enormous expense.[121]

As early as 1929, Lasserre had asked Maurain to find a young man thirty-five to forty years old who could be groomed to replace him. The understudy "would truly have one day a beautiful domain." Lasserre emphasized that he was working for his successor, whom he wanted to know and train. "The task is crude and one must be firm, but satisfactions are to be had."[122] By 1936, having established the Tamanrasset

118. PIPG, Lasserre to Maurain, 6 July 1933.

119. PMN, Cicely M. Botley to Philippe Wehrlé, 12 Apr. 1935; Wehrlé to Dubief, 27 Apr. 1935. Jules-Gaston-Henri Carde was a colonial administrator who rose to become governor general of Algeria from 1930 to 1935 (*DBF*, s.v. "Carde").

120. Philippe Wehrlé, "La participation de l'Office national météorologique à l'année polaire internationale 1932–33," *La météorologie* 77 (1934), 1–29.

121. Faillettaz and Robert Bureau, "Les enregistrements d'atmosphériques à Tamanrasset (Hoggar) au cours de l'année polaire," *CRAS* 199 (1934), 376–78; Faillettaz, "Un nouveau procédé d'enregistrement des atmosphériques pour la prévision des orages," ibid., 1647–48; idem, *La radiation solaire et le trouble atmosphérique à Tamanrasset (Hoggar)* (diss., University of Paris, 1937); Bruyère, *Rapport de la mission météorologique française sur les travaux effectués à Tamanrasset pendant l'année polaire* (Paris, 1941).

122. PIPG, Lasserre to Maurain, 6 Oct. 1929.

Observatory and having been forced to reintegrate the bête noire Petitjean, Lasserre had had enough. The world of geophysics scrambled to take advantage of his imminent retirement. The man who became master of Lasserre's empire was Algeria's greatest scientist.

Roots of Excellence

Lasserre's successor, Jean Coulomb, was born in Blida, Algeria, where his father was a teacher of Greek and Latin at the local collège and his mother was a leading light of the Ecole primaire supérieure—the modernist middle school for girls. Coulomb attended his father's school and then the lycée at Algiers, situated some fifty kilometers from Blida, where in 1910 his mother had become a teacher (his father moved to Algiers in 1911). He lived in a perfect French cocoon, quite isolated from the local Arab population. Coulomb, whose talents lay in mathematics, finished secondary school at Marseilles, where his mother was transferred in 1920 (his father followed in 1921). He sat for the most prestigious national examinations and succeeded brilliantly, gaining entry to both the Ecole normale supérieure and the Ecole polytechnique. He made the usual choice for an aspiring scientist in attending the former instead of the latter.[123]

Coulomb picked up his licence in 1925 and graduated from the Ecole normale in 1926, after which he spent a year as a reserve officer and then second lieutenant teaching at Metz.[124] Following military purgatory, Coulomb obtained a scholarship from the Fondation Commercy at the Sorbonne's Faculty of Sciences and began predoctoral research by asking advice of Elie Cartan, professor at the Collège de France. Coulomb's friend Yves Rocard, who had been a class ahead of him at the Ecole normale, was then préparateur for Marcel Brillouin at the Collège. Brillouin had turned to devote his attention to geophysics, specifically tides and seismology. When for reasons of his own Rocard decided to find another patron, Brillouin advertised for a mathematician replacement. Coulomb, desiring a different master, opted for the open position, which, however, had been devalued in rank.[125] The new pupil got along well with his teacher and progressed to become Brillouin's assistant. Brillouin proposed that Coulomb write his thesis on

123. Jean Coulomb, interview with author, 28 Feb. 1989; Coulomb, letter to author, 7 Sept. 1989; Coulomb, *Titres et travaux scientifiques de Jean Coulomb* (Algiers, 1941).

124. Coulomb, *Titres.*

125. Rocard recalls that the assistantship was downgraded to the lower-paying position of préparateur temporaire; he joined the Société la radiotechnique as a radio engineer, and the devalued assistantship went to Coulomb (Dominique Pestre, *Physique et physiciens en France, 1918–1940* [Paris, 1984], p. 239).

the propagation of seismical waves, and he successfully defended it in 1931. Charles Maurain sat as a member of Coulomb's jury.[126]

Just as he was finishing his doctorate, two positions materialized at the geophysical institute of the University of Clermont-Ferrand and its associated observatory at Puy-de-Dôme. Neither the directorship (opened by the retirement of Emile Mathias) nor a post of météorologiste-adjoint was a plum. Jean Dufay, for one, then at the physics laboratory of the University of Lyons, preferred to wait for an opening at the local observatory rather than move to direct a minor institute.[127] The leading candidate for the directorship was Pierre Weiss's student Louis Néel, then twenty-seven years old and a graduate of the Ecole normale; Gaston Grenet, already at Puy-de-Dôme, would be promoted to the meteorology position.[128] Néel, accomplished in both mathematics and experiment (although lacking in geophysical experience), accepted the directorship. He arrived in Clermont, took the measure of his job for several weeks, and then fled in terror. He explained his reasons in a long letter to Maurain. He was really a physicist interested more in experiment and the general laws of nature than in observation and the accidental form of the globe; borrowing from a military metaphor, he enjoyed trying to "subdue nature in its last positions, . . . to work out theories and find crucial experiments that allow one either to invalidate them definitively or to give them a beginning of probability." Néel did not enjoy administration, and he found himself unable to put up with "the total nonexistence of a library, the complete intellectual isolation in his [*sic*] specialty, the number of researchers understandably limited, and finally the unwholesome regard [*l'allure un peu morte*] that the Institut de physique du globe occasions in the Faculty at Clermont." He would have put up with it all had he been able to continue his research into magnetism, but Clermont's resources simply did not permit it. He was not "*the right man in the right place.*"[129] The central authorities' next choice to direct Clermont was fifty-nine-year-old Joseph Lacoste, an atmospheric physicist then *professeur sans chaire* at Strasbourg. Lacoste and his wife descended on Clermont to view the situation firsthand.[130]

126. Interview with Coulomb, 28 Feb. 1989; Coulomb, *Titres*. In his autobiography, Yves Rocard is uncharitable toward Coulomb's later administration of geophysics, although he is silent about the early years (Rocard, *Mémoires sans concessions* [Paris, 1988], pp. 213–14).

127. PIPG, Dufay to Maurain, 25 May 1931.

128. Ibid., Director of Higher Education to Rector of Clermont, 23 July 1931.

129. Ibid., Néel to Maurain, 15 Oct. 1931.

130. Ibid., Rothé to Maurain, 28 Oct. 1931; Rothé to Cavalier, 15 Oct. 1931, for Lacoste's vita; Charles Mathias to Maurain, 31 Oct. 1931.

While Lacoste dithered about moving south, Maurain sought to install Coulomb in Lacoste's position at Strasbourg. The dean of geophysics sent a copy of Coulomb's thesis to Rothé at Strasbourg, hoping to persuade the senior seismologist that Coulomb would be a useful acquisition. Rothé responded with skepticism and hostility. He wrote to Maurain that Coulomb's work was "interesting but it has a purely mathematical character; it is even a thesis in mathematical sciences, and under these conditions I have strong apprehensions about the way that it [presumably his institute] would view replacing Mr. Lacoste." Rothé wanted an experimenter and an observer, not a mathematician with theoretical inclinations; lack of observational material in the thesis indicated how Coulomb approached seismology.[131] But Lacoste placed too many conditions on his appointment, and the director of higher education, Jacques Cavalier (a metallurgical chemist and a normalien), finally decided to let him go in favor of Grenet. What Cavalier regretted most was not the loss of Lacoste in Clermont but the loss of Coulomb in Strasbourg. When Rothé learned about Grenet's appointment, he quickly suggested Charles Bois, one of his seismologists, for the junior post.[132] The post was to be Coulomb's, however.

Already at the end of August 1931, Marcel Brillouin had written to Maurain to suggest Coulomb for the directorship at Clermont. Coulomb spoke to Néel about the situation there and then affirmed his desire to be considered for the post. A mitigating factor was his wife's roots in Auvergne. A stronger argument for a move derived from Coulomb's wife's teaching position at Chartres, which had turned Coulomb into a weary commuter. Coulomb presented himself to Grenet, a man of about his own age, who saw no reason to oppose what the central authorities favored. Coulomb installed himself in Clermont. Over the next four years he published more than a score of communications in seismology, a number with the Academy of Sciences but most in the Clermont institute's house organ. Coulomb and Grenet worked well together. In 1938 the Academy of Sciences awarded them the Victor Raulin prize in geophysics.[133] There was enough time for Coulomb to join his old schoolmates from the Ecole normale in founding the mathematical circle known as Bourbaki. His contribution, limited to the early years of the circle, concerned drafts of texts on the theory of integration and special functions. Coulomb worked together with two

131. Ibid., Rothé to Maurain, 2 Nov. 1931.

132. Ibid., Cavalier to Maurain, 24 Nov. 1931; Lacoste to Director of Higher Education, 23 Nov. 1931 (copy); Rothé to Maurain, 28 Nov. 1931.

133. Ibid., Brillouin to Maurain, 21 Aug. 1931; Coulomb to Maurain, 17 Oct. 1931; Coulomb, interview with author, 28 Feb. 1989. PIPG, Grenet to Maurain, 20 Jan. 1932. Coulomb, *Titres*.

other members of the circle who were posted to Clermont-Ferrand, René de Possel and Szolem Mandelbrojt. As his mathematician colleagues progressively abandoned reference to the real world, Coulomb withdrew from the enterprise.[134]

Maurain's protégé put in a short tour of duty under Grenet and then looked to replace Albert Lasserre as director of the geophysical institute in Algiers upon the latter's retirement in 1936. Coulomb sent his vita to the ministry of national education around the end of February.[135] Coulomb received Algiers — in his native land — while a directorship of lesser importance nearby, that of the meteorological observatory at Tunis, went to Rothé's protégé Charles Bois.[136] The minister of foreign affairs noted with satisfaction in February 1937: "With Bois in Tunis and J. Coulomb at Algiers, North Africa will be well provided for."[137]

Crown of Glory

Coulomb found the institute populated with young scientists and overseen by the retired Lasserre, who had contrived to receive the title of honorary director. The new man set about to make Algiers second only to Paris in the world of French geophysics. Even though his interests were in seismology, he decided not to dismantle the weather service, which was highly valued by the local authorities, and to continue to collaborate with disciplinary cousins in astronomy and meteorology throughout North Africa. What was most important, he "began to ask for a bit of scientific research (actinometry and atmospheric electricity) from those of my staff who seemed equal to the task." Coulomb had the right administrative touch: "It doesn't seem to be going badly. All told, the people are friendly and I've been able to dispose of the few questions that seemed delicate."[138]

In reorienting his institute, Coulomb faced a heritage not untypical in the world of geophysical observatories. He had a mandate both to provide practical information and to undertake original research. Administratively he was tied into the university system, and although the

134. Liliane Beaulieu, *Bourbaki: Une histoire du group de mathématiciens français et de ses travaux (1934–1944)* (diss., University of Montreal, 1990), pp. 183–86.

135. PIPG, Coulomb to Maurain, 1 Feb. 1936, where Coulomb declines to accept a temporary post in Japan until he is sure that he will be considered as Lasserre's successor; 20 Feb. 1936.

136. Ibid., Bois to Maurain, 29 Jan. 1935. Bois completed an uninspired doctorate under Rothé: Bois, *Recherches sur les séismographes destinés à l'inscription de la composante verticale du mouvement du sol* (diss., University of Strasbourg, 1933).

137. PIPG, Minister of Foreign Affairs to Maurain, 10 Feb. 1937.

138. Ibid., Coulomb to Maurain, 4 June 1937.

affiliation with pure learning carried a certain measure of prestige, it deprived him of access to other sources of funds. He reported on his peculiar situation to Maurain late in 1937. His request to the Algiers university council for a modest increase in funding for his meteorological service had summarily been refused. Coulomb explained the position of the university rector, the specialist in French literature Pierre Martino: "The Institute of Meteorology and Geophysics had, according to him, a dual character: that of a scientific institute and that of a meteorological service. The University could fund scientific work but could not commit itself to the regular requirements of meteorological positions or to the Tamanrasset Observatory." Coulomb decided that he would press the matter by recalling the general status that had been accorded geophysical institutes in France. Furthermore, "I would insist on symmetry with the Astronomical Observatory, which here avoided being reduced to a service of the [governor general's] General Secretariat (this would not perhaps be bad for funding, but it would no longer be possible to do anything except routine meteorology, which would depress me)." And to argue his case effectively before local bureaucrats, Coulomb needed to have a permanent voice on the university council, not, as was the tradition, a consultative presence during debates concerning his institute. In fact, he wrote, according to a law of 1877, astronomical observatory directors were members of university councils ex officio, and geophysical observatories possessed the juridical status of their astronomical sisters.[139]

The best way to demonstrate that his institute was designed for scientific research, Coulomb continued in his letter to Maurain, was to produce science. He sought "with all in my power to shake up the routine [meteorology]," and his program to send the meteorologists back to basics led one of his staff, Lucien Petitjean, "to accuse me of transforming him into a pedagogue." Coulomb pushed ahead in actinometry and atmospheric electricity. With Maurain's support, Coulomb counted on receiving a grant from the Caisse nationale de la recherche scientifique. Two large projects absorbed his attention. The first concerned making sense of the meteorological and magnetical data that had accumulated over the preceding decades; its logical outcome would be publication of a general climatology for Algeria. Before such a climatology could be produced, however, it would be essential to obtain reliable data from the southern part of Algeria. And so Coulomb's second project concerned renovating and expanding the observatory set up by Lasserre in Tamanrasset.

Coulomb expended a great deal of effort to staff and maintain the

139. Ibid., 4 Dec. 1937.

Tamanrasset Observatory and to publish the data that it collected. Soon after arriving, he learned that Queney would receive his own former post at Clermont-Ferrand. This set in motion the usual scramble to fill a subordinate position. Coulomb, however, was concerned that any staff reorganization permit Jean Dubief to take temporary leave of Tamanrasset and finish his licence in residence at Algiers. Like his predecessor, he inspected the facilities of the southern station firsthand. By late summer 1938 he could report to Maurain that the Tamanrasset observers were forging ahead on a number of fronts. The senior men there, Dubief and Queney's replacement Louis Arbey, took magnetical and meteorological observations of all sorts. Arbey was a highly recommended graduate of the Algiers faculty whom Coulomb had schooled in measuring solar radiation and charged, in a distant future, with producing a thesis in the general area of atmospheric transparence.[140] Dubief, as an old hand, readily volunteered to test theoretical propositions emanating from Algiers.[141] The men erected a hilltop shed to study temperature inversions. "I have often had to brake Dubief, who has too much ardor and would easily undertake too many things. With Arbey, who is much cooler, the two make an excellent team."[142] Late in 1938 Coulomb proposed to send a new observer, a certain Kriéger, to study atmospheric electrification produced by the collision of sand particles in the desert wind.[143]

Galloping across the desert sands with a modified Villey-Bouty mirror electrometer (such was the instrument that Coulomb had in mind for his atmospheric-electrical observer) would be the stuff of romance and legend. The mountain of data accumulating in Algiers, however, was an administrative nightmare. Making sense of it would be an exacting and, in effect, a stupefying task. There would be no funds to engage a real scientist for the job. Coulomb looked around Algiers and found an aspiring playwrite, the very antithesis of an imperialist adventurer, who, having recently concluded that "there is nothing here for one who wanted to learn, educate himself, or better himself," was by all accounts slowly starving to death. He signed on Albert Camus as a temporary assistant in meteorology.[144]

140. Ibid., 4 June 1937; 21 Aug. 1938; 14 June 1937 for Arbey's origin; 4 Dec. 1937 for Arbey's training before being sent south; 14 Dec. 1937, for Arbey's research orientation.

141. Ibid., Queney to Maurain, 2 Feb. 1937, reporting Dubief's observations at Tamanrasset confirming the theoretical calculation of winds that Queney had provided in his dissertation.

142. Ibid., Coulomb to Maurain, 21 Aug. 1938.

143. Ibid., 21 Aug. 1938 and 20 Dec. 1938.

144. Herbert R. Lottman, *Albert Camus*, trans. Marianne Véron (Paris, 1978), pp. 179–

Camus began in November 1937. He spent all of a working day setting up climatological card files for the data that had accumulated over twenty-five years from 400 meteorological stations in Algeria. This completed, he began a study of atmospheric pressures in all of North Africa. Comparing recordings made at neighboring stations resulted in the identification and correction of numerous errors. Coulomb was delighted at Camus's progress. The literary assistant possessed "a broad level of culture very much higher than that of people I supposed I would be able to hire." Meteorologist Lucien Petitjean, who directly supervised Camus's analysis of the pressures, described the meticulous and fatiguing nature of the work. For twenty-seven years' data Camus plotted curves of barometrical pressures at 121 stations; he then calculated tables of successive sums for monthly pressures and mean values (determined by using a planimeter). This allowed Petitjean to trace out isobars of deviations from the mean. Petitjean estimated that the final corrected barometrical curves would take Camus around four hundred hours to complete.[145]

Camus worked at Coulomb's institute until September 1938, constructing averages from the meteorological data that represented, as he noted in his journal, only "an arbitrary slice of reality." These reflections of the natural world, however, rose to occupy a central place in Camus's writings. The physical climate is omnipresent in his novels and essays. For example, near the end of a piece from 1948 on the aftermath of the war he writes: "Nature . . . contrasts her calm skies and her reasons with the madness of men. Until the atom too catches fire and history ends in the triumph of reason and the agony of the species." The earliest surviving notes of the text that would become *L'étranger*, written in December 1937, focus on the weather – rain in winter. Then comes a passage about a man who, despite having a promising future before him, worked in an office; Sunday he rose late and looked out the window to see sun or rain. "Thus the entire year. He waited. He waited to die."[146]

Coulomb waited for no man. The major tune-up he gave to the Algiers institute made such an impression on his metropolitan patron that after scarcely a year on the job, he received from Maurain a list of open-

80; Camus, "Summer in Algiers [1936]," in Camus, *The Myth of Sisyphus and Other Essays*, trans. Justin O'Brien (New York, 1955), pp. 104–13, on p. 104.

145. PIPG, Coulomb to Head of Central Service of Scientific Research (Paris), 29 June 1938, application to renew a grant, to which is appended "Rapport sur les travaux effectués par M. Camus, aide-techique," from which the quotations attributed to Petitjean are taken.

146. Lottman, *Camus*, p. 181; Camus, "Helen's Exile," in *Myth*, pp. 134–38, on p. 137.

ings in the hexagone. Two plums were in the offing: Rothé's directorship at Strasbourg and Maurain's Paris lieutenantship, then under Ernest Labrouste. Coulomb replied to his master:

> In principle the work at Strasbourg would interest me more than what I must do in Algeria. But I do not despair of making the latter interesting, and perhaps when that happens I would prefer my present situation. (Regarding courses, the Faculty of Sciences would like me to involve myself, so that it is not what would attract me to Strasbourg.) On the other hand, I believe that there is no position comparable to that of Mr. Labrouste as far as the possibilities of work without the usual complement of administrative responsibilities (you will excuse this cynicism), and I will be happy if you still feel, when the time comes, that I might be a candidate.[147]

Coulomb was not interested in exiling himself to France's second center of science, Strasbourg. Once he had mastered the situation in the colony, nothing short of Paris would do. He watched as Maurain suffered bouts of ill-health.[148] When Coulomb was en route to Tamanrasset, he learned about the death of Maurain's son in an Alpine accident; he wrote eloquently to express his sympathy.[149] He lost no opportunity to stick a knife in the back of a potential rival. Charles Bois told Coulomb that Tunis was an impossible posting and that Maurain had advised him to anticipate a transfer to France in a few years; Coulomb wrote to Maurain that Bois's position was indeed not promising, although "unfortunately Mr. Bois seems to me to lack tact, and I hardly see how he can be helped."[150]

By early 1939, however, tact and metropolitan connections could help none of the North African researchers. The impending war had closed in. Of Coulomb's three junior staff, Kriéger was mobilized, Dubief was on leave from Tamanrasset, and Arbey had been transferred to the astronomical observatory in nearby Bouzaréah. Coulomb implored Maurain to find him a new man to fill Arbey's shoes; otherwise he would have to advertise the position locally, "even though this method has hardly succeeded during previous vacancies." The end of August found Coulomb and his family in France. Mobilized, he returned to Algeria by way of England and joined the military meteorologists at Oran. His wife and three children arrived in Algiers several weeks later. Dubief

147. PIPG, Coulomb to Maurain, 14 Dec. 1937.

148. Ibid., 8 Nov. 1938, where Coulomb comments on Maurain's cancellation of a trip to Syria: "I hope that it would not be for reasons of health, but only because of the alarms of September, that you would have canceled going there. It would not be any less of a disappointment for Mrs. Maurain, with whom you were counting on beginning your retirement in this way."

149. Ibid., Coulomb to Maurain, 12 Jan. 1939.

150. Ibid., 25 Apr. 1939.

was mobilized at his post in Tamanrasset. Only Petitjean and his assistants remained at the geophysical institute, predicting weather in secret for the French National Meteorological Office. The institute lost its radio equipment to the army and henceforth had no means of contacting Tamanrasset. Lasserre suffered the ultimate insult: In Coulomb's absence Petitjean was given direction of the institute's shell.[151] "Return Coulomb to us, have him mobilized in Algiers," Lasserre cried to Maurain, for Coulomb, "with admirable ardor and very broad competence," had given the institute "a new spirit." After Maurain pulled strings to have Coulomb returned to the observatory, Coulomb – transferred to Algiers – decided to stay in the army as a meteorologist, while apparently retaining his director's title.[152] He headed the principal military meteorological station in Algiers, leaving Petitjean in charge of the derelict institute. Fealty found its true reward. After lecturing at the Algiers agricultural institute in 1940, Coulomb was tapped to succeed Maurain as director of the Paris geophysical institute in 1941.

The impression should not be left that Coulomb did nothing but climatology while in the colonies. These particular interests figured along with systematic work in a number of additional areas, and in each area the geophysicist strove to arrive at *general* laws. Coulomb found time to gratify his passion for mathematical physics applied to seismical waves and to exploit a number of lesser themes.[153] Apotheosis was the expected result. Having prospered under the German occupation, Coulomb became a dominant force in the Fourth Republic. His chest sagged under the weight of medals, and his appointment book filled with meetings of one or another council of elders. The Academy of Sciences admitted him as a member in 1960.

The Algeria of Jean Coulomb and Albert Camus was an enclave of French culture in an expanse of Arab land. Physics and astronomy at Algiers were exemplary, exuding Parisian essences. The ensemble of exact sciences there compared favorably with those of provincial locations in the hexagone. Algeria's rise to scientific prominence derived from the persistent labors of administrators who engineered research projects and brought them to a published conclusion – Charles Trépied, François Gonnessiat, Albert Lasserre, Lucien Petitjean, and Jean Coulomb.

151. Ibid., 25 Apr. 1939 and 1 June 1939, quotation from the latter; Lasserre to Maurain, 21 Sept. 1939 and 8 Oct. 1939.

152. Ibid., Lasserre to Maurain, 8 Oct. 1939, for the quotation, and Coulomb to Maurain, 23 Nov. 1939; A. Lasserre and J. Dubief, *Carte de la déclinaison magnétique sur le Sahara pour 1938,0* (Algiers, 1940), pp. 5–6, where Coulomb signs the preface as the institute's director.

153. See Coulomb, *Titres*, which, prepared in anticipation of Coulomb's triumphal return to Paris, includes a summary of his research career.

Their epistolary efforts on behalf of the French civilizing mission were minimal, and one sees no colonial stamp on the fruit of their work. By serving the scientific interests of Parisian masters, they incarnated and spread metropolitan values.

5

Tunisia and Morocco: The Antebellum Satrapies

The Eastern Regency

The generation following the founding of the Third Republic saw the cancer of French presence spread from beachheads in Algeria to the entire coastal quadrant of northwest Africa. The more immediate gratification of French ambitions lay in Tunisia, the farthest removed of the Turkish dependencies. Tunisian cities were liable to naval bombardment, and it was a small matter to ferry troops in from nearby Algerian garrisons. When France moved against Tunisia, however, it did so with the hindsight of nearly two generations of rule in Algeria. There being a spiritual attraction to the indigenous rulers of Tunisia, who themselves had been disciplined to accept foreign supervision first from Constantinople and then from European bankers, France recognized the beylic regime as regents and undertook to instruct them as compradores.

Introducing Tunisians to Racine and Descartes took place in the context of a sophisticated system of higher learning. By the time the French took control of Tunisia in the early 1880s, Tunisians enjoyed a system of higher education that had already been apprised of European developments. At the pinnacle of the educational pyramid was the Mosque of the Olive Tree, Djemaà Ez-Zitouna, which had dispensed Islamic wisdom for a millennium. Ez-Zitouna assumed its modern university character in 1842, when Ahmed-Pacha Bey provided a charter establishing thirty chairs; professorial salaries were augmented and assistant professorships established in 1870. Five years later the Tunisian minister Khéreddine set statutes of conduct at Ez-Zitouna in great detail, in areas ranging from staff appointments, lesson plans, and teaching methods to diplomas, grades, and student dossiers. The university—there is every reason to provide it with such a designation—drew instruction from recent developments in the eastern Mediterranean, in Beirut and Cairo, toward which it looked for inspiration in the form of teachers and learned treatises. The universe of the mind in Islamic Tunis, however, had definite limitations. In principle any publication by a university professor could be censored by the state. The subjects of instruction were limited to theology, grammar, rhetoric, and Islamic law.[1]

European learning percolated to Tunisian shores during the middle of the nineteenth century. Various military missions between 1831 and 1854 dispensed elementary technical education.[2] A French-language military school, modeled (it is claimed) on the Ecole polytechnique, operated at Bardo with beylic patronage from 1855 to 1867.[3] The most significant infiltration of foreign ideas occurred on the eve of the French conquest. The minister Khayr ad-Din founded the Collège Sadiki (giving it the name of the reigning potentate, Sadok Bey). It dispensed secondary instruction in Arabic, European languages and natural sciences, and athletics to 150 pupils. Graduates of the Sadiki could then go on to acquire traditional instruction at Ez-Zitouna, and a number traveled to Paris for additional study at the Ecole centrale des arts et manufactures. The modernist part of the curriculum was overseen by Zéphyrin Aymond, the French professor of mathematics and physics who had previously taught at the Bardo military school.[4] The French conquest could have been expected to provide impetus for identifying compradorial talent. Gallic policy was not of a mind to encourage independent scientific thinking. During the first decades of the twentieth century, the Sadiki declined precipitously in enrollment and élan.[5]

1. Mohamed Lasram, "L'enseignement supérieur musulman à la Mosquée de l'olivier, ou Grande Mosquée, à Tunis," in *Congrès de l'Afrique du Nord tenu à Paris du 6 au 10 octobre 1908, compte rendu des travaux*, ed. Charles Depincé, 2 vols. (Paris, 1909), 2:144–73; Mahmoud Abdel Moula, *L'Université zaytounienne et la société tunisienne* (Tunis, 1971) [diss., University of Paris, 1971], pp. 79–85. The medieval background in Tunisia is discussed in Ahmed Chérif, *Histoire de la médecine arabe en Tunisie* (med. diss., University of Bordeaux, 1908). For a detailed discussion of the present state of knowledge about Islamic institutions of higher learning, especially in medieval times, see George Makdisi, *The Rise of Colleges: Institutions of Learning in Islam and the West* (Edinburgh, 1981).

2. L. Carl Brown, *The Tunisia of Ahmad Bey, 1837–1855* (Princeton, 1974), pp. 283–94.

3. Mustapha Kraïem, *La Tunisie précoloniale*, 2 vols. (Tunis, 1973), 2:176–81; Mongi Smida, *Khéreddine: Ministre réformateur, 1873–1877* (Tunis, 1971), pp. 291–93.

4. Abdel Moula, *Université zaytounienne*, pp. 85–86; Kraïem, *Tunisie précoloniale*, 2:195–201; Smida, *Khéreddine*, p. 303, for Aymond's name and pedigree.

5. According to Abdel Moula, *Université zaytounienne*, p. 116, by 1907 the Collège Sadiki had only seventy-five pupils; the number of Islamic Tunisian professors fell to seven in 1910 and three in 1919. According to Tunisia, Direction générale de l'instruction publique et des beaux-arts, *L'oeuvre scolaire de la France en Tunisie (1883–1930)* (Bourg, 1931), pp. 63–71, by 1930 the Sadiki had nearly two hundred pupils in its senior section; nine teachers had French diplomas, and six Arabic teachers came from the "Great Mosque" (Ez-Zitouna). See also Paule Laberge, *Politiques scientifiques au Maghreb: L'im-*

Higher education was left to Ez-Zitouna and the French university in neighboring Algiers.

Following the French pattern of cultural imperialism, a mission of scientific exploration arrived shortly after the military declared a protectorate over Tunisia in 1881. The commission's head, physician and botanist Ernest Cosson, employed three geologists to chart the new land. One of the three, the military veterinarian Philippe Thomas, had previously served in Algeria; in 1885 Thomas discovered a large deposit of commercially exploitable phosphates in southern Tunisia, along the border with Algeria. Cosson died in 1889, the commission expired, and the Tunisian landscape received no synthetic appraisal. The Tunisian public works administration delayed mining phosphates until 1899. In 1902 the ministry of public instruction in Paris commissioned Thomas to publish a geology of the protectorate; two volumes had appeared at the time of Thomas's death in 1910.[6]

The history of advanced instruction under the Tunisian regency is written in private initiative. Foremost among educational innovators were Catholic missionaries from Algeria, who arrived in 1880 to establish the Collège St-Louis in Carthage. Subjected to curricula of French lycées, students were forbidden to speak anything but French; the school soon moved to Tunis, and after a bit more than a decade of operation there under the name Collège St-Charles, it took a secular turn by adopting the name Lycée Carnot.[7] Muslim intellectuals followed their religious rivals. Responding to interest among modernist professors at Ez-Zitouna and Sadiki, the French resident general René Millet permitted establishment of an Islamic society with the aim of reconciliating tradition and science. The society took its name from the Muslim historian Ibn Khaldoun and promptly offered instruction in arts and sciences. The Khaldouniyya's library contained recent Arabic periodicals from Beirut and Cairo, such as *al-Hilâl* and *al-Mouqtataf*, which discussed such controversial ideas as Darwinian evolution. In scientific matters, the society's lecturers turned toward practical certification. A

---

plantation du système scientifique dans les sociétés maghrébines de 1830 à 1980 (diss., University of Montreal, 1987), pp. 66–67, 73–77.

6. Eusèbe Vassel, "L'auteur de la découverte des phosphates africains et la mission d'exploration scientifique de la Tunisie," Institut de Carthage, *Revue tunisienne* 5 (1898), 401–22, on pp. 414–16; Georges Millot and Philippe Morin, "Centenaire de la découverte des phosphates d'Afrique du Nord par Philippe Thomas," *La vie des sciences* [Paris, Academy of Sciences, *Comptes-Rendus*, série générale] 3, no. 1 (1985), 71–79.

7. Kraïem, *Tunisie précoloniale*, 2:192–93; Dwight L. Ling, *Tunisia: From Protectorate to Republic* (Bloomington, 1967), pp. 92–94; Tunisia, Direction générale de l'instruction publique et des beaux-arts, *Oeuvre scolaire*, pp. 58–63.

diploma program for land surveyors came in 1905. The Khaldouniyya's principal function remained, however, to popularize science for students at Ez-Zitouna, the neighboring classical university.[8]

The French invaders invested Tunisia with permanent institutions to facilitate exploitation, and among these was the Colonial School of Agriculture at Tunis. Created in 1898, the school spawned an agricultural experimentation station in 1908 and a Botanical Service in 1913. The man in charge of all the incarnations of Tunisian agricultural climatology from 1898 until the late 1930s was Félicien Boeuf. As early as 1906 he applied Mendelian genetics to breed better cereal plants. During the course of his tenure, Boeuf published five doctoral theses in his house organ, the *Annales du Service botanique et agronomique.*[9] By the 1920s he became interested in climatology, and he encouraged Władysław Gorczyński to measure solar radiation in Tunisia. Boeuf published Gorczyński's actinometrical measurements for Algeria as well as for Tunisia, and he permitted his *Annales* to serve as a forum for Gorczyński's climatological speculations.[10]

Though styling himself a patron of science, Boeuf could not compete with the premier scientific institution in Tunisia, the Pasteur Institute. It emerged in 1893 as one of Resident Millet's practical concerns. The institute's first director, Adrien Loir, resigned in 1902. Loir's veterinarian adjoint took over until the appointment of Charles Nicolle shortly thereafter. Nicolle renovated the institute with the help of Resident Stephen Pichon; in 1909 he and his colleagues announced to the world that fleas were the carrier of bubonic plague, a discovery that later garnered a Nobel prize. Nicolle's forays into public health, like biologist Henri Heldt's into physical oceanography, drew on laboratory analyses. But it was a French professor at the Collège Sadiki, Gaston Ginestous, who pioneered exact sciences in the Tunisian protectorate.[11]

8. Abdel Moula, *Université zaytounienne*, pp. 103–6; Mohamed Lasram, "L'enseignement supérieur musulman: La Khaldounia à Tunis," in *Congrès de l'Afrique du Nord*, 2:173–80; Laberge, *Politiques*, pp. 218–26.

9. Boeuf, "Trente-huit années d'expérimentation agricole en Tunisie," Tunisia, Direction des affaires économiques, *Annales du Service botanique et agronomique* 12–13 (1935–36), 3–31; Laberge, *Politiques*, p. 133. On nineteenth-century Arabic intellectual life in Beirut and Cairo, see Adel A. Ziadat, *Western Science in the Arab World: The Impact of Darwinism, 1860–1930* (London, 1986).

10. *Annales du Service botanique et agronomique* 3 (1925), 5–87, featuring Boeuf's general remarks on solar radiation and Gorczyński's data.

11. Yves Châtelain, *La vie littéraire et intellectuelle en Tunisie de 1900 à 1933* (Paris, 1937), pp. 237–49, for the Pasteur Institute and the physical analyses of ocean water in the Salammbô oceanographical station; Augustin Bernard, "La Station océanographique de Salammbô," *Annales de géographie*, 1926, pp. 468–69. On the overwhelmingly biological focus of Heldt's station, see Heldt, *Rapport sur l'organisation, l'activité et les travaux de la*

Although climatology found interest among the agricultural botanists, the exact sciences emerged from the Tunisian meteorological service, founded in 1885 by the resident's educational authorities. The first five weather stations came under the control of Algerian and metropolitan meteorologists, and others were added by various Tunisian services. Between 1889 and 1900 the meteorological service was passed from one portfolio to another in the resident's secretariat, while its part-time director Ginestous sought to standardize the incoming climatological data. Ginestous's first synthesis appeared in 1903, and it earned him a Parisian doctorate. Then he published occasional meteorological and geophysical observations in the house organ of the local literary and ethnographical society, the Institut de Carthage.[12] By 1921 Ginestous ran a reporting network of 179 stations. No theoretician, the French meteorologist aspired to instruct colonial administrators. He labored for more than thirty years to document the climatical regimes of the protectorate, turning his hand to such practical problems as aviation and evaporation.[13]

On the eve of his retirement late in the 1920s, Ginestous added seismological and magnetical recording to his central observatory in Tunis.[14] Ginestous was, however, no Albert Lasserre, and he did not investigate general questions in geophysics. Ginestous's successor Victor Lacroux, who arrived from Paris to take over direction of the meteorological observatory late in 1929, fared no better. Lacroux took notes on Lasserre's organization of weather prediction. More than five years later Lacroux had yet to activate his magnetical recorders.[15] When La-

Station océanographique de Salammbô depuis sa création (1924–1931) (Paris, 1931) [Station océanographique de Salammbô, Bulletin, no. 24], pp. 119–23.

12. Ginestous, *Etude sur le climat de la Tunisie* (Tunis, 1903), where the preface discusses the history of the meteorological service; Ginestous's doctorate of nearly the same title, for 1906, recorded in Châtelain, *Vie littéraire*, p. 248. A lecture by Ginestous, where he reports on an earthquake of 24 January 1909, is noted in *Revue tunisienne* 16 (1909), 167.

13. Ginestous, *Conférences de météorologie faites aux élèves de l'Ecole coloniale d'agriculture de Tunis* (Tunis, 1922), p. 3, for the extent of the meteorological network; Ginestous, *Etude climatologique du golfe de Tunis en vue de l'aéronautique et de l'aviation* (Tunis, 1925); PIPG, Ginestous to Maurain, 28 Dec. 1927, where he outlines a project to measure evaporation.

14. PIPG, Ginestous to Maurain, 23 Nov. 1928, where he describes the magnetical apparatus that he is ordering from France and the seismological equipment already in place.

15. Ibid., Lacroux to Maurain, 16 Dec. 1929, reporting his arrival in Tunis; Lasserre to Maurain, 13 Feb. 1930; Maurain to Colonel Ferri (Florence), 3 Mar. 1931 (copy); Lacroux to Maurain, 16 Mar. 1932, where Lacroux has chosen Moghrane (sixty kilometers from Tunis) as his site for the terrestrial magnetism recorders; ibid., 15 Jan. 1935, where

Victor Lacroux (*left*) prospecting at Raz el Djebel, Tunisia, for underground water by means of short-wave echoes, 1932. Scientists in North Africa depended on the French army for logistical support and protection. (Courtesy of Institut de physique du globe, Paris)

croux died suddenly of a heart attack in the middle of 1936, the director of public works in Tunis asked Charles Maurain to find a replacement. The new man would best be an aide-physicien attached to Maurain's institute; he could hope for a salary as large as Fr 53,760, to which would be added liberal vacations and free lodging.[16]

Charles Bois received Lacroux's post. Bois, then forty-six years old, had parlayed an agricultural diploma into a position at the National Meteorological Office in the years after the war. In 1922 he moved to Edmond Rothé's institute as stagiaire and earned a number of university certificates and diplomas, ending with a doctorate on vertical seismometers; he was at home in every branch of geophysics. Bois arrived early in 1937. He was "quite satisfied" with what he found: "My coworkers evidently lack scientific culture, but from the technical point of view they are rather informed about their service, and they are very devoted. I can have a bit of time for my personal work once I am familiar with

---

Lacroux laments that he is unable to install the recorders because the administration of the protectorate will not construct a building for them.

16. Ibid., Director of Public Works (Tunis) to Maurain, 18 Aug. 1936. On Lacroux's death, see Lasserre to Maurain, 29 Aug. 1936.

the service, which is rather complex." Bois was visited by a meteorological delegation from Algiers, and they discussed measures to avoid duplication of effort.[17]

Bois continued Ginestous and Lacroux's program to measure Tunisian rainfall. Although he was a functionary in the public works portfolio of the protectorate, and not in agriculture, water remained the most important question on the colonial agenda; public works did not respond to internal military priorities, since the Tunisians, unlike their counterparts to the west, abandoned armed struggle against the French invaders during the period before the Second World War. Bois wrote to Maurain in 1938, referring to a paper he wanted to give at the French congress of learned societies:

> My work here is very interesting, but very official. It assumes an increasingly technical character. The note that I am preparing for you is the first element of a work that must allow evaluation of the onset of floods in the wadis. The methods now used have led to large deforestation in Tunisia. The results of calculation have been, in certain cases, one-tenth the actual case. Practical result: bridges washed out.
>
> Elsewhere, I have been named a member of the Permanent Secretariat of the Agricultural Hydraulic Commission, where numerous proposals are refined and some are quickly brought into being. The interest here is mostly technical, economic, and social. These various questions nevertheless provide points of departure in the domain of scientific research.[18]

Bois did not mind working as an engineer. He did mind being paid substantially less than his predecessor. He had begun in 1937 at thirty-three thousand francs, while Lacroux had begun at forty-nine thousand. His indemnity for having a doctorate had been suppressed. Taking various deductions into account, Bois received as much in Tunis as he had been paid in Strasbourg ten years previously. Bois wanted to return to France in the near future because on his salary he would be unable to send his children there for higher education. His repatriation would not be easy. He had entered Rothé's service at Strasbourg as only an ingénieur-géophysicien. He followed Rothé's advice in studying all the physical sciences as he prepared for his licence; his doctoral dissertation was so unspectacular that he had not been included on the official government waiting list for a position of maître de conférences.[19]

17. Ibid., Bois to Maurain, 24 Jan. 1935 and 29 Jan. 1935, where he applies for the post of météorologiste-adjoint in Algiers; Lasserre to Maurain, 26 Jan. 1935, where Lasserre reserves all posts of this kind for internal candidates; Edmond Rothé to Jacques Cavalier, 15 Oct. 1931, for Bois's vita and competences; Bois to Maurain, 22 Apr. 1937, for the quotation.

18. Ibid., Bois to Maurain, 24 Jan. 1938.

19. Ibid., 15 Apr. 1939, for the salary figures and Bois's future plans, and 21 Apr. 1939.

Rain dominated Bois's research to the exclusion of other enterprises. In 1939 he met the Coulombs at Tozeur, near the Algerian border. There he and Jean Coulomb discussed how to set up a Tunisian magnetical station, promised since the time of Ginestous. Albert Lasserre, Coulomb's predecessor at Algiers, who was still in harness, proposed to erect a joint Algerian-Tunisian station at the border of the two countries. Bois would supply his three Mascart recorders, the Algerians everything else. The Tunisian protectorate evidently saw no need of impractical learning in its various civilizing missions. As Bois wrote to Maurain in 1939, the Tunisian public works administration "does not want to provide for the maintenance of the [magnetical] station, reserving its funds for works having an immediate technical application." The protectorate then made it entirely impossible for Bois to conduct his own research by forcing him to teach physics and chemistry at the local Lycée Carnot.[20] A small country did not need a full-time rainman. Tunisia, falling under the administrative shadow of the powerful French presence to the west, acted as a buffer against currents emanating from the rest of the Arab world. Lacroux had reminded Maurain in 1930: "Algiers is the beautiful European city where you will find yourself once again in France. Tunis is the beginning of the East."[21]

As dean of the Paris Faculty of Sciences, Charles Maurain was eclectic about encouraging young men overseas in their search for a doctorate. He could be, however, a demanding patron. Louis Montlaur, a physicist working in the Tunisian agricultural service, lived under Maurain's authority for nearly a decade. Montlaur's research interests focused on the relationship between climate and genetics.[22] Through the 1930s he worked on a geophysical thesis for Maurain, insofar as his precarious health allowed such extracurricular exertion. After seven years he applied for the post of meteorological director in Tunis, vacated by Lacroux—the appointment won by Charles Bois. Montlaur remained bitter about the outcome. It was the only post to which he could aspire in Tunisia, he complained to Maurain. He continued to work on a thesis, however, and to publish on topics of concern to Tunisia, such as soil evaporation.[23] For his part, Bois entered into friendly relations with Montlaur, whose interest in meteorology had to do mostly with its applications for mildew protection.[24]

20. Ibid., 15 Apr. 1939; 21 Apr. 1939 for the quotation; 5 Nov. 1939.

21. Ibid., Lacroux to Maurain, 27 May 1930.

22. Montlaur, "Le climat et les besoins physiques de la plante: Moyens de les comparer," *CRAS* 199 (1934), 464–66.

23. PIPG, Montlaur to Maurain, 6 Jan. 1937. As early as 1930 Montlaur sought advice and encouragement from Maurain (ibid., 30 Oct. 1930).

24. Ibid., Bois to Maurain, 22 Apr. 1937.

By June 1937 Montlaur had a first draft of his "second" thesis subject about rainfall in Tunisia and had inaugurated a Tunisian section for the French Physical Society. The thesis plan he sent to Maurain a month later concerned the acclimatization of a strain of wheat in Tunisia. The first part of the thesis provided a history of agricultural meteorology; the second part presented the analyses Montlaur would carry out. Only one quarter of the third part was given over to meteorology. Maurain was not impressed with Montlaur's first two parts. "Too many neologisms, the history much too long, nothing original," were some of his private comments.[25] He urged Montlaur to present the data that he had worked up over the preceding six years. Montlaur was glad to do whatever Maurain suggested, because a completed thesis was his ticket back to France:

> You know our isolation in Tunisia. We are far from any university, and those who, like me, only know faculties from passing examinations at them feel the extent to which they are handicapped. Also, my greatest desire would be to save my son and my daughter from the difficulties of a similar situation, and this is one of the reasons why I would be happy, in a few years, to be called to carry on my work in France.[26]

Montlaur sent him a revised manuscript two years later.[27] Charles Maurain, as dean, gave Montlaur permission to print up the dissertation (a step paving the way for a defense) on 22 May 1940; the German invasion brought everything to a halt. Georges Déjardin, Montlaur's old physics professor at Lyons, extended the possibility of a defense there, but Montlaur finally held forth at the Sorbonne in 1942, with the retired Maurain presiding.[28]

The Sherifian Empire of Science

The land in Africa closest to Europe, Morocco, was the last part of the continent to come under nominal European control. To the European imperialists, with their visions of vast plantations and enormous

25. Ibid., Montlaur to Maurain, 4 June 1937; 12 July 1937; comments attached to Montlaur's letter of 12 July 1937, on which Maurain noted "wrote 19 July."

26. Ibid., Montlaur to Maurain, 21 July 1937.

27. Ibid., 14 June 1939.

28. Montlaur, *Relations entre les éléments météorologiques et la croissance d'une lignée pure d'un blé* (diss., University of Paris, 1942), also in Tunisia, Direction des affaires économiques, *Annales du Service botanique et agronomique*, 1939–40, pp. 3–139. In Yale University Library's copy of the dissertation, Montlaur dates the preface from June 1939; the tipped-in title page was printed in 1941 and dated for a defense in 1942. The complicated history of the defense may be read in a papered-over note on p. 6 of the dissertation.

deposits of mineral wealth, Morocco was severe, savage, and unremarkable. The Atlas Mountains received less attention than more distant ranges; the littoral off to either side of Gibraltar – too close to serve as ports of call for distant voyagers – hosted small European enclaves of traders and expatriates.

Late in the nineteenth century, Western learning came to a number of Moroccan centers in the form of medical missionaries. European savants were welcomed by the reigning sultans, just as foreigners had been welcome for a thousand years at the world's oldest surviving university, located at Fez. New practices existed alongside traditional medical, pharmacological, and psychiatric doctrines; European-style hospitals coexisted with traditional Moroccan ones. The sultan's regime encouraged these contacts. The generation following the Spanish-Moroccan War of 1860 saw hundreds of Moroccans studying engineering, medicine, and science in Europe and Egypt. As long as Morocco remained independent, its peoples assimilated new learning confidently and unselfconsciously.[29]

Moroccan talent and accomplishment in the exact sciences received a mixed evaluation by French scholars. Gaston Delphin, incumbent of the public chair of Arabic at Oran, reported in detail on the remarkable persistence of a medieval tradition of astronomy and mathematics, and he noted the introduction of modern astronomical instruments (including a Dolland equatorial telescope) at the court of the emperor under the French military mission headed by Jules Erckmann from 1877 to 1883. Delphin concluded that although the Arab tradition had ossified, there was in Morocco "innate talent" for astronomy; he could foresee Arabs' quickly assimilating modern methods.[30] But forty years later Henri Paul Joseph Renaud observed about the exact sciences:

29. On traditional medicine, see Nozha Ibn Lkhayat Zouggari Houssani, *Politique médicale et production scientifique des médecins au Maroc, 1912–1987* (diss., University of Montreal, 1989), pp. 27–54. For the technical instruction of a handful of Moroccan students, see Jacques Caillé, "Les Marocains à l'Ecole du génie de Montpellier (1885–1888)," *Hespéris* 41 (1954), 131–45; Jean-Louis Miège, *Le Maroc et l'Europe*, 4 vols. (Paris, 1961–63), 3:222–23. For the Engineering School at Fez, established in 1844 and running at least through the 1860s, see James Albert Paul, *Professionals and Politics in Morocco: A Historical Study of the Mediation of Power and the Creation of Ideology in the Context of European Imperialism* (diss., New York University, 1975), p. 83; see pp. 84–88 for complements to Caillé's treatment, p. 231 for the slow extinction of traditional medical institutions by their Western counterparts. See Laberge, *Politiques*, pp. 65–66, for the modernizing urge.

30. Delphin, "L'astronomie au Maroc," *Journal asiatique* 17 (1891), 179–201, on pp. 200–201. In his book, *Fas, son université et l'enseignement supérieur musulman* (Paris, 1889), Delphin urged that professors from the University of Fez be installed in Tlemcen, Algiers, and Constantine to counteract Algerian decadence in Muslim-based higher education (p. 112).

> If they were practiced by Muslims, one sees, it was for quite specific goals and in an almost exclusively religious context. Contrary to the Greeks, theoreticians and philosophers who applied this turn of mind to contribute to the progress of science as no other people did until the modern epoch, the Arabs only saw mathematics, geometry, and astronomy from their utilitarian side. It would be unjust to say that they did not produce applications, but their contribution is modest. . . . If from time to time the patience of researchers was rewarded by happy discoveries, these have done nothing until now to change the opinion of competent specialists regarding the finally rather mediocre quality of what we call Arab science. . . . What was perpetuated in Morocco until the early years of this century was not an attenuated memory of the age of great observers, compilers of tables, authors of star catalogues; it was only the amateurish routine practice [*petite pratique routinière*] of the science of calculating the hour of the day and the construction of instruments for this determination.[31]

And in 1942, in the course of an examination of Moroccan astronomical practice, Renaud appealed to the "profound decadence" of Moroccan savants to explain why mosques were oriented so poorly toward Mecca.[32]

According to French teachers of mathematics and science in Moroccan secondary schools, the malaise had its roots in Moroccan culture. Georges Bonjean, a physics teacher at the Collège Moulay-Youssef who had tried unsuccessfully to obtain a doctorate in geophysics from the Sorbonne, summarized the situation in 1939. Physics instruction required the talents of an ethnographer, an orientalist, and a physicist. Bonjean's students followed his lectures slavishly, without understanding the meaning of words. They dressed in traditional garb and perceived the world in "prelogical" ways. Students learned rules of calculation as a "moral imperative but apply them rather ineptly" to new situations. They invested instruments such as microscopes with mysterious parts hidden from view; they had no sense of the "requirements of precision" in measurement.[33]

Medicine, a technology whose adepts receive state patents allowing them to anticipate an honored and comfortable career, carries with it a

31. Renaud, "L'enseignement des sciences exactes et l'édition d'ouvrages scientifiques au Maroc avant l'occupation européenne," *Hespéris* 14 (1932), 78–89, p. 88.

32. Renaud, "Astronomie et astrologie marocaines," ibid. 29 (1942), 41–63, on p. 63. The question of mosque orientation had been set by George Sarton in the early 1930s (*Isis* 20 [1933], 262–63; 24 [1935], 109–10). It is an open question whether the standards of astronomy in French faculties at, say, Poitiers or Caen differed much from those of traditional savants in Moroccan cities.

33. Bonjean, "L'enseignement scientifique au Collège musulman," *Sémaine pédagogique de Rabat*, 1939, pp. 244–49; a companion critique by a mathematician follows Bonjean's text.

learned doctrine that furnishes parameters for daily life. Since antiquity, the doctrine has been informed by ideological prejudice and practical experience. Healers are naturally suited as the cultural avant-garde of a hostile foreign power. Whether affiliated with Christian missions – those of the Franciscans and Protestant denominations – or military ones, medical men spread European values. The political or religious responsibilities of these early physicians, men such as Fernand Linarès and T. D. Churcher and Robert Kerr, deprived them of the leisure to inventory the countryside.[34] Such inventory, a part of which depended on the instruments and methods of the exact sciences, began through the efforts of diplomats and private adventurers.[35]

The diplomats reported reliably on the Moroccan climate. Beginning in the 1860s, French consuls on the coast and less frequently in the interior published meteorological observations, just as German consuls did later at Tangiers and Casablanca. The most famous of the private adventurers, Charles Eugène, vicomte de Foucauld, traveled through Morocco in the early 1880s and evoked climatological interest in his two constituencies – the French army and the Catholic church.[36] The lessons to be learned from the new land were not lost on European academics. German meteorologists tried to organize a network of stations among German residents. Marburg *ordinarius* of geography Theobald Fischer focused special attention on Morocco, publishing a climatological treatise and supervising the work of students and colleagues – for example, Karl Knoch and Heinrich Jung – on the same subject. But it was the French military that first commanded climatology and all related enterprises in the exact sciences.[37]

34. The physician-diplomats and their role in conquering Morocco are admirably treated in Paul, *Professionals and Politics in Morocco*, pp. 100–130. An essential account is René Cruchet, *La conquête pacifique du Maroc et du Tafilalet* (Paris, 1934), focusing exclusively on the medical imperialists.

35. In P. Boudy et al., eds., *La science au Maroc* (Casablanca, 1934), see Charles Marin, "Les travaux astronomiques, géodésiques et de nivellement au Maroc," pp. 19–40; and Georges Roux, "Les travaux de météorologie et de physique du globe au Maroc," pp. 41–56.

36. For a revisionist appraisal of the adventurers, see Abderrahmane El Moudden, "Exploration et pénétration: Un siècle d''itinéraires' dans le couloir de Fès-Taza (1805–1911)," *Hespéris tamuda* 22 (1984), 103–18; a photograph of Father de Foucauld and Marshal Lyautey riding through the Sahara in 1905 appears in Paul Lesourd, *L'oeuvre civilisatrice et scientifique des missionnaires catholiques dans les colonies françaises* (Paris, 1931), p. 160.

37. The heroic explorers at the end of the nineteenth century – German, English, and French – are discussed by Louis Gentil in his *Maroc physique* (Paris, 1912), pp. 26–34. See also Theobald Fischer, *Wissenschaftliche Ergebnisse einer Reise im Atlas-Vorlande von Marokko* (Gotha, 1900) [*Petermanns Mitteilungen*, suppl. no. 133]; idem, *Meine dritte Forschungsreise im Atlas-Vorlande von Marokko im Jahre 1901* (Hamburg, 1902) [*separatum*

When the army disembarked at Casablanca in the west and entered Oujda in the east, in 1907, it found that there were no reliable maps of the country to be conquered. As French forces moved inland, military geodesists carried out a triangulation. After following in the train of professional killers for seven years, the western and the eastern triangulators met at Taza; the junction failed to close by only about 400 m in latitude, 60 m in longitude, and 10 m in altitude. The French geodesists in Morocco, wrote one of their historians, often had to collect their instruments and take cover under hostile fire, "thinking only of fighting and of rejoining the column from which they had been separated."[38] What the geographical service of the army had conquered would not be relinquished to civilian hands. For the next thirty years the army continued to map Morocco at great expense and with increasing precision. While the enterprise helped to direct troop movements, it produced little of scientific value. Natural knowledge came from the pen of an academic adventurer, Louis Gentil.

Gentil was born in Algiers in 1868. In a scientific vita published late in life, he credited his early interest in Algerian geology to Montpellier professor of mineralogy Jacques Curie, the discoverer (with his brother Pierre Curie) of piezoelectricity, but this association may refer to nothing more than correspondence or a short visit.[39] In 1891 Gentil became préparateur in the Algiers School of Sciences, where he began his university-level studies, and at about this time he worked as a cartographer for the Algerian geological service. With two papers in inorganic chemistry under his belt, in 1894 Gentil became préparateur at the Collège de France for the mineralogist and vulcanologist Ferdinand André Fouqué. His master and Alfred Lacroix gave Gentil the task of analyzing volcanoes in the Atlas Mountains. The study opened into a comprehensive examination of North African geology. Gentil issued a stream of descriptive papers, culminating in a doctorate for an exhaustive survey of the geology of the Tafna basin. His inventory of North Africa resonated with Lacroix's enthusiasm for overseas mineralogy, especially in French spheres of influence. Gentil rose to become chargé de conférences in 1899, maître de conférences in 1905, professeur-

---

from *Mittheilungen der geographischen Gesellschaft in Hamburg*, 18]; and Karl Knoch, *Die Niederschlagsverhältnisse der Atlasländer* (diss., University of Marburg, 1906).

38. Marin, "Astronomique," p. 27.

39. Gentil, *Titres et travaux scientifiques* (Paris, 1922), p. 25. It seems that Curie did publish a study of Algerian volcanic rocks and then suggested to Gentil that there would be interest in a study of the Tafna basin, but cause and effect here are far from clear (Gentil, *Esquisse stratigraphique et pétrographique du bassin de la Tafna* [*Algérie*] [Algiers, 1902], pp. 1–2).

adjoint in 1912, and professor of physical geography in 1919, all at the Sorbonne.

Gentil's numerous publications – dull, primitive, and pretentious although comparable to those of many among his Parisian colleagues – were not the motor of his brilliant career. It was in his role as an enthusiastic apologist for French imperialism that he won the hearts and minds of the metropolitan educational bureaucracy. He wrote in 1922: "My entire career was transformed by this desire: to traverse and know Morocco, and perhaps to contribute to the best of my ability to French expansion up to the Atlantic. . . . Without Morocco, the attempt to create, by the collaboration of our civilization and Arabo-Berber elements, a marvelous field of action for young French energies risked being ephemerous and dangerous. All the storms began on this somber point." Gentil set out to chart the storms that had been observed by Charles de Foucauld. "Without waiting for the French penetration that our policies had us glimpse, I set off across the Maghreb, resolved to work with all at my disposal toward a scientific understanding of this country and to begin there the work of peaceful conquest that France would be led to undertake."[40]

With the Anglo-French accord of 1904, according to which Great Britain gave France a free hand in Morocco, Gentil decided to travel inland to see those parts of the country not bound to the sultan's rule. (He was fluent in Arabic from the time of his youth.) Gentil went along and across the High Atlas Mountains, reporting all manner of geological curiosity. He published his observations and then returned in 1907, when he traveled to Marrakech in the company of a medical doctor, Emile Mauchamp. Popular sentiment in Marrakech ran against the European imperialists and their puppet, the Moroccan sultan. Mauchamp perished in a riot eight days after he and Gentil arrived in the city. Gentil recovered the body of his companion and traveled north to join General Louis-Hubert-Gonzalve Lyautey, who had invaded Oujda at the head of Algerian troops. Gentil looked over the geology of northern Morocco and generally made himself useful to one or another invading French army, for example, by locating waterholes. He was a regular visitor to Morocco thereafter, lending his hand and voice to the French army and serving as an occasional councillor of the minister of war in Paris. His descriptive summary, *Le Maroc physique*, appeared in 1912, the year Lyautey became resident general. Lyautey promptly named Gentil as his scientific adviser. The Parisian geographer worked with the local French agricultural and public works offices to "pacify" the country. He won the Delesse prize of the Academy of Sciences in 1917

40. Gentil, *Titres*, pp. 25–27.

for his "first-rate role in the *scientific conquest of Morocco.*"[41] The conquest achieved, Gentil passed his torch to physical geographers who would be permanently situated in the new French protectorate.

To read Gentil's own account of his career is to think that he traveled in Morocco as a Berber, carrying little more than a notebook under his coarse gown. By the time he received Lyautey's patronage, however, he had risen to head a major scientific commission financed by a pleiad of metropolitan institutions. Early in 1914, the Société de géographie of Paris met to discuss Gentil's operation, which drew funds from both public and private coffers:

Accounts received

Société de géographie	Fr 12,000
Société antiesclavagiste	12,000
Museum d'histoire naturelle	10,000
Maison Mirabeau	10,000
Consortium of banks	25,000
Société de l'Afrique occidentale	5,000
Ministry of foreign affairs	2,000

Funding requests to come

Ministry of foreign affairs	Fr 8,000
Ministry of public instruction	50,000
Academy of Sciences	50,000
Protectorate of Morocco	25,000
Museum d'histoire naturelle	10,000
Various	25,000

The initial projections ran to Fr 244,000 for a six-year term. In addition to Gentil, who served as the mission's geologist-mineralogist, there were places for a zootechnician-agronomist, a zoologist, a botanist, a forester, and a sociologist-ethnographer. The staff would have all their costs covered and would receive Fr 10,800 per year for dividing their time equally between Morocco and France; Fr 40,000 would be reserved for publishing the mission's results. Gentil's patrons declined all responsibility in the event of "accident, death, or violence."[42]

Learning in France has never been separated from fealty to the state, and the state has traditionally expressed its authority by force of arms.

41. Ibid., p. 36, citing Charles Depéret's judgment in the Academy's *Comptes-Rendus*, emphasis in the original.

42. PSG, colis 91/4371, "Mission d'études scientifiques au Maroc organisée par la Société de géographie et la Société antiesclavagiste de France." The typescript, a carbon copy, bears many corrections by hand. It serves as a report on the mission's first year, 1912–13, as well as a projection for the future.

Professor Gentil, in his enthusiastic support for the French invasion of Morocco, is remarkable only in the accident of his timing, which propelled him to the heights of the Sorbonne. Neither his record of accomplishment nor his administrative experience was sufficient, however, to open the gates of the Academy of Sciences. He solved no burning scientific issues, formulated no general propositions, and made use of no great body of existing knowledge. In his publications one finds no trace of his *life* as an imperialist. His conventions and his concepts are those of a metropolitan savant, in just the way that his contemporary, ethnographer Alfred Le Chatelier, viewed Moroccan peoples through spectacles polished at the Collège de France.[43]

During his reign as resident general in Morocco (1912–25, except for a short term as minister of war during the First World War) Lyautey employed several hundred physicians in pacification campaigns.[44] Having risen to the Académie française in 1912, he sought to endow his conquest with the attributes of high culture that would synthesize information useful in subjugating the protectorate and would also advertise the intellectual superiority of the conquerors. By 1920 Lyautey could point to the usual colonial establishments for dealing with local diseases, customs, crops, and languages. In this universe of French learning, the exact sciences were limited to meteorology. As the weather was important for agronomists, mariners, aviators, and government engineers, men in all these offices collected climatological information. There was a rudimentary meteorological service in the colonial agriculture portfolio, which emerged shortly after Lyautey's arrival; a well-appointed military meteorological service; and a maritime meteorological service at Kénitra, set up at the initiative of a naval aviator in order to forecast gales.[45]

43. See Morocco, Direction générale de l'instruction publique, des beaux arts et des antiquités, *Historique (1912–1930)* (Rabat, [1930]), pp. 185, 194–95, for Gentil's creation of an Institut des recherches scientifiques (as a service of "economic scope") under the administration of the Sherifian government, which from 1914 to 1921 "was maintained on the budget *pour mémoire.*" See also Paul, *Professionals and Politics in Morocco*, pp. 136–38; and Christophe Charle and Eva Telkès, *Les professeurs du Collège de France: Dictionnaire biographique, 1901–1930* (Paris, 1988), s.v. "Le Chatelier." Le Chatelier was behind a scientific mission sent by France to Morocco in 1904; ethnographical in scope, the mission served to prepare the ground for an eventual French invasion.

44. Paul, *Professionals and Politics in Morocco*, pp. 124–28; Oberlé and H-P-J. Renaud, "La pénétration pacifique par le médecin au Maroc depuis 1908," France, Service de santé militaire, *Revue* 77 (1922), 229–61.

45. See Jacques Liouville, *Rapport général sur l'Institut scientifique chérifien* (Rabat, 1921), for the state of science in Morocco generally, p. 39 on meteorology; and Georges Roux, "Introduction," Société des sciences naturelles du Maroc, *Mémoires*, no. 39, 1934 [Mémoire annuel consacré à la physique du globe et à la météorologie au Maroc, 1932],

A military governor like Lyautey predicated pacification on strong central organization, and he made knowledge march to the drums of his legionnaires. In 1920 he consolidated existing centers of the humanities to create the Institut des hautes études marocaines. Georges Hardy, then director of public instruction in Morocco, stated the institute's mission at the time of its creation. Science had arrived in Morocco to "the sound of the fusillade," he noted. The time had come to "apply the methods of French *science* to the understanding of special situations," meaning Morocco. French learning would radiate the authority of the French presence, since it was important "to present the results of our research in a form as clear as possible, without useless clouds, in the French manner."[46]

After the humanities came nature. In 1921 Lyautey breathed life into the Institut scientifique chérifien, a scientific institution placed under the auspices of the Academy of Sciences in Paris that would supervise research in all areas of natural and exact sciences. The administrator of public instruction assumed responsibility for funding it, although the institute's activity was supervised by a council of "economic application," the majority of whose seats went to chiefs of Lyautey's principal colonial services.[47] It would be an extraordinary opportunity for a science bureaucrat, given that science found no representation in Moroccan institutions of higher education: the Sherifian Institute would have direct access to government funds.[48]

Following Lyautey's scheme for the Sherifian Institute, the natural sciences fell under military tutelage. One of Lyautey's military physicians, Jacques Liouville, who had served as a marine biologist on his uncle Jean Charcot's first expedition to Antarctica and who married a daughter of politician Pierre-Marie-René Waldeck-Rousseau's, was called to direct the institute and also its oceanographical service.[49] The

---

pp. 3–11, pp. 4–5 for the military and naval meteorologists during the first two decades of the twentieth century.

46. Hardy, "Discours prononcé par M. G. Hardy, directeur de l'enseignement, à la séance d'inauguration du congrès des hautes études marocaines le 9 juin 1920," Institut des hautes études marocaines, *Bulletin* 1 (1920), 4–9; *DBF*, s.v. "Hardy."

47. Liouville, *Rapport général*, pp. 5–7.

48. In 1924–25 the budget of the Sherifian Institute for scientific personnel ranked seventh in greater France (Fr 250,000). Ahead of it were the metropolitan ministries of public instruction and agriculture, the Paris Academy of Sciences, the Pasteur Institute, the Edmond de Rothschild Foundation, and the Curie Foundation (Harry Paul, *From Knowledge to Power: The Rise of the Science Empire in France, 1860–1939* [Cambridge, 1985], p. 285). On the absence of local competitors, see Morocco, Direction générale de l'instruction publique, des beaux arts et des antiquités, *Historique*, pp. 18–19.

49. PIPG, Liouville to Charles Maurain, 1 Apr. 1939 (written 16 Apr. 1939), where

latter, which united both the biological and the physical side of the discipline, was intended to be placed at the disposal of Albert I, prince of Monaco, although Liouville operated a personal laboratory and institute branch office in Paris. Alexis Rolland, a military pharmacist who had obtained a licence ès sciences and had studied with Louis Gentil in Paris, became chief of a new geological service; he also served as the institute's secretary, handling much of its paperwork. The largest of the institute's sections, meteorology, fell to Adrien Jury, who had headed the military meteorological service. Lyautey was able to reassign some of the maritime meteorological functions to Jury's charge, but he could not pry the gale watchers away from the public works portfolio – a singular failure due in some measure to rivalries between the army and the navy.[50] Jury came to Liouville highly recommended by Colonel Emile Delcambre, in the French National Meteorological Office. Initially, Jury had the services of only a sergeant-archivist at his central office.

The military provided muscle and organization, but Lyautey sensed that before prestige radiated from the departments of his institute, there had to be an apparently dispassionate mechanism for circulating the institute's new knowledge to the scientific world. He looked to Algeria and borrowed its solution: he created a Society of Natural Sciences to attract civilian researchers and to supervise publication of memoirs deriving from the institute's various laboratories and observation posts. The society's founding members were delighted at the prospect of regular funding for a scientific journal.[51] Like their metropolitan colleagues, they found nothing aberrant about pursuing science while shackled with military fetters.

Director Liouville aired his thoughts on the new scientific institute soon after its foundation. His organization formed the counterpart in natural sciences of the Institut des hautes études marocaines. Both institutes were "dominated by the desire to see clearly into the enigmas of

---

Liouville notes his relation to Charcot and Waldeck-Rousseau and also his monograph on whales (resulting from the Charcot expedition), which garnered him awards from Scandinavian institutions (Liouville, *Cétacés de l'Antarctique* [*baleinoptères, ziphiidés, delphinidés*] [Paris, 1913]). A man of many talents, Liouville was attracted to the historical side of his specialty (Liouville, "La chirurgie conservatrice au XVIIIe siècle," *La tribune médicale*, 1905, pp. 583–86, 600–603, 614–15). See also Laberge, *Politiques*, p. 203 n. 120.

50. Liouville, in his *Rapport général*, p. 42, notes that the maritime meteorological service would retain "its independence and its misery." See Laberge, *Politiques*, pp. 158–61, on the institute through 1940.

51. André Théry, the founding president, reflects on the symbiotic relationship between society and institute in his "Historique de la création de la Société des sciences naturelles du Maroc," in *Volume jubilaire de la Société des sciences naturelles du Maroc, 1920–1945* (Rabat, 1948), pp. 401–3.

a new country, by this need to put order in the chaos of problems that confront our mind, in short, by this thirst to understand for understanding's sake, which is the honor of our Western culture and of which disinterestedness is the most noble trait." The Sherifian Institute had scientific research as its mandate. What was science, Liouville asked? It could only be, he answered, what went on in the French science faculties. To call all learning by the term *science*, he went on, would be a "regrettable confusion." He heaped ridicule on humanists who vested scientific authority in their work by speaking about language and philology laboratories or by announcing the advent of experimental novels. These figures of speech flattered immature science students, "but it is all just an image born of confused expression." Nor did Liouville feel it suitable to speak about scientific exploration, investing intrepid adventurers in the mantle of pure thought. The reign of science in Morocco would be austere and classical.[52]

Liouville was a medical doctor and so, by his very words, outside the scientific enterprise. The center of gravity of the preceding discourse lies in the realm of politics: the Sherifian Institute took shape to provide France with another vote in the new International Research Council and its affiliated international scientific unions. Liouville wrote that the institute's creation,

> providing Morocco with an organ of disinterested research although by definition placed at the service of the economic progress of the protected country, permitted Morocco to take its place among the nations called to sit in the International Research Council, founded at the end of the war by the Allies and afterwards destined to receive representatives of neutral countries attracted by its prestige.

The privilege of membership was restricted to nations that had made "a sacrifice to Science by establishing on national soil those institutions dedicated to scientific research." To be a member of the international council signified that a nation had "left childish instruction to attain its intellectual majority." Liouville crowed that Morocco had joined the council as a full voting member ahead of all other French protectorates in North Africa and Indochina. He promptly placed himself and Louis Gentil in charge of the various national Moroccan committees attached to the international body.[53]

Liouville's expansive claims for the *scientific* independence of Mo-

52. Liouville, *Le Maroc et la recherche scientifique contemporaine* [Rabat, 1922], pp. 2–7.

53. Ibid., p. 18. This paved the way for Tunisia's joining the council in the middle 1920s. See also Georges Perrier, "La coopération internationale en géodésie et en géophysique," Bureau des longitudes, *Annuaire pour l'an 1928* (Paris, [1928]), pp. C.1–C.120, on pp. C.48 and C.118.

rocco found little echo during the 1920s. The naval meteorological office at Kénitra and then at Rabat continued forecasting gales by collaborating with Portuguese homologues in the Azores. The naval operation received a civilian engineer as its head in 1923, and in 1928 it was absorbed by the Sherifian Institute.[54] Climatological inventory continued to dominate the scientific agenda of Liouville's meteorologists, who divided their time between work of a general nature and military work conducted under the eye of Emile Delcambre.

Meteorological chief Adrien Jury had the services of an assistant, Lieutenant Georges Marius Joseph Dedebant, a graduate of the Ecole polytechnique, who succeeded to head the service upon Jury's promotion in 1928.[55] The practical side of their work concerned weather prediction for air transport and rainfall charts essential for agriculture.[56] Research of a general nature involved extending Norwegian ideas about fronts from polar air masses to the interface between cool Atlantic trade winds and hot Saharan air currents. There, Jury and Dedebant classified weather conditions by type and specified directional perturbations to Atlantic anticyclones; weather in Morocco depended not only on fronts but also on the air masses pushing them.[57] Dedebant went on to finer things in the metropolis. With a colonial directorship to his credit and having been promoted to captain in the air force, he spent 1929 preparing for the three certificates that would garner him a licence ès sciences; he moved to Maurain's geophysical institute in Paris to complete the laboratory work associated with one of them.[58] Dedebant rose to become director of the cartographical section in the National Meteorological Office, and he ended up in the late 1950s at the La Plata Observatory in Argentina.[59]

Moroccan science depended on Liouville's negotiations with metropolitan ministries — foreign affairs, colonies, war, the navy, and public

54. Roux, "Travaux de météorologie," pp. 43–45.

55. VAA. "Dedebant," service record, and entries in the annual *Etat des officiers de l'armée de l'air*, published by the ministry of the air.

56. Roux, "Travaux de météorologie," pp. 45–47; Jury and Dedebant, *Etude sur le régime des pluies au Maroc* (Rabat, 1924) [Société des sciences naturelles du Maroc, Mémoires, 9].

57. Jury and Dedebant, *Les types de temps au Maroc* (Paris, 1925) [Office national météorologique de France, Mémorial, 13]. In the preface, Delcambre insists on the close convergence between Norwegian and French meteorology.

58. PIPG, Dedebant to Maurain, 11 Sept. 1929, where Dedebant also provides his address in Paris.

59. Beginning in the 1930s, Dedebant was widely known in Latin America. See Dedebant and André-Jules-Armand Viaut, *Manual de meteorología para pilotos*, trans. Luis Morandi (Montevideo, 1938); and La Plata, Observatory, *El observatorio astronómico de La Plata en el octogésimo aniversario de su fundación* (La Plata, 1966), p. 40 (in a contribution by Simón Gershánik).

works. Dedebant's departure for Paris coincided with the grand interministerial commission that set out a new agenda for geophysics in the French colonies. As one result of the subsequent discussions, foreign affairs decided to confide Moroccan air navigation to Delcambre's Paris office, while remanding the rest of meteorology to the Sherifian Institute. The reorganization provided Morocco with a new Geophysical and Meteorological Service, by *arrêté viziriel* in 1933. While the army continued to issue flight bulletins, Liouville followed his institute's new geophysical mandate by creating an observatory near Berrechid, some forty kilometers south of Casablanca on the main road to Marrakech. The Averroès Observatory, as it was called in honor of Morocco's greatest Islamic natural philosopher, had provision for research programs in seismology and gravimetry and for extending preliminary studies of solar radiation and terrestrial magnetism.[60]

Algeria provided a model for French aspirations in Morocco, and in geophysical matters Liouville looked to Algiers for direction. In May 1930 he wrote to Albert Lasserre, director of the geophysical and meteorological institute there, for details about the organization of Lasserre's operation so that he could bring Morocco into line with it.[61] By 1931 he had appointed naval lieutenant Georges Roux as director of the geophysical and meteorological service, which, despite its name, was not entirely demilitarized. In November of that year, Roux outlined an ambitious research program. He envisaged hiring a chief geophysicist who held a doctorate or who had passed the agrégation; three junior geophysicists drawn from the ranks of licenciés; an agronomist (ingénieur agronome); ten meteorologists and thirteen assistants, who would be assigned tasks of calculation; and a complement of menial support staff. His personnel would be divided among the head office at Casablanca (and at Berrechid, when the new site was ready) and regional stations to be erected at Ifrane in the foothills of the western Atlas, at Kénitra, and at Taroudannt on the south coast. The projected cost of this scientific empire ran to over 6 million francs.[62]

Roux received encouragement from Liouville. Roux "has worked a great deal in Morocco," Liouville explained to Charles Maurain in 1933, "and I am pleased with his services, something I have by no means failed to indicate on his evaluations, where I thank the navy for having

60. Roux, "Travaux de météorologie," pp. 47–54. Roux initially spelled it *Averrhoes*, which then changed orthography several times.

61. PIPG, Lasserre to Maurain, 16 June 1930, where Lasserre describes Liouville's letter of the preceding month.

62. Ibid., Liouville to Secretary General of the Protectorate of Morocco, 9 Dec. 1931, plus appendix dated 26 Nov. 1931 (copy).

sent me this distinguished, modest, and enterprising officer."[63] Liouville then hired Maurain's former student Jean Debrach as a geophysicist and soon installed him at the Averroès Observatory to supervise a Mainka seismograph.[64] Georges Bidault, licencié ès sciences from Paris and a graduate of the Ecole supérieure d'électricité, came to Casablanca as a meteorologist. Liouville promptly encouraged Bidault to obtain a certificate in geophysics under Maurain. By the middle 1930s Roux had picked up Pierre Schindler as a meteorologist.[65]

Into the late 1930s, Liouville continued to emphasize the importance of pure science. He admonished the cabinet of the French protectorate that, unlike British meteorological aims in Africa, which were practically and regionally oriented, the "interest of French science in Africa" was to "simplify methods by which general problems can be treated whose solution concerns the African continent as a whole."[66] In 1936 he urged Roux to publish "articles in the periodical press on disinterested scientific subjects." He continued: "You know the price that I attach to aiding officers on active duty to renew the French tradition of having the navy participate in works of pure science."[67]

The geophysical observatories in Morocco—the administrative seat at Casablanca, the major installation of Averroès, and the station at Ifrane—were indifferently supplied with equipment. Debrach complained in 1933 that "the experiments I intend to carry out here have been delayed by the total absence in this service of all measuring apparatuses and by funding restrictions."[68] In 1936 he blamed poor facilities for lack of progress toward finishing his thesis: "I have found here no matériel, no laboratory. Interrupted at the beginning, it [the thesis research] found itself subsequently slowed down by difficulties with apparatus and above all by the slowness with which I was able to obtain the indispensable equipment, all the way down to galvanometers."[69]

63. Ibid., Liouville to Maurain, 28 June 1933.

64. Ibid., 26 Oct. 1931 and 5 Nov. 1931 for Liouville's maneuvers to bring in Debrach. At the time, the general secretariat of the protectorate had declared a freeze on all new positions. See Debrach to Maurain, [1929], for Lieutenant Debrach's intention to begin work on a thesis in geophysics after his military service. Debrach took until 1937 to place his seismological station in order (Debrach, "La séismologie à l'Observatoire Averroès," *Comptes-Rendus du 71e Congrès des sociétés savantes* [Paris, 1938], 75–78).

65. PIPG, Bidault to Maurain, 11 Oct. 1932, with Liouville's favorable note on the proposal in a postscript; Schindler to Maurain, 5 Apr. 1935.

66. Ibid., Liouville to Chief of the Diplomatic Cabinet of the Resident General, 31 July 1936 (copy).

67. Ibid., Liouville to Roux, 20 Jan. 1936 (copy).

68. Ibid., Debrach to Maurain, 4 Jan. 1933.

69. Ibid., 17 Nov. 1936.

The struggling geophysical service received a setback when Roux's sympathetic director Liouville decided to retire; suffering bouts of dengue fever and enduring chronic problems with his sciatic nerve, he sought to divest himself of administrative chores.[70] His impending exit brought about a new regime in Moroccan science. The Sherifian Institute would henceforth be governed by a committee with representatives from both the protectorate and the metropolis. Roux anticipated that naturalists would dominate things, and he feared as a result that the Franco-Moroccan committee would view geophysics as a first cousin to the descriptive sciences. He urged Charles Maurain to join the committee and defend the interests of his discipline.[71]

In Roux's correspondence with Maurain, there is no trace of concern about scientific independence for the Moroccan protectorate. From our perspective we see that the Franco-Moroccan committee had been struck to rein in an unruly charge and generally to subjugate Moroccan research to metropolitan desires; a majority of the committee's voting members came from the hexagone. In the universe of French overseas science, however, such centralizing tendencies were welcomed rather than despised. Geophysicist Roux certainly did not mind being tethered more firmly to the metropolis, where his salvation lay.

And the metropolis sent out secular missionaries of science. From June to October 1937 the atmospheric physicists Etienne Vassy and Arlette Vassy (née Tournaire) measured the absorption of light at Ifrane. The Vassys, supported in the late 1930s by the Clément-Félix Foundation of the Academy of Sciences, were experts on atmospheric ozone and spectroscopy. Morocco offered attractions to the husband-and-wife team, not the least of which was peace and quiet. Early in 1937 they saw Morocco as a safe haven from the strikes and controversy that had convulsed Léon Blum's Popular Front government and had made daily life difficult for young researchers at the beginning of a career. The Vassys' involvement with French overseas science was, in the scheme of things, peripheral.[72]

70. Ibid., Liouville to Maurain, 28 June 1933, for the fever and the nerves; 1 Apr. 1939 (written 16 Apr. 1939), for Liouville's appointment as honorary director of the Sherifian Institute upon his retirement.

71. Ibid., Roux to Maurain, 18 Dec. 1937.

72. See *CRAS* 205 (1937), 1289 and 209 (1939), 923–24 for the support of the Clément-Félix Foundation; Vassy and Vassy, "Méthode expérimentale pour la comparison de l'absorption de la lumière par la basse atmosphère et par l'atmosphère totale," ibid. 206 (1938), 1893–95; idem, "Variations journalières de la température moyenne de l'ozone atmosphérique," ibid. 207 (1938), 1232–34. Mme Vassy explained her motivation for the Moroccan research in an interview with the author, 19 Feb. 1989. Both Vassys wrote dissertations on upper-atmospheric physics under Charles Fabry and Eugène Darmois (Etienne Vassy, *Sur quelques propriétés de l'ozone et leurs conséquences géophysiques* [diss., Uni-

Arlette Vassy in trousers beside a transformer erected by the Moroccan Electrical Company to provide power for spectroscopical lamps, 1937. She and her husband had come to Ifrane to study atmospheric ozone, a research area in which French physicists were leaders. (Courtesy of Arlette Vassy)

If Morocco held attractions for the elite of Parisian physics, men and women permanently situated in the protectorate saw things in a different light. Roux's staff melted away—there were five resignations from a total of eighteen employees in 1937 alone—for their salaries ranged far below the colonial norm.[73] Not only were salaries low but the positions remained insecure. Writing to thank Maurain for entering him on the government waiting list for the position of aide-physicien, Jean Debrach underscored that he would try to remain in Morocco to complete his thesis and related work on solar radiation and its absorption by the

versity of Paris, 1937]; Arlette Vassy, *Sur l'absorption atmosphérique dans l'ultraviolet* [diss., University of Paris, 1941]). For a description and photographs of Ifrane, see Georges Roux, "Une station météorologique de montagne au Maroc: La station d'Ifrane," Société des sciences naturelles du Maroc, *Mémoires*, no. 44, 1937, pp. 61–63.

73. PIPG, Roux to Maurain, 18 Dec. 1937.

atmosphere; his position was far from ideal, and he anticipated moving to a provincial French observatory in the future:

> I would not think of leaving Morocco, at least for now, if our administrative situation were stable, but this is not the case. I am still in fact employed by the protectorate under contract, the same as all my colleagues at the Scientific Institute; furthermore, I have just signed a rider to this contract, extending its validity until the end of the year 1937.

Staffing the Sherifian Institute on a permanent basis had been under discussion for a long time, Debrach emphasized. Referring to political turmoil at home, he offered that the "present crisis" in France was responsible for his poverty.[74]

Debrach labored on his doctoral dissertation and continued to supervise seismology at the Averroès Observatory (seismology came to constitute his "second" thesis).[75] The war cut short his progress. Late in 1939 he informed Maurain that he had been mobilized and consequently had failed to complete the dissertation, which in any case would have been difficult to print under wartime conditions. He would nevertheless send Maurain two notes for the *Comptes-Rendus* of the Paris Academy. As night closed in on science in France, Debrach had reason to be happy. Following a pattern common to geophysical institutes in the Maghreb, he planned to marry an assistant at the Sherifian Institute.[76]

The geophysical service, starved for operating expenses, found itself a potential beneficiary of one hundred thousand francs given to the Sherifian Institute by the new, bilateral Franco-Moroccan research committee. The grant, similar in structure and amount to funding authorized by the ministry of foreign affairs for French Canada, went entirely to support French scientists who desired to carry out research in Morocco. Announcing the new program to Charles Maurain, Georges Roux emphasized that French geophysicists who received funding

74. Ibid., Debrach to Maurain, 17 Nov. 1936. Roux commented to Maurain in 1938 that Debrach had been in the Moroccan geophysical service for seven years and was still paid as an *auxiliaire* at sixty-four francs per day (Roux to Maurain, 30 Dec. 1938). The franc devalued dramatically in the late 1930s, falling to 39 francs to the dollar in September 1938 (René Sédillot, *Histoire du franc* [Paris, 1979], p. 166).

75. See PIPG, Debrach to Maurain, 17 Nov. 1936, for the thesis on solar radiation; 9 June 1939 for the second thesis on seismology.

76. Ibid., 11 Nov. 1939. At the Algiers geophysical institute, Paul Queney married Miss Thoumazou, an observer there (Albert Lasserre, "Historique de l'Institut de météorologie et de physique du globe de l'Algérie," introduction to Paul Seltzer, *Le climat de l'Algérie* [Algiers, 1946], pp. vii – xxiii, on p. xxii).

would have to bring in all their own equipment.[77] Apparently none availed himself of the invitation to follow in the steps of the Vassys.

Morocco's scientific establishment compared unfavorably with Algeria's, but it stood head and shoulders above Tunisia's. The military, retaining a strong hand in Morocco, supported much of Sherifian science; Tunisia, a small and relatively homogeneous nation submitting peacefully to French rule, charged civilian science administrators with routine work relating to agriculture. The scientific discourse produced by each protectorate, nevertheless, can be segregated by neither military nor civilian attributes. Reflecting the latest word emanating from the hexagone, pure learning constituted a microcosm of metropolitan sensibilities.

77. PIPG, Roux to Maurain, 11 Oct. 1938. French Canada is considered in part 4 below.

III Science and the Celestial Host

The Third Republic, by harrying Catholic schools, contributed to the missionary zeal of the Jesuits, who spread abroad what was proscribed at home. In the beginning it was a policy born of expediency. Jesuit presence overseas antedated the Republic, and Parisian mandarins frequently enlisted Catholic support in pacifying newly acquired lands. During the 1920s, French politicians used grants-in-aid to assume effective control of important Jesuit educational and scientific institutions. In the end, Jesuit operations survived the republic that had compromised their authority.

Essential to the Jesuits' proselytizing progam was an inventory of God's laws in the natural world. In their view, constant devotion to the stars, to meteorology, or to seismology produced an affirmation of faith and a tale of edification for converts. Through their activity, the Jesuits also promoted French political aims. Sent by French superiors to conquer new land for the cross, they distinguished their nation from its government. In their view, no civil regime could alter the fact that France remained a profoundly Catholic country.

With a major university and a number of observatories, China hosted the largest concentration of French Jesuit scientists. With a university and a small observatory, Lebanon came a respectable second. Madagascar fell behind in last place. Each research complement followed the scientific progress of its fellow missionaries around the world; occasionally scientists transferred from one location to another. The Jesuits adapted to the special features of their environment: the various imperial, republican, provincial, and imperialist regimes of greater Shanghai; the imperial and then the mandated territory of Lebanon; the kingdom and then the colony of Madagascar. In each case they obtained autochthonous patronage and French civilian funding.

By around 1930 each religious establishment had ceded much control to political chieftains in Paris or to their colonial factotums. The Jesuits agreed to the resulting condominium reluctantly, although they took delight in receiving support from a regime that had repressed them in the hexagone. By considering China, Lebanon, and Madagascar together, we see how government ministries in Paris absorbed remarkable institutions built by French cultural imperialism.

6

China: Missionaries to the Middle Kingdom

A Chinese Trinity

In the seventeenth and eighteenth centuries, the Society of Jesus acted as a key scientific liaison between China and Europe. The Jesuit presence at Beijing continued well after it had been extinguished at Rome in 1773, the Vincentians succeeding the Jesuits in China only by a decree of 1785. The Jesuits returned to China in 1842, when Claude Gotteland and François Estève landed near Shanghai to start up a mission. Following the tradition of previous centuries, Gotteland and Estève brought with them a number of astronomical instruments, including a respectable lunar telescope, although they were unable to accomplish more than longitude determinations and incidental observations. Their site of operations was Zikawei, a suburb of Shanghai named after Hsü Kuang-ch'i, a high mandarin baptized by João da Rocha in 1603. A generation passed before the mission was established securely enough to sustain a permanent observatory.[1]

When France entered China during the middle of the nineteenth century, metropolitan attention focused on securing territory by force of arms. Scientific observations were limited to meteorological reports sent by naval physicians to metropolitan academicians.[2] The French government left science to private initiative, and among all private missions to China the Jesuits at Zikawei, who answered to the Jesuit Province of Paris, had the experience, the resources, and the conviction to carry out research in physics and astronomy.

The first systematic meteorological observations at Zikawei began in

1. Henri Cordier, "L'Observatoire de Zi-Ka-Wei," *La nature* 33 (1905), 179–81. Hsü's remains lay in a sepulcher maintained by the Jesuits. Joseph Dehergne, *Répertoire des jésuites de Chine de 1552 à 1800* (Rome, 1973) [Bibliotheca instituti historici s.i., 37], p. 223.

2. For example, the weather reports of *médecin major* Champenois, taken between September 1840 and February 1841 at Teng-Haï, in PMN.

1872 under the supervision of Augustin Colombel. The following year Colombel constructed a permanent observatory building in the mission's garden. By 1874 Henri Le Lec took charge of meteorological recording, following the thrice-daily regime advocated by the Montsouris Observatory in Paris; Marc Dechevrens began magnetical recordings in a wooden shed according to William Sabine's instructions. Le Lec succeeded Colombel as director of the observatory in 1875, only to be replaced in turn by Dechevrens in 1876. Dechevrens set the finest geophysical observatory in East Asia on its feet.[3]

Dechevrens was born at Chêne, Switzerland, near Geneva. Both he and his brother attended the Collège de Belley in France at the height of the Second Empire; both brothers declared themselves for the Jesuits while in their teens. Dechevrens became a physics teacher at the Jesuit school in Vannes, and in 1872, at age twenty-seven, he received the chair of physics at the elite Jesuit Collège Vaugirard in Paris. The following year he was called to the Zikawei Observatory. He devoted his early years to terrestrial magnetism, installing photographical recorders and seeking to understand periodical fluctuations in the earth's magnetical field. Magnetism was to remain a lifelong interest, but circumstances led him to become an agent of imperialist design.[4]

From the time of Dechevrens's arrival at Zikawei, the English newspaper *Shanghai Courier* carried a daily report of Jesuit meteorological readings. In 1879 a typhoon devastated the China coast. Dechevrens published a monograph on the destructive gale, based on observations at Zikawei and Kobe, Japan; he also signaled the imminent approach of another storm. His expertise persuaded the International Chamber of Commerce – a council of imperialist merchants in the International and French settlements – to invest Zikawei with a typhoon warning service; making use of a large semaphore erected on the Bund (the Shanghai waterfront), the service fell under Robert Hart's Chinese Maritime Customs. Beginning in 1882 Dechevrens lobbied for a telegraphical meteorological service to serve the entire China coast; the project received funding from the various imperialist enterprises and govern-

3. *Vregille*, pp. 493–507; Henri Gauthier, "L'Observatoire de Zi-Ka-Wei," *RC* 6 (1918–21), 137–46; Jean Brunhes, "Les travaux des jésuites à l'Observatoire de Zi-Ka-Wei près Chang-hai (Chine)," *Revue d'histoire des missions* 1 (1924), 26–38; Joseph Dehergne, "Zi-Ka-Wei: L'observatoire des cyclones," *La météorologie*, no. 4, 1976, pp. 174–88; Yen Lin-San and Ma Chung-Liang, "The Foundation and Development of the Xu-jia-hui Observatory (1872–1950)," *China Historical Materials of Science and Technology* 5, no. 2 (1984), 65–72 (in Chinese).

4. See Henri Gauthier, "Un maître de physique du globe," *Etudes* 178 (1924), 272–91, for extended discussion of Dechevrens's scientific work.

ments, and in 1896 the Jesuits finally obtained the privilege of sending out telegraphical typhoon warnings without charge.[5] It was a small step from running a weather service to taking responsibility for assuring the time of day, which followed in 1884.

In the early 1880s a French naval lieutenant reported eloquently about Dechevrens and his observatory. According to the officer, Dechevrens had persuaded a large Chinese shipping concern to connect Zikawei telegraphically with Shanghai so that the French Jesuits might transmit their storm warnings directly to the English administrators who controlled the harbor. The Zikawei Observatory had in fact become, according to the words of Dechevrens, "the central observatory of China," and it received financing from Chinese and English insurance companies. To accentuate the fact that the service was under French control, Dechevrens insisted on using metric measures instead of English ones.[6]

Dechevrens labored at Zikawei until 1887, when illness recalled him to the Isle of Jersey. There he erected a meteorological observatory. The data that he had accumulated in China served as the basis of his subsequent theoretical writings. While at Zikawei in the middle 1880s he pioneered measurements of rising columns of air in cyclonic formations and presented a concise description of vertical circulation at the interface of cyclones and anticyclones. In his view, cyclones arose at higher atmospheric levels; beneath them were air masses with higher temperatures, and around the cyclonic axis circulated air with lower temperatures. Since the theory included a temperature inversion along the vertical axis of a cyclone, it permitted Dechevrens to predict the arrival of anticyclones by observing altocumulus and cirrus clouds in the upper atmosphere. He wrote in 1888:

> The high temperature of the low parts of the vortex increases the capacity of the air for water vapor; furthermore, the lowering of pressure and the horizontal and vertical movement of the air accelerate evaporation at the earth's surface, resulting in high humidity, which counters the loss of heat by radiation. This humidity gives rise to abundant condensations and precipitations

5. See Louis Froc, *Observatoire de Zi-ka-wei: Signaux aux marins* (Zikawei, 1909), p. 2, for the date of the telegraphical warnings cable. On the Chinese Maritime Customs, an international civil service during the nineteenth and twentieth centuries, see Lester Knox Little's introduction to *The I. G. in Peking: Letters of Robert Hart, Chinese Maritime Customs, 1863–1907*, ed. John King Fairbank, Katherine Frost Bruner, and Elizabeth MacLeod Matheson, 2 vols. (Cambridge, Mass., 1975), 1:3–34.

6. PAN, Marine, 3JJ/403, G. Sandem [?], "Notice sur les typhons des mers de Chine et du Japon, l'Observatoire de Zikawei," received in the Dépôt de la Marine in September 1883. The author cites a letter written by Dechrevens.

> when the altitude and expansion of the humid masses result in a lowering of air temperature.

The innovation is characteristic of Jesuit geophysics in China: it is on the cutting edge of European research, with no discernible trace of having been conceived on the periphery of a French empire.[7]

When Dechevrens returned to Europe, he was replaced for several months by Bernard Ooms until a permanent director arrived. The new man, Stanislas Chevalier, was born into a large and devoutly Catholic family in the Vendée. From the time of his early childhood he wanted to be a missionary. His academical progress was slow. He failed to obtain a baccalauréat upon leaving secondary school in 1871. Independent study provided him with the elements of science (and a licence), and from 1877 to 1881 he taught mathematics, chemistry, and physics at several Jesuit collèges. The disestablishment of Jesuit secondary schools by Jules Ferry drove Chevalier, along with his collège, which was preparing students for the Ecole navale at Brest, to the Isle of Jersey. Two years in the English Channel persuaded Chevalier to set off for China. After studying Chinese for five years, he became chief meteorologist at Zikawei. Chevalier continued the programs set in place by Dechevrens, but he was no meteorologist at heart. He brought into being the short-lived Shanghai Meteorological Society; in his free moments he surveyed the Yangtze River and published an atlas of China.[8]

In the middle 1890s the Zikawei meteorological establishment was pushing at the seams of the small structure that housed it. Chevalier failed to obtain the confidence of his superiors at a time when they were planning for a new observatory compound on ground adjacent to the mission. In 1891 the Jesuit hierarchy had Bernard Ooms replace him for a short time as director.[9] Then, in 1897 Chevalier was finally relieved of his scientific charge and sent to carry out pastoral work in the countryside; he elected this assignment rather than serve a new scientific chief at Zikawei.[10] His replacement, Louis Marie Froc, who had arrived in China on the same day as Chevalier, oversaw the observa-

7. Gauthier, "Maître," for summary and citation; *Vregille*, pp. 536–46, for an extensive account of Dechevrens's meteorology; Gisela Kutzbach, *The Thermal Theory of Cyclones: A History of Meteorological Thought in the Nineteenth Century* (Boston, 1979), pp. 137–38.

8. J. de Lapparent, "Le R. P. Stanislas Chevalier (1852–1930)," *Lettres de Jersey* 46 (1937), 323–79; A. Salvini, *Cent ans sur le fleuve Bleu: Une mission des jésuites* (Shanghai, 1942), pp. 148–69. Chevalier's hydrographical surveying did not prevent the Germans from repeating his work: see Arthur Kniep, *Der Yang-tzï-kiang als Weg zwischen dem westlichen und östlichen China (Eine hydrographisch-verkehrsgeographische Studie)* (diss., University of Strasbourg, 1904).

9. See Cordier, "Observatoire," for the substitution by Ooms.

10. Salvini, *Cent ans*, p. 154.

A Jesuit meteorologist taking measurements on a garden plinth at the Zikawei Observatory ca. 1900. The observatory's central tower, destroyed by fire before the First World War, was replaced by a skeletal frame. (Courtesy of Shanghai Meteorological Bureau)

tory's move to new quarters. Two years of construction, in 1899 and 1900, produced a magnificent two-story structure surmounted by a tower that measured thirty-three meters. A separate granite pavilion housed magnetical instruments, and a cupola was provided for a 10-cm equatorial telescope. Zikawei's first seismographs – horizontal Omori pendulums – came as gifts from a visiting Japanese delegation in 1904.[11]

Froc, the son of an armorer in Recouvrance, near Brest, entered the Society of Jesus in 1879 at age twenty. Four years later he arrived in China, finding his way to the Zikawei Observatory in 1885. After two years there he returned to Europe for further study. He attended the Catholic faculty at Angers and sat at the feet of physicist Edouard Branly in Paris. Ordained priest in 1892, he shipped out to Zikawei in 1894 with the view in mind of replacing Chevalier as director. Froc was

11. *Vregille*, pp. 498–503.

an able, if uninspired, administrator; twenty-five years of service at the helm of the Zikawei Observatory (in the 1910s he went on sick leave for a number of years) brought him membership in the Legion of Honor.[12]

In a long letter of 1904 written to a German professor, Froc discussed his operation in detail. The Zikawei Jesuits assumed responsibility for meteorology in continental East Asia, Froc began. All the other weather services, including the British one at Hong Kong and the German one at Tsingtao, as well as the military ones attached to each foreign fleet in Shanghai harbor, were local in scope. Zikawei received 150 bulletins daily from sixty stations: 14 from Japan and Formosa, 11 from Siberia, 6 from Indochina, 6 from the Yangtze valley, and 11 from the China coast, among others. It took one Jesuit the better part of a day to make sense of them and prepare the forecast. The second part of Zikawei's mandate, terrestrial magnetism, provided work for Froc himself every day of the year. The time service required astronomical observations during 132 nights in 1903. To this charge was added the honor of receiving guests, from German royalty to curiosity seekers; Froc reported 2,103 visitors in 1903. To help him with these tasks, Froc had a Jesuit colleague, Joseph de Moidrey, and five Chinese assistants. The latter "would never be more than aides of a lower degree; it is useless to think of giving them one of the services to direct." Because of overwork and understaffing, Froc was unable to install a photographical Mascart electrometer for measuring atmospheric electricity (it had remained in its box since 1887), and he could do nothing with the data obtained from the Omori seismographs.[13] Froc's situation at the helm of Zikawei geophysics improved through the creation of separate observatories for astronomy and terrestrial magnetism.

Astronomy had attracted Froc's attention for some time. He visited the Paris Observatory in 1894, and upon returning to Zikawei he informed Auguste Fraissinet, the observatory's secretary in Paris, about his plans to determine the geographical location of Zikawei. He hoped to acquire more instruments and establish a real astronomical observatory, the first French one in China.[14] Beginning in 1895, Stanislas Chevalier dunned the French and international settlements in Shanghai

12. André Bonnichon, *Le R. P. Louis Froc de la Compagnie de Jésus, "Le père des typhons"* (n.p., 1927); Carroll Lunt, ed., *The China Who's Who 1925 (Foreign)* (Shanghai, 1925), s.v. "Froc"; Henri Gauthier, "Le 'père des typhons,'" *RC* 6 (1918–21), 457–59.

13. See Chant., Froc to Provincial Superiors, 20 Feb. 1904, which provides a transcription of the letter to an unnamed German science professor; and "L'Observatoire de Zi-ka-wei," *RC* 1 (1903–5), 648, for the royal German visit. According to Paul Mellon, in 1904 Wilhelm II sent the Zikawei Jesuits Fr 50,000 for their weather service ("L'Observatoire de Zi-Ka-Wei," ibid., pp. 357–60).

14. PO, "Zikawei, 1895–1896," Froc to Fraissinet, 15 Feb. 1895.

for money to purchase an equatorial telescope. The two administrations, together with large shipping companies active in Shanghai, came through with Fr 30,000.[15] Chevalier asked the French navy to contribute Fr 20,000 toward the cause, which it declined to do.[16] When Froc took over Chevalier's position, he persuaded the provincial superiors in France to grant more than Fr 120,000 for a twin equatorial refractor from Gautier in Paris (with each objective 40 cm in diameter) and to install it on a hill, Zô-Sè (in Pin-Yin, She Shan), twenty-five kilometers southwest of Zikawei.[17]

Zô-Sè had better observing conditions than Zikawei, but it was eight hours distant from the central observatory.[18] As a result, Jesuit astronomy necessarily fell under independent management. The natural candidate for directing the new telescope was Robert de Beaurepaire de Louvagny, the Jesuit called in to install it. De Beaurepaire, a product of the Collège Ste-Geneviève and the Ecole polytechnique, had gone to work as a civil engineer on the Panama canal in 1883. Yellow fever forced him to return to France, his health destroyed. The brush with death may have caused him to recall the teachings of the Jesuits at Ste-Geneviève and the experiences of an uncle who had been a Jesuit missionary in China, for in 1885 he presented himself to the Society of Jesus. The new recruit was soon put to work at St-Joseph's College in Trichinopoly, India. There de Beaurepaire conceived a love for astronomy. Having returned to France on leave, he was drafted to set the Zô-Sè telescope in order. He knew nothing of the Chinese language; his indispensable asset, in view of the three hundred boxes of equipment that accompanied him, was his cousin Georges-Gaston-Servan, comte de Bezaure, who was the French consul general in Shanghai. De Beaurepaire arrived late in 1898 and stayed through 1901. He then returned to French India.[19]

Having lost a trained mathematician and engineer, Froc found him-

15. *Vregille*, p. 499; PAN, F17/3827, Ministry of Foreign Affairs to Ministry of Public Instruction, 4 Sept. 1895, reporting the settlements' grant of Fr 20,000 and the shipping companies' grant of Fr 10,000 toward Chevalier's equatorial.

16. PAN, F17/3827, Chevalier to Navy, early 1895; Navy to Ministry of Public Instruction, 22 May 1895.

17. *Vregille*, pp. 498–99; de Lapparent, "Chevalier," pp. 344–45; Chant., Froc to Provincial Superiors, 20 Feb. 1904, for the total cost of the Zô-Sè Observatory; "Le Pèlerinage de N.-D. auxiliatrice à Zô-Cé," *RC* 1 (1903–5), 81–95, for a description of the Christian enclave of Zô-Sè in the nineteenth century.

18. Chant., Froc to Provincial Superiors, 20 Feb. 1904, for the time of transit. As late as 1933 the trip from Zikawei to Zô-Sè involved train, carriage, and one and one-half hours on an airboat navigating along canals (A. Le Palud, "L'Observatoire de Zikawei," *Le Journal de Shanghai*, supplement, 24 Dec. 1933).

19. See *Le révérend père Robert de Beaurepaire de Louvagny, un vaillant fils de Saint-Ignace*

Constructing the large dome at Zô-Sè (Courtesy of Shanghai Meteorological Bureau)

self forced to rely on local talent for directing the new astronomical observatory. Stanislas Chevalier was waiting in the wings. For the preceding four years he had been mapping the Yangtze River. His river atlas having received prizes from the Paris Academy of Sciences and the Société de géographie, only a recall to France would have denied him the directorship. Chevalier remained in charge of Zô-Sè for twenty-three years. During this time his principal astronomical contributions were a survey of the equatorial skies and a long study of solar rotation. His early colleagues were Antoine Weckbacher and the observatory's mechanic, Brother Joseph Aguinagalde (both of whom had ac-

(Chartres, 1917), pp. 48–50, on China; the date of de Beaurepaire's arrival is given as the end of 1899, which seems to be in error by a year.

Constructing the large dome at Zô-Sè (Courtesy of Shanghai Meteorological Bureau)

companied de Beaurepaire to China in 1898); later he worked with the Japanese Jesuit Paul Tsutsihashi, Pierre de Mathan, and his successor Louis Gauchet, Zô-Sè's double-star man.[20] Under Chevalier's direction, Zô-Sè produced a stream of publications, both with the Jesuit presses in Shanghai and with North Atlantic periodicals. Broadcasting Jesuit devotion to science was inextricably linked with propagating the faith. Chevalier prefaced a summary of his solar observations:

> To hope that science will bring men an era of universal peace would be illusionary; science is powerless to raise man above his passions, powerless to inspire in him the virtues of justice and charity. But science may help him see in nature the Higher Author of nature. No purely human science is closer to God than astronomy. In revealing to our astonished gaze, in the immensity of the skies, the order and harmony that reign among all the celestial globes, astronomy allows us to touch the finger as well as the infinite wisdom, the infinite greatness and the omnipotence of the Creator.[21]

20. De Lapparent, "Chevalier"; Salvini, *Cent ans*, p. 159, for Weckbacher and Aguinagalde in 1898.

21. Chevalier, "Préface," Zô-Sè, Astronomical Observatory, *Annales* 8 (1912), i, dated December 1914.

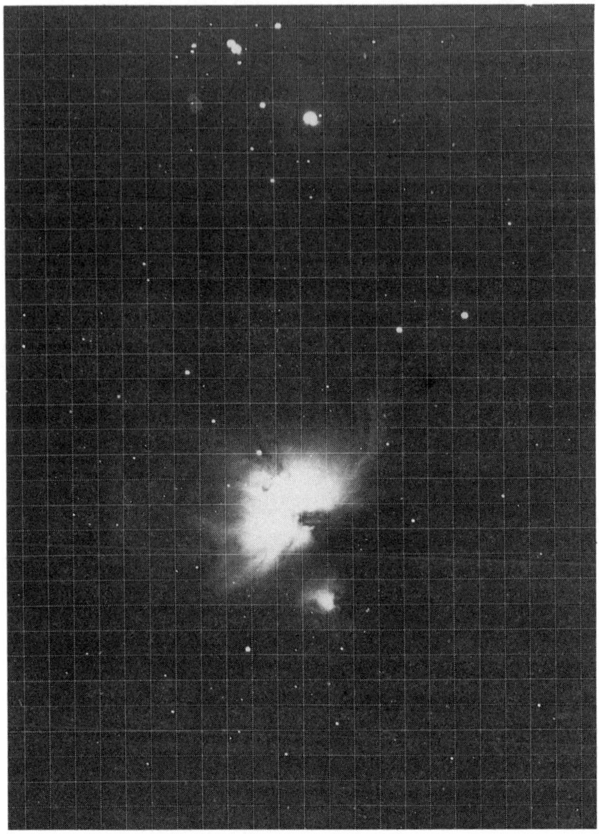

Photograph of the Orion nebula, taken by the double refractor at Zô-Sè on 30 November 1902. The superimposed grid was used in exposures made for the *carte du ciel*, a survey of the heavens supervised by Paris astronomers. (Courtesy of She Shan Observatory, China)

Chevalier's double refractor remained one of the jewels of French astronomy for more than thirty years.[22]

As Louis Froc had noted in 1904, magnetical measurements became a perpetual headache for the Jesuits. They were not only delicate and time-consuming but also highly sensitive to locally generated electromagnetical fields. The principal culprit for producing such fields at the turn of the century was the electrical tramway.[23] The French and Inter-

22. SMB, Pierre Lejay to Louis Dumas, 1 Feb. 1928, where Lejay emphasizes that the Zô-Sè telescope is surpassed in greater France only at Paris, Meudon, and Algiers.

23. The Jesuit observatory at Manila, for example, had to close its magnetical observatory in 1904 due to interference from new streetcar lines (HJA, "Manila, 1908–1923," Miguel Saderra-Masó to E. F. Pigot, 2 Apr. 1909).

national settlements of Shanghai, modern administrations, implemented a tramway network; their long history of supporting Jesuit science at Zikawei led to their providing funds — five thousand taels and twenty-five thousand francs in 1905–6 — for a new magnetical observatory. The Jesuits first thought to add terrestrial magnetism to Zô-Sè's charge, but they decided on a more distant location after learning that so-called vagabond currents could disturb magnetical instruments twenty kilometers from a radiating source. They chose Lukiapang, a small Christian community eighty kilometers away, which was linked to Zikawei by train and canal. Construction of the buildings took two years, during which time the Jesuits suffered robbery (apparently condoned by the local mandarin) and incompetence on the part of the builders. Lukiapang began recording in July 1908; the Chinese government strung a direct telephone line from Zikawei to the new observatory.[24] The man placed in charge of Lukiapang was Joseph de Moidrey, another scientist who had arrived in connection with the Zô-Sè telescope.

De Moidrey spent more than twenty years at his Lukiapang post, directing it to become the premier French station for terrestrial magnetism. An indication of his supremacy comes from a letter he sent to Charles Maurain in 1922, in which he asked for Paris's magnetical readings after 1914 and proposed an exchange of publications; de Moidrey indicated that he had all the magnetical readings from St-Maur and Val-Joyeux from between 1888 and the First World War.[25] Maurain answered by describing the changes that had occurred in French geophysics since the war. Terrestrial magnetism was the province of Maurain's institute, and he promised to send de Moidrey the magnetical readings that had been made in France since 1914, the publication of which was imminent. De Moidrey expressed disappointment at Maurain's torpor: "It is regrettable that you are unable to continue the exchange after 1914. Chinese or Japanese observers who come here must naturally think that France is out of the picture, and so they must go to England or Germany. That is what happened for a distinguished pooh-bah [*fourniture notable*] whom it is best not to identify. How can one say anything to the contrary to this man?"[26]

In the eyes of the Jesuit magnetician de Moidrey, torpor was a serious failing. He gave voice to his credo in an essay of 1919 entitled "What Is the Use of a Magnetical Observatory?" The question might be asked of

24. Zikawei, Observatory, *Observations magnétiques faites à l'Observatoire de Lu-Kia-Pang, année 1908* 1 (1911), 1–31; cf. *CIES*, pp. 275–76.

25. PIPG, Maurain to de Moidrey, n.d., answering de Moidrey's letter of 29 Mar. 1922; Maurain to de Moidrey, 12 Mar. 1923.

26. Ibid., de Moidrey to Maurain, 3 Feb. 1923.

Lukiapang Observatory, ca. 1910, looking southwest from the church. To the right of the two-story observatory building, at the center, is the meteorological mast. (Courtesy of Jesuit Archives, Hawthorn, Victoria, Australia)

Lukiapang Observatory, ca. 1910, Zu-tseng and his pupil beneath the trees. The small equatorial is at the right, next to the hut for absolute measurements; the meteorological mast is at the center. (Courtesy of Jesuit Archives, Hawthorn, Victoria, Australia)

"all research in pure science," de Moidrey began. The answer was readily apparent: sooner or later applications stemmed from the disinterested pursuit of knowledge about the natural world. These kinds of endeavor — solar physics, expeditions to distant places — invoked enthusiasm and curiosity, and "the Company of Jesus has never shrunk from directing its men and its resources toward purely theoretical studies." Among significant phenomena worthy of theoretical studies in geomagnetism, de Moidrey signaled the observations of Willem van Bemmelen, director of the Royal Magnetical and Meteorological Observatory at Batavia, regarding regular magnetical pulsations. Regularities of nature such as these were "riches to the greatest glory of God."[27]

The Zikawei Observatory and its two dependencies attracted foreign scientific travelers, and they inspired imitators. The most successful of the Zikawei spin-offs took shape following the labors of Edward Francis Pigot, an Irish Jesuit physician who arrived in Shanghai in 1899 at age forty-one. He grew interested in solar observations, which he carried out while studying Chinese. Failing health sent him to Australia, where he worked at both the Sydney and Melbourne observatories while teaching physics and experimenting with wireless transmission at St. Ignatius' College in Riverview, a suburb of Sydney. In 1904 he returned to Zikawei, only to ship out for Riverview three years later as a dying man. Pigot regained his health, and he conceived a desire to replicate in Australia what the French Jesuit scientists had done at Zikawei. Of all the activities he had seen, only seismology — which required no magnetical isolation and for which the instruments were relatively inexpensive — was within his reach. Jesuit seismology, Pigot wrote to an aunt who he hoped would help finance his observatory, was "a standing refutation of the Church's supposed antagonism to modern scientific progress."[28] Over the next decades Pigot built the most productive geophysical and astronomical observatory in early twentieth-century Australia.[29]

In his early years, Pigot valued support from his former colleagues at Zikawei to the extent that he wondered whether Riverview could become a branch of the Chinese operation. Henri Gauthier opposed such

27. De Moidrey, "A quoi peut servir un observatoire magnétique?" *RC* 6 (1918–21), 201–5; on Van Bemmelen see *ER*, pp. 108–18.

28. HJA, box 6, Pigot to Aunt Kate, [1911].

29. D. J. K. O'Connell, "Father Edward Francis Pigot, S.J.," *Studies* 41 (1952), 189–96, 323–32; *CIES*, pp. 79–81; *ER*, pp. 78–80. Pigot's first seismometer in 1908, a Wiechert inverted pendulum, was donated by a wealthy member of the upper house of Parliament from Sydney; his second and third seismometers — a Wiechert and a Mainka — and the cellar to house them came courtesy of a £270 legacy from a relative (HJA, box 6, "Notes, etc. by Fr. Pigot").

an arrangement, even though he thought it would be an honor for Zikawei to publish Riverview's scientific bulletin. Nor did Gauthier think that the time was ripe for an Asian counterpart to the Jesuit Seismological Association, organized for North America by Frederick L. Odenbach at St. Ignatius College in Cleveland. He did find merit in the thought of cooperation among Zikawei, Riverview, and a North American Jesuit observatory.[30] Pigot soon became master in his own house. By the middle of 1911 the Zikawei Jesuits were asking him about analyzing seismograms and using the time-distance curves of Emil Wiechert's Göttingen school.[31]

The French Jesuits readily asked advice of a junior man in Australia because their notion of prestige transcended chauvinism. Making sense of the open book of nature required consulting commentaries and issuing reports in a number of languages. Meteorological reports circulated in English—the lingua franca of the China coast—but matters of theory and instrumentation came from German texts.[32] The Jesuits were more than willing to cooperate with German colleagues in Asia. In 1902 Chevalier traveled to Tsingtao and advised the German governor, Oskar von Truppel, about founding an observatory there.[33] Zikawei hosted the German geophysicist Georg Angenheister, director of the Apia Observatory on Samoa, and readily made use of Angenheister's expertise in setting up a Wiechert inverted-pendulum seismometer. Angenheister was "very friendly and very devoted, not to speak of his expertise," Henri Gauthier wrote to Pigot in 1909. "He put the instrument in perfect shape, and it has been working for three weeks."[34] Angenheister's successor Ludwig Geiger looked to the Jesuits for succor. Late in 1913 he anticipated raising Zikawei with Apia's soon-to-be-completed radio transmitter. Radio contact would allow him to refine Apia's position on the globe. The new precision would contribute to

30. HJA, "Zikawei, 1901–1928," Gauthier to Pigot, 15 Jan. 1911. On Odenbach's observatory, see *Vregille*, pp. 550–54. Odenbach proposed a network of North American Jesuit geophysical observatories in 1909; by 1910 a dozen Wiechert 80-kg astatic, horizontal seismometers had been ordered or set in place (James Bernard Macelwane, *Jesuit Seismological Association, 1925–1950: Twenty-fifth Anniversary Volume* [St. Louis, Mo., 1950], p. 78).

31. HJA, "Zikawei, 1901–1928," Gauthier to Pigot, 7 May 1911. On the curves, see *CIES*, pp. 46–51.

32. Ernesto Gherzi chose Carl Mainka's *Physik der Erdbebenwellen* (Berlin, 1923) as his fundamental seismological handbook (HJA, "Zikawei, 1901–1928," Gherzi to E. F. Pigot, 4 May 1924).

33. De Lapparent, "Chevalier," p. 550; *CIES*, p. 262.

34. HJA, "Zikawei, 1901–1928," Gauthier to Pigot, 1 May 1909. On Angenheister at Zikawei, see *CIES*, pp. 79–80.

Ernesto Gherzi at Zikawei (Courtesy of Shanghai Meteorological Bureau)

theoretical seismology: "Our entire calculation of the innermost core of the earth will be modified."[35]

The pattern of recruitment for Zikawei, where a Jesuit was targeted years before he was given scientific responsibility, continued into the twentieth century. The career of the observatory's wireless expert Ernesto Gherzi recalls elements of the careers of other Jesuit geophysicists, such as Louis Froc and Edward Francis Pigot, and it also reveals the extent to which the Zikawei administration bent over backwards to stay on good terms with its civilian benefactors.

Gherzi was the son of a hotel owner in San Remo. When the tourist trade declined in the Italian resort following an earthquake, Gherzi's father lost his properties, taught languages in a local technical school, and then died of pneumonia. The family went to live in a fifth-floor apartment. Gherzi, sent to a Jesuit school in Monaco, had to decide at age seventeen between the Society of Jesus and the Italian navy. His brother had already signed up with the Salesians at Turin, and so he went to the French Jesuits on the Isle of Jersey. He shipped out to Zikawei in 1910 as a temporary physics teacher. Three years later he returned to Europe to recover from amoebic dysentery. After spending the war on Jersey and in England, he set off again for Zikawei in 1920 as Louis Froc's assistant.[36]

35. Göttingen, Niedersächsische Staats- und Universitätsbibliothek, HW 22, Geiger to Hermann Wagner, 18 Nov. 1913.

36. SJQ, Gherzi, "Autobiographie."

Froc initially placed Gherzi with the seismographs. Among Zikawei's four earthquake monitoring devices, only the Wiechert astatic pendulum was working. Froc and Gherzi decided to let the two horizontal Omori pendulums rest in peace and to renovate the Galitzin machine by converting its photographical recorder to a visual system.[37] Gherzi ordered a vertical Wiechert from the German firm of Spindler and Hoyer, and with the prestige of the Jesuits behind him, he anticipated setting up a seismological department at the Beijing Observatory.[38] Gherzi also took responsibility for radio storm warnings and meteorology in general. He proceeded, as he wrote in English to Pigot in 1924, in an intellectual vacuum:

> I found here myself completely alone, without any real practical help (politesse does not change anything in science). . . . To speak sincerely, French ways, especially now after this war, would *tue* you, as I have seen, since 20 years they do with English speaking people. I am myself not a Frenchman, but Italian, and having accepted this sort of life, I am quite happy here. . . . But I dare to say, their different ways, would surely be trying to you—(not for a stay of 6 months).
>
> Now as you know, I am here number nth . . . not 2°, neither 3°. . . . So I have no authority on the subject. For my experience I have always found a practical opposition (notwithstanding good words) to any new scientific research.[39]

The full measure of Gherzi's isolation from his Zikawei colleagues lies in his apparent ignorance of Pigot's time in China.

Froc returned to Europe on furlough in 1926, leaving Gherzi alone in charge of twelve Chinese assistants, who were paid by the Jesuit Province of Paris, and a string of inland observers, paid by the Chinese Maritime Customs.[40] But he had no administrative authority. A French Jesuit (probably Henri Gauthier) took charge until Stanislas Chevalier emerged from retirement.[41] The French consulate in Shanghai would not allow the directorship to go to Gherzi, a foreign national. Zikawei

37. HJA, "Zikawei, 1901–1928," Froc to Pigot, 5 Jan. 1921; Gherzi to Pigot, 4 Jan. 1921.

38. Ibid., Gherzi to Pigot, 19 Feb. 1922 and 1 Apr. 1922.

39. Ibid., 4 May 1924, ellipses and English text by Gherzi.

40. SJQ, Gherzi, "Autobiographie." Gherzi reports that a consortium of shipping companies came to supply 20–30 percent of the cost of the Zikawei assistants; all other expenses at Zikawei and inland were covered by the French Settlement and the Chinese customs.

41. In Zikawei, Observatory, *Cinquante ans de travail scientifique* (Paris, [1928]), a profusely illustrated brochure, Froc retains the title of director, Gauthier is *directeur honoraire*, and Chevalier is *directeur suppléant*.

fell to a brilliant young physicist, Pierre Lejay, who left Gherzi alone in his meteorological tasks.[42]

Meteorology meant predicting the weather. In his early work on seismology, Gherzi tried to track typhoons by watching the displacements of his seismographs.[43] With time the weather service absorbed much of his attention. Beginning in 1926 he published a yearly monograph on typhoons, which constituted his report to the Shanghai General Chamber of Commerce; the reports accounted for the chamber's financial support of the Zikawei weather service, and they featured Gherzi's synthetic observations and speculations about tropical weather. In his report for 1933 Gherzi applied the theory of fronts to typhoons. He concluded that typhoons were not Margules-type cyclones driven by potential energy due to temperature differences in a revolving air mass; rather, typhoons drew their energy from latent heat set free by condensation of water vapor, by the vertical circulation of heat in the air mass, and by the residual work done in the system as the air mass was pumped from high-pressure to low-pressure areas.[44] These are the writings of a dispassionate meteorologist. There is nothing, except perhaps scrupulously straightforward prose, to suggest that the analysis came from a French Jesuit situated at an imperialist outpost in China.

In 1928 Gherzi received two hundred weather reports daily; his synthetic predictions, analyzed with the help of Chinese coworkers, went out to seventy-five semaphores controlled by the Chinese Maritime Customs. When tracking typhoons, Gherzi used Zikawei's wireless to broadcast storm warnings in English, Russian, and French.[45] The following year saw civil aviation come to China, and with it came the need

42. SJQ, Gherzi, "Autobiographie." Gherzi mentions that this is the reason why he was later passed over as effective director in favor of a "young *scholastique*" with no advanced training in exact sciences – Louis Dumas. See Henri Jalabert, *Jésuites au Proche-Orient* (Beirut, 1987), pp. 321–22, for Dumas at Zikawei; and idem, *Le souvenir de nos morts* (Beirut, 1983), p. 139. In fact, at the end of 1938 the French Concession refused to allow Gherzi to direct the semaphore tower on the Bund; his Italian nationality was at issue (SMB, Dumas to French Concession, 5 Dec. 1943, where Dumas records his rise in the observatory hierarchy and the municipal objections to Gherzi's promotion).

43. See Brunhes, "Jésuites," pp. 34–35, for Gherzi's belief that seismograms would be able to provide warning of approaching tropical storms; see also Benjamin F. Howell, Jr., *An Introduction to Seismological Research: History and Development* (Cambridge, 1990), p. 132. We shall see in part 4 that Gherzi's belief was also voiced by French seismologists on Martinique.

44. Gherzi, *Typhoons in 1933* (Shanghai, 1934), p. 6.

45. Chant., "Notes prises après et pendant une conversation avec le P. Froc," 26 Nov. 1928.

Deciphering incoming telegrams at Zikawei, a task preliminary to issuing daily meteorological bulletins. The telegraphical network, extending south to Saigon, east to Manila and Tokyo, and north to Vladivostok, for the most part served maritime commerce. (From Zikawei, Observatory, *Cinquante ans de travail scientifique* [Paris, (1928)])

for regular wireless forecasting.[46] By 1931 Gherzi was broadcasting daily weather reports on 24-m and 42-m bands, with transmitters provided by the Italian navy. His authority extended from Vladivostok to Guam and Singapore.[47]

Gherzi described his internal regime in detail. In the 1920s he directed fourteen senior Chinese staff, who ranged in age from young Joseph Koé Tsang Zié to octogenarian François-Xavier Seng Fang Bing. One man, Tsiang Wen Da, compiled the daily meteorological charts; Bang Seng Ghi recorded radio data, Ze Tsing telephone data. Koé Tsang Zié prepared national synopses by summarizing reports from stations of missionaries and the Chinese Maritime Customs. Seng Fang Bing took charge of barometrical and heliometrical measurements, while Ziing Fang handled the seismographs. Deaf Dang Lizen supervised the local thermometer, barograph, wind, and rainfall rec-

46. Ernesto Gherzi, "The Scientific Work of the Catholic Church at the Zikawei Observatory in Shanghai," Instituto Português de Hong Kong, *Boletím*, July 1950, pp. 45–57, on p. 52.

47. M. Burgaud and B. [*sic*] Gherzi, "L'annonce des typhons dans les mers de Chine par l'Observatoire de Zi-ka-wei," *Journal de la Marine marchande* 13 (1931), 1917–18, 1965–67, 2013–17.

Recording atmospheric ozone at Zikawei, 1902. From the turn of the century, Chinese staff supervised many meteorological measurements. (Courtesy of Shanghai Meteorological Bureau)

ords. Zi Ping Wen oversaw all the mean calculations and took care of Chinese correspondence. Sickly Zi Ming Kwei was available to help where needed. For their exacting and fatiguing labor, the Chinese assistants received on the order of fifty dollars per month.[48]

Despite the strictures of Louis Froc in 1904, both Zikawei and Zô-Sè offered Chinese geophysicists and astronomers the possibility of working and publishing. The earliest man to author a treatise was Pierre Hoang (Huang Po-lu), an expert on Chinese law who grew interested in historical observations of earthquakes and eclipses in China. The Zikawei Jesuits published several of his compendia and produced posthumous editions of his manuscripts.[49] François Lou published a de-

48. SMB, Gherzi, "Travail des siésangs de la météorologie," n.d. Gherzi notes that the "coolies" employed at the semaphore on the Shanghai waterfront received minimal wages of forty dollars per month and that office staff had to be paid better. The amount here may be in Mexican silver dollars, the coin of the imperialist realm in Shanghai (see *CIES*, p. 255). In the following pages it is unknown whether unqualified dollar amounts are in Chinese, U.S., or Mexican currency.

49. Hoang, *A Notice of the Chinese Calendar and a Concordance with the European Calendar*

scriptive brochure about Lukiapang in both French and Chinese.[50] Tsai Chang-tche published observations of stellar occultations in the *Journal des observateurs.*[51] Kao Kiun (Kao Chin) spent two years as Chevalier's assistant and helped write up a lunar atlas; then he looked into the early astronomical writings known as the *Chou Pei Suan Ching.*[52] By the middle 1930s Kao was an astronomer at the Nanking Observatory, run by the Academia Sinica.[53] There he continued Hoang's and his own interest in early Chinese astronomy by publishing a revised concordance between the Julian calendar and Chinese calendars; he also began working with a Hale spectroheliograph.[54]

The internal side of Jesuit scientific activity, assimilating young men to French research norms, found a complement in the radiation of Jesuit authority beyond their observatories. During the first half of the twentieth century, Jesuit physicists and astronomers projected their expertise throughout China. National stature derived in large measure from their extensive network of weather reporting by telegraph and radio. Ernesto Gherzi in fact proposed that the League of Nations finance a national Chinese meteorological service comprising forty first-class stations under direct Jesuit supervision spread across nine provinces.[55] Zikawei also sought a commission from Chiang Kai-shek to draw a new map of the country. In collaboration with Henri Roussilhe,

(Zikawei, 1885; 2d ed., 1904), of special value for its long introduction; idem, *Catalogue des tremblements de terre signalés en Chine d'après les sources chinoises (1767 avant J. C.—1895)*, 2 vols. (Shanghai, 1909–13), the second volume appearing posthumously, ed. Jérôme Tobar and Henri Gauthier; idem, *Catalogue des éclipses de soleil et de lune, relatées dans les documents chinois et collationnées avec le canon de Th. Ritter v. Oppolzer* (Shanghai, 1925), apparently edited by Louis Gauchet; idem, "Du sauvetage du soleil ou de la lune dans une éclipse," *RC* 1 (1903–5), 76–77, where Hoang provides a description of the traditional Chinese eclipse ceremony; "Un anniversaire," ibid., 241–42, for a photograph of Hoang.

50. François Lou, *Lu-kia-pang: Observatoire magnétique* (Zikawei, 1918), plus a companion Chinese volume, n.d., in Chant.

51. Tsai, "Occultations d'étoiles observées à Zô-Sè," *Journal des observateurs* 10, no. 6 (1927), 72.

52. Kao, "Etude sur une observation d'étoile rapportée par le Tcheou-pei," *BUA*, no. 5, 1922–23, pp. 1–16. Kao's commentary concerned the *Chou Pei*'s identification of the pole star. See Joseph Needham, *Science and Civilisation in China*, vol. 3, *Mathematics and the Sciences of the Heavens and the Earth* (Cambridge, 1959), 19–25, which, however, does not note Kao's treatment.

53. "Le rayonnement de l'Aurore en Chine et à l'étranger," *BUA*, no. 33, 1935–36, pp. 8–12.

54. *Academia Sinica with Its Research Institutes* (Shanghai, 1929), p. 33; *The Academia Sinica and Its National Research Institutes* (Nanking, 1931), pp. 63–72; Kao, *Conversion des dates chinoises en jours de la période julienne* (Nanking, 1932) [Academia Sinica, National Research Institute of Astronomy, Monographs, 1].

55. SMB, Gherzi to F. J. M. Bourdrez (Nanking), 14 May 1937.

the naval hydrographer who had risen to become general secretary of the International Union of Aerial Photography, the Jesuits wanted to send French aviator surveyors across China, province by province, in a project that would cost tens of millions of dollars and take more than a decade to complete.[56]

Neither proposal materialized, possibly because the Chinese resented a Jesuit infrastructure that seemed to offer little place for Chinese initiative. Until the late 1920s, when political and economic circumstances drove them to be more circumspect, the Jesuits generally avoided professional contact with Chinese counterparts. They conducted the longitude project, for example, without official Chinese participation. The French consul general at Shanghai reminded Louis Froc in 1923 about the difficulties such a collaboration would entail: "If the Chinese could be excluded, that would only be for the better, since you know all the objections they have raised and continue to raise about foreign radio installations."[57]

Chinese astronomers and physicists nevertheless sought out Jesuit counterparts. Wong Wenlay, at the Beijing Geological Survey, asked in 1921 about an exchange of scientific publications and about the possibility of assistance in setting up a seismograph. Li Yng Piao, a physicist at the University of Amoy, turned to Maurice Burgaud in 1926 for advice about meteorological recordings, and he sent the son of his university's president to Zikawei for "theoretical and practical" instruction. Tchou Koantchal, a graduate of the Ecole polytechnique and the Ecole des ponts et chaussées and a member of the physics section of the Academia Sinica, sought close relations with the Jesuits on astronomical and physical matters. In 1931 he asked about Pierre Lejay's gravimetrical techniques, wondered whether Edmund dc la Villemarqué could help his Beijing group place their small telescope in order, and hoped to learn how to receive European time signals.[58] In recognition of this growing scientific respect, Zikawei's director Pierre Lejay entered the Academia Sinica in 1933 as an associate member. When he accepted this honor, Lejay expressed hope that political turmoil would not compromise Chinese research.[59]

At a time when China was convulsed by war and revolution, Jesuit science underwent major restructuring. Stanislas Chevalier's successor at Zô-Sè in the 1920s was Louis Gauchet, who lasted less than a year

56. Ibid., Lejay to Chiang, 23 Aug. 1933.

57. Ibid., Consul General to Froc, 24 Feb. 1923.

58. Ibid., Wong Wenlay to Froc, 17 Mar. 1921; Li Yng Piao to Burgaud, 21 Jan. 1926; Tchou Koantchal to Burgaud, 28 Feb. 1931; Jean A. Bussière to Burgaud, 2 Feb. 1930, with Tchou's qualifications.

59. Ibid., Lejay to Li Shou Houa [?], 13 May 1933.

View from the Zô-Sè Observatory, 1951, shortly before the Jesuits' expulsion from China by the communist regime (Courtesy of Cantilliaceusis Domus Sessionum et Exercitiorum, Les Fontaines, Chantilly)

before returning to Europe; in Gauchet's place came Edmund de la Villemarqué, a former French naval officer. The guard changed at a time of external dislocation. Fighting in 1923 prevented Joseph de Moidrey from contacting colleagues in Beijing.[60] The threat of civil war in 1927 inspired the Zô-Sè astronomers to send all their delicate apparatus and records to Zikawei, where Gherzi took up arms to defend the central observatory against the Chinese communists.[61] By 1933 the Lukiapang magnetical operation, operated by the aged de Moidrey, had moved to Zô-Sè, a hilltop location removed from the path of advancing armies.[62] But as would happen with the Jesuit observatories on Madagascar and in Lebanon, Zô-Sè itself became a military objective. In 1937 the observatory narrowly escaped destruction by bombs and suffered through

60. PIPG, de Moidrey to Maurain, 8 Oct. 1923.

61. E. de la Villemarqué, "Préface," Zô-Sè, Astronomical Observatory, *Annales* 17 (1932), i – iii; SJQ, Gherzi, "Autobiographie."

62. Emile Joulia, "Trois observatoires français d'outre-mer: Ksara, Ambohidempona, Zi-ka-wei," Toulouse, Société d'astronomie populaire, *Bulletin mensuel*, no. 197, 1939, pp. 122–39, on p. 136; Dehergne, "Zi-Ka-Wei," p. 184, which gives 1929 as the date of the move; SMB, Lejay to President of the Caisse des oeuvres d'intérêt public [French Concession of Shanghai], 1 Oct. 1933, where the Zô-Sè magnetical operation is mentioned as being in service since the beginning of the year. The Zikawei administration had planned the move from 1927 (ibid., Lejay to Father Provincial, 21 Feb. 1927).

a battle that saw a thousand men killed; it offered shelter to hundreds of famished refugees, while the magnetical recorders kept measuring the pulse of the heavens.[63]

The heavens withheld blessings from other institutions. The president of the Chinese Physical Society, in a circular letter sent out from Kunming in 1940, put the situation succinctly: "With the fall of our larger cities, our better-equipped universities and other institutes of learning have suffered tremendously. Instances of the actual destruction of these institutes are too numerous to be cited; and wherever they have not been destroyed, the conditions have become such as to render their normal functioning utterly impossible."[64]

Adept at coexisting with the ascendant temporal power, the Jesuit physicists and astronomers persisted at their posts. The Japanese occupied Shanghai in 1937; in 1940 they took over direct administration of the area around Zikawei. The Jesuits, under protection from Rome and Vichy, were deprived of their meteorological network but still went about their business unmolested.[65] Director Pierre Lejay was clear about his observatory's good fortune:

> When hostilities began, the observatory was in an excellent position. Many publications appeared over the preceding years, and our reputation in international scientific circles was without question spreading. Doubtless Zikawei is not one of the institutes of first rank whose researches follow the path to great discoveries, but among the secondary observatories that contribute to general progress, we are in a very good position, and in a great number of areas our work must be taken into account.

He continued: "The consequences of the war have been in the past much less than could have been foreseen." Zikawei and Zô-Sè were out of communication with each other and with the outside world, but "the whole has continued to work normally, and new research at least as interesting as the old research has been undertaken." The scientific staff were overworked and taxed with emotions, but they continued to produce results unflaggingly. Most important, local funding had been found to make up for missing government grants.[66]

63. PIPG, Burgaud to Maurain, 18 Nov. 1937, where Burgaud acknowledges receiving the Benjamin Valz prize of the Academy of Sciences for his magnetical surveying; 30 Dec. 1937, for the number killed and the virtually uninterrupted recording. SMB, "Rapport sur la situation financière de l'Observatoire de Zikawei" [1938], for ten thousand dollars' worth of damage to Zô-Sè by Japanese bombs.

64. PIPG, Y. H. Woo, circular letter of 1 Apr. 1940.

65. SMB, Louis Dumas, "Rapport sur l'activité de l'Observatoire de Zikawei en 1940," 7 July 1941, where the only consequences of the Japanese invasion are a lack of outside communication and a momentary loss of constant-frequency electrical power.

66. Ibid., Lejay, "Situation de l'observatoire," Oct. 1938.

Financial records provide mute testimony to Jesuit tenacity. Their annual budget amounted to between $20,000 and $30,000. The Jesuit hierarchy in France contributed $6,000 of this amount.[67] The French navy, a major patron, provided about Fr 10,000. It expected to receive in return useful weather predictions twice daily and "general theories about the atmosphere in the Far East"; it advised naval vessels to transmit meteorological data to Zikawei and to calibrate their instruments with the Jesuits when in port.[68] The French Concession provided $1,500 in the late 1920s, a grant that rose to $3,000 in 1933; the money was drawn on the Bank of Indochina. The government of French Indochina directly contributed several thousands of Mexican dollars. The French ministry of foreign affairs gave Fr 10,500 through its Shanghai consul general, and the undersecretariat of state for scientific research contributed Fr 20,000 for equipment in 1937. Airlines and shipping concerns provided more than $2,700.[69]

As the preceding tally suggests, the Jesuits continually demanded more. Director Lejay asked the French Concession for $10,000 in 1933 and the ministry of foreign affairs for Fr 25,000 in 1939; Louis Dumas dunned Indochina for $3,000. Jesuitical eloquence prevailed. The ministry of foreign affairs gave U.S. $1,298.70 in 1939, "a large increase in the sums previously granted." By 1941 the ministry was contributing Fr 42,900 ($18,000), the lion's share of the observatory's budget; interim director Dumas promptly asked that it be raised to Fr 60,000.[70]

The optimistic vision of scientific life under Japanese administration late in 1940 is revealed in an exceptional grant of $5,000 by the French Concession for the Jesuits' libraries. Of this amount, $1,000 went for purchasing and binding scientific periodicals, whose very list testifies to a desire to remain aware of the latest word in physics and astronomy: the *Revue de physique et de radium*, *Terrestrial Magnetism*, *Radio Engineering*, the *Journal des observateurs*, the *Astrophysical Journal*, the *Astronomical Journal*, the British astronomical *Monthly Notices*, and the *Journal of*

67. Ibid., Lejay to Dumas, 15 Mar. 1936.

68. Ibid., French Naval Forces in the Far East, "Recherches de renseignements météorologiques par les bâtiments des F.N.E.O.," 5 Oct. 1928.

69. Ibid., Shanghai Consulate General of France to Froc, 28 Oct. 1929; Consulate General to Bursar of the Kiang-nan Mission (Verdier), [1933]; Governor General of Indochina to Lejay, 6 June 1935, where two thousand dollars is promised for 1936; Consulate General to Lejay, 2 Dec. 1933; Undersecretariat of State for Scientific Research to Service des oeuvres in the Ministry of Foreign Affairs, 26 June 1937; Lejay to unknown, 20 Mar. 1937; French Concession to Lejay, 4 June 1937.

70. SMB, Lejay to Shanghai Caisse des oeuvres d'intérêt public, 1 Oct. 1933; "Rapport sur les activités de l'Observatoire de Zi Ka Wei en 1938"; Dumas to Shanghai Consulate General, 20 Dec. 1940; Consulate General to Lejay, 14 Feb. 1939; Dumas to Consulate General, [1941].

Edmund de la Villemarqué observing at Zô-Sè. The telescope, a double-barreled refractor mounted at the turn of the century, is currently in use. (Courtesy of Cantilliaceusis Domus Sessionum et Exercitiorum, Les Fontaines, Chantilly)

the Franklin Institute.[71] By this time, when all Chinese universities in Shanghai had been closed or had moved to another location, Dumas believed that the Japanese were more kindly disposed to the church than were the communists on the other side of the lines. The invaders allowed the Zô-Sè astronomers to continue their work; Dumas had traveled by automobile to Zô-Sè and had secured a supply of coal for the hilltop observatory. He thought the Japanese wanted to return the French Concession to Vichy France. He concluded that the new administrative regime was an incompetent extortion racket in which the price for municipal services increased by a factor of ten.[72]

Dumas's benign interpretation of Japanese imperialism was not shared by everyone on his staff. Ernesto Gherzi, for one, spent a month on leave in Japan to recover his health, but he returned from the visit

71. Ibid., Dumas to French Concession, 26 Nov. 1940.

72. Chant., Dumas to Lejay, 15 Nov. 1940. Dumas was at this time interim director of Zikawei and Zô-Sè, a position that he held through the war. He became rector of the Université l'Aurore in 1946. Expelled from China in 1952, he relocated in the medical faculty at the Université St-Joseph in Beirut. A sniper's bullet cut him down in 1975 (Jalabert, *Jésuites au Proche-Orient*, pp. 321–22).

more depressed than ever. The Italian national had not been prevented from traveling to a friendly country.[73] His nationality earned him no premiums, however, at Zikawei. He published a climatological atlas of East Asia in 1944, but the Japanese refused to allow it to circulate.[74]

The Academy's Envoy in China

Pierre Lejay, who in 1940 received word about his observatory from Louis Dumas, was born in a wealthy suburb of Toulon, the haunt of officers stationed at the nearby naval base. His father rose to the rank of rear admiral during the First World War; two of his four brothers also served as naval officers, the younger one going down with his submarine during the Second World War. Lejay's biographer emphasizes that from his family Lejay retained "an acute sense of duty and discipline, and the taste for distant voyages." Lejay succeeded brilliantly in his studies at the Jesuit collège of St-Louis-de-Gonzague on the rue Franklin in Paris, receiving a double baccalauréat in philosophy and elementary mathematics in 1914. Then sixteen, he wanted to join the Jesuits. His father, feeling that he was a bit young to make such a decision, persuaded him to prepare for the naval entrance examinations at the Lycée St-Louis. But in 1915 there was no entrance competition, so Lejay completed two certificates at the Sorbonne, in general mathematics and higher astronomy. In October 1915, after the summer vacation, the seventeen-year-old Lejay entered the Jesuit novitiate at Canterbury. Soon thereafter the French state claimed him as conscript. He served at the front, where he was wounded, and finished the war as an officer.[75]

The decision to become a Jesuit is, at least for the Jesuits, a mystery — the result of vague intimations and inspirations. It is certain that Lejay knew and rejected the privilege, pretension, and corruption that attended on those who owed their life to the state. Just as certainly, he —

73. SJQ, Gherzi, "Autobiographie." Apparently before 1940, Gherzi implored Lejay, then in France, to have Benito Mussolini intercede with Japan on Zikawei's behalf.

74. Dehergne, "Zi-Ka-Wei," p. 187. SMB, Gherzi [?], "Aperçu du travail météorologique fait à Zi Ka Wei durant les années de guerre," 8 Jan. 1945.

75. Pierre Tardi, "Notice nécrologique sur le R. P. Pierre Lejay," *CRAS* 247 (1958), 1265–70; Henri Milloux, "Notice sur la vie et les travaux de Pierre Lejay (1898–1958)," Académie des Sciences, Paris, *Notices et discours* 4 (1957–62), 416–22; François Russo, "The Reverend Pierre Lejay, S.J., 1898–1958," *Jesuit Science Bulletin* 37 (1959), 80–87. Details of Lejay's career, especially his progress in the Society of Jesus, have been relayed to me by Père Robert Brunet, S.J., Bibliothécaire des Fontaines, Chantilly, France. Brunet and Russo ("Lejay") have Lejay preparing to enter the Ecole navale by attending the Lycée St-Louis. It is possible that this refers to the Maison St-Louis on Jersey, which until 1900 housed a collège for this purpose.

like his predecessor Stanislas Chevalier — was one of the boys for whom the Jesuits had set their traps.

The postwar years found Lejay acquiring the scientific tools of his trade and the research sensibilities particular to French scientists. He took a diploma in radioelectricity from the Ecole supérieure d'électricité in 1919. After a year of literary studies at Canterbury, he received permission from his Jesuit superiors to finish a Sorbonne licence in 1921. In 1922–23 he studied philosophy at the Jesuit establishment on Jersey. From 1924 to 1926 he read theology in Paris, while at the same time serving as an assistant at the Paris Observatory in the meridian department and the time service.[76] Lejay was a bright hope in the exact sciences.

Lejay married his technical facility in electricity with his more abstract interest in the heavens to study electrical storms in the atmosphere and especially their long-distance propagation. He wrote a doctoral dissertation under Charles Maurain, who provided him access to the observatory on the Pic-du-Midi, whence came his data.[77] Beginning in 1923, Lejay worked with Maurain on designing his instruments.[78] Notes of Lejay's appeared in the *Comptes-Rendus* of the Academy of Sciences, courtesy of General Gustave Ferrié, who was a close friend and colleague of Maurain's from the early days of French aviation.[79] Maurain used his influence with the military aeronauts to provide Lejay with the kites and wire that were necessary for his research and then steered the Jesuit to a grant of two thousand francs from the Caisse des recherches scientifiques.[80]

Early in 1923, Ferrié identified Lejay as an ideal candidate for one in a series of ambitious missions hatched by the geophysicists attached to the general staff of the French army.[81] In charge of French wireless transmission, General Ferrié had distributed amplifiers to outposts of

76. These details from Robert Brunet. Lejay was ordained a priest in 1926, the same year he received a doctorate from the Sorbonne. On Lejay's scientific progress, see Lejay, *Notice sur les travaux scientifiques du R. P. Pierre Lejay* (Paris, 1946).

77. Lejay, *Les perturbations orageuses du champ électrique et leur propagation à grande distance* (diss., University of Paris, 1926).

78. PIPG, Lejay to Maurain, 20 Dec. 1923 and 19 Apr. 1924.

79. Ibid., 9 May 1924; Lejay, "Sur un electromètre à lampe triode et son application à la mesure du gradient électrique de l'atmosphère," *CRAS* 178 (1924), 1480–82; idem, "Sur l'emploi des lampes à plusieurs électrodes en électrométrie," ibid., 2171–73.

80. PIPG, Maurain to Director of the Technical Service of Aeronautics (Issy-les-Moulineaux), 1925, draft; Lejay to Maurain, 11 Mar. 1925; Maurain to Lejay, 14 Mar. 1925.

81. SMB, Ferrié to Froc, 28 Feb. 1923, where Ferrié has Lejay become familiar with the mission's astronomical and wireless instruments, to be discussed below.

French civilization, and with a suitably rigged antenna these enabled Paris to raise a number of stations around the world. Ferrié had offered his new technology to French physicists and astronomers on a number of occasions. In 1913–14, he used his Eiffel Tower broadcaster and receiver to work with colleagues from the United States Naval Observatory in determining the longitude difference between Paris and Washington, and in 1919 he performed a similar operation between Paris and Brussels.

The geodesical determinations became precedent for a global measurement of longitudes based on radio signals. Early in 1919, the Bureau of Longitudes identified a Northern Hemisphere radio triangle connecting Paris, Shanghai, and the region of San Francisco; a related operation would connect Paris with New Zealand to study the global propagation of radio waves and the possible influence of the earth's rotation upon them. By the summer of 1919 Algiers was included as one of the fundamental vertices of the longitude determination (Algiers in fact replaced Paris as the principal French vertex).[82] The United States Naval Observatory was glad to collaborate on the project, but the American astronomers wanted to reduce their own data independently. After proposing to add in a handful of English-language stations (and after astronomers at Lick Observatory declined to join the project), the Americans settled on San Diego for the third fundamental vertex.[83] In

82. PO, carton 3, II-C, Bureau of Longitudes, "Projet de circulaire," Feb. 1919; Baillaud to Ferrié, 15 Aug. 1919 (draft). Emile Picard, Benjamin Baillaud, and Gustave Ferrié, "Sur un projet du Bureau des longitudes relatif à la détermination d'un réseau mondial de longitudes et de latitudes," *CRAS* 168 (1919), 1074–76, where Algiers, Zikawei, and an observatory in the region of San Francisco, California, are identified as the main stations. Charles Edward Adams at Wellington asked Baillaud to send information about the global longitude operation to the secretary of state for colonies in London; the latter apparently did not supply funds for Adams to participate (PO, carton 3, II-C, President of the Bureau of Longitudes to Minister of Public Instruction, n.d.).

83. PO, carton 3, II-C, summary of a letter from the U.S. Naval Observatory, 1 June 1919, where Greenwich, Washington, Paris, San Francisco, and Shanghai are proposed as the fundamental vertices and secondary stations are offered in Panama, Alaska, Hawaii, and Manila. See SMB, Gustave Ferrié, "Projet de programme de l'opération préliminaire de longitudes que l'on se propose d'effectuer à la fin de 1923," 22 Jan. 1923, where Lick Observatory is proposed as the third vertex. The San Diego Naval Air Station (North Island) was eventually chosen as the third observatory. Captain Frank B. Littell (a veteran of the 1913–14 Paris-Washington longitude measurement) and his second, John C. Hammond, were sent there from Washington by the naval authorities; Colonel Charles Mailles of the French army's geographical service joined them. See F. B. Littell, J. C. Hammond, C. B. Watts, and P. Sollenberger, "World Longitude Operation of 1926: Results of Observations at San Diego and Washington," *Publications of the United States Naval Observatory* 12 (1929), 527–93 (titled "Appendix"), which also details discussions about the operation in the International Astronomical Union and the International Geodesical and Geophysical Union from 1922 to 1926.

1921 the French parliament awarded Fr 250,000 to the operation, the money intended for purchasing three identical, portable meridian circles and other equipment.[84]

Zikawei had previously received radios courtesy of Ferrié and the army.[85] Having directed a wireless weather service for the China coast since 1914, the Jesuit fathers at war's end could pick up signals from San Diego and Bordeaux.[86] Even with circumstances favoring a transoceanic collaboration of church and state, Parisian mandarins took years to decide how they would deal with the Jesuits. As late as November 1925, Ferrié saw three possible arrangements. France could occupy itself officially only with Paris and Algiers, letting the Jesuits take charge themselves of Shanghai and having the Americans operate their station at San Diego independently, a course that would cost Fr 45,000; France could send an astronomer to San Diego and radio apparatus to Shanghai, at a cost of Fr 100,000; or France could send one astronomer to San Diego and two to Zikawei, which would run to Fr 190,000.[87] France settled on the most expensive scenario, although some money was saved by engaging the Jesuit Lejay, who cost the state nothing.

Two French scientists went out to China for the operation – Lejay and the director of the Nice Observatory, Gaston Fayet. Fayet arrived in September 1926 with his own astrolabe telescope. He remained at Zikawei until December, sharing responsibility for the celestial observations with a team under Edmund de la Villemarqué, who used the Zikawei meridian telescope. Lejay was charged with receiving the hourly time signals and verifying the observatory's clocks, which were placed in a special, isolated room.[88]

Lejay provided the project – an exercise for fine-tuning a technology useful to the military – with an abstract rationale. Shortly after arriving in Shanghai, he published a note about the longitude operation. He began by recalling a few points from "the relativity of time and space," although he cautioned his readers that he would not go into Einstein's theories. The problem that preoccupied him was essentially astronom-

84. PO, carton 3, II-C, Commission des appareils, session of 19 Oct. 1921.

85. De Lapparent, "Chevalier," p. 352, where Chevalier agreed to participate in the longitude project "following an exchange of views with Mr. Benjamin Baillaud."

86. On the early Jesuit wireless operation see Gauthier, "L'Observatoire de Zi-ka-wei," p. 12; and "The Story of Siccawei Observatory: Fifty Years of a Great Work," *North China Daily News*, 15 Dec. 1923, Christmas supplement.

87. PO, carton 3, II-C, Ferrié to Baillaud, 14 Nov. 1925.

88. For the responsibilities and instruments, see *Coopération de l'Observatoire de Zi-ka-wei à la révision internationale des longitudes* (1927) [Zô-Sè, Observatoire, Annales, 16], a collective work under the direction of Stanislas Chevalier. The theoretical justifications are in Ernest Esclangon, "Sur la précision des observations méridiennes et des déterminations de longitude," *Annales de l'Observatoire de Strasbourg* 1 (1926), 373–404.

Ernesto Gherzi with antenna used at Zikawei for the longitude operation of 1926. With the end of the Jesuit regime at Zikawei, Gherzi moved to Macao. He then began a new career in Montreal, where he established a seismological observatory at the Collège Jean-de-Brébeuf. (From *Coopération de l'Observatoire de Zikawei à la révision internationale des longitudes* [1927] [Zô-Sè, Observatoire, Annales, 16])

ical, but the results would relate most directly to geophysical issues. The three stations participating in the exercise would each measure the positions of two thousand stars over the course of two months. Through radio communication, each station would know its location on the globe to within three meters. Such precision allowed for testing Alfred Wegener's theory of continental drift. Lejay informed his readers about the current opinion vindicating isostasy, according to which the earth's crust floated in equilibrium on substrata. Wegener's theory provided interesting explanations in geology, but evidence for it remained inconclusive. If it turned out that relative longitudes did not remain fixed, however, "our understanding of the interior of the earth, and perhaps even of the condition of humanity, will have taken a small step." Having gone from Einstein to Wegener to humanity at large, Lejay used geophysics to inspire confidence in divine purpose. Science might be cultivated more for evoking mystery than for explaining things, Lejay offered, but it had a role to play in dispelling doubt about the grandeur of mankind which had arisen at the end of the nineteenth century. The doubt festered in Pierre Eugène Marcelin Berthelot's "blasphemous exaltation" and Henri Poincaré's pessimism. Science was no stranger to the well-meaning anguish of the times, Lejay emphasized. The coming generation would recognize that even though science was neither an ultimate goal nor a final explanation, it hardly constituted a danger. Science was "a need; it is good to realize simply that science is one of the most glorious parts given to man in unfolding the Creation that continually advances."[89]

Pierre Lejay spent a year at Zikawei, working with the director there to analyze the data of the longitude operation. He returned to France for his fourth year of theology at the advanced seminary of Fourvières, which had just come home to Lyons after surviving in English exile (at Canterbury and then at Ore Place, near Hastings) for a quarter-century. In 1929, to prepare for profession of the three vows, he took his third year as a novice (the tertianship) at Paray-le-Monial in Burgundy; then he installed himself in Ste-Geneviève, the elite Jesuit science collège at Versailles, to continue scientific work in anticipation of returning to Zikawei.[90] Lejay soon found a metropolitan partner with whom he would thereafter be associated.

89. Lejay, "Le problème de l'heure," *BUA*, no. 13, 1926–27, pp. 16–24. This seems to be one of the first references to Wegener (called "Wegeneq" by an apparent typographical error) in China; cf. Yang Jing Yi and David Oldroyd, "The Introduction and Development of Continental Drift Theory and Plate Tectonics in China: A Case Study in the Transference of Scientific Ideas from West to East," *Annals of Science* 46 (1989), 21–43, where Lejay makes no appearance. Testing Wegener's theory to justify the great expense on scientific grounds was also uppermost in the mind of French mandarins. See Georges Perrier, "L'Académie des sciences, le Bureau des longitudes et les grandes missions scientifiques," Bureau des longitudes, *Annuaire pour l'an 1933* (Paris, [1933]), pp. C.1–C.20, on pp. C.18–C.19.

90. See Alain Guichard, *Les jésuites* (Paris, 1974), p. 112, for Fourvières. Lejay's trajectory is according to Robert Brunet. Fourvières and Ste-Geneviève are in Gabriel Le Bras, *Les ordres religieux*, 2 vols. (Paris, 1979–80), 2:622, 647. It is probable that after his year at Zikawei, and perhaps well before then, Lejay was targeted by the Jesuit authorities as the next director of the observatory there. Paul Duclos indicates the targeting, but is vague about when it began, in his notice on Lejay in *Dictionnaire du monde religieux dans la France contemporaire*, vol. 1, *Les jésuites*, ed. Jean-Marie Mayeur and Yves-Marie Hilaire (Paris, 1985), p. 177. Lejay is listed as a member of the Zikawei staff for 1928, and all of his publications from 1923 to 1929, including his doctoral dissertation of 1926, are credited to Zikawei in Zikawei, Observatory, *Cinquante ans de travail scientifique*, a profusely illus-

Three Jesuits (Pierre Lejay, Stanislas Chevalier, and Maurice Burgaud) at a small meridian telescope, 1926 (From *Coopération de l'Observatoire de Zikawei à la révision internationale des longitudes* [1927] [Zô-Sè, Observatoire, Annales, 16])

The man was Fernand Holweck, then attached to the Radium Institute at Paris as its finest instrumentalist. Born into a Catholic Alsatian family, Holweck had graduated first in his class at the Ecole de physique et de chimie industrielles in Paris. From there he fulfilled his military obligations under Ferrié's command at the Eiffel Tower. He became involved in Ferrié's earliest determinations of longitude differences by radio. During the war, Ferrié had Holweck develop range finders and carry out research into submarine ultrasound. Demobilized, Holweck returned to a berth at the Radium Institute (which he had first occupied briefly in 1913 as Marie Curie's assistant) and in 1922 completed a doctorate under André Debierne on long-wavelength X-rays. In the course of his research, he developed a new vacuum pump. This knowledge of emptiness he used to invent portable and reliable radio emitters, again in the service of General Ferrié.[91]

trated brochure printed by Lejay at the end of 1928. SMB, Lejay to Henry [?], 26 Dec. 1928, for the booklet.

91. Georges Perrier, "Fernand Holweck (1890–1941)," *Bulletin géodésique*, nos. 69–76, 1941–42, pp. 39–64.

Enter Lejay, who had been looking into irregularities in the contacts of astronomical pendulum clocks and who (he recalled later) wanted to design a reliable and portable gravimeter for use in China. He first thought to amplify the precision of a free pendulum by using a registering chronograph developed by the firm of Belin; he would travel with the pendulum and relay its oscillations to Shanghai by short-wave.[92] This was nothing other than a project of Gustave Ferrié's, for since 1921 Ferrié had been working on electrical registration of pendulum gravimeters, marked with radio time signals.[93]

The idea of a quartz pendulum came to Lejay, and Belin's engineers showed him a quartz filament produced by Holweck, who intended it to serve as a time standard. Lejay considered having the filament oscillate in a vacuum as a free pendulum. But, undoubtedly with the model before him of Emil Wiechert's inverted-pendulum seismometers, Lejay decided to anchor the filament's base and let its top oscillate. He telephoned Holweck, whom he did not know, and the two promptly met to elaborate the instrument. Over the course of weeks the physicists refined the design of their gravimeter. To take into account the thermal sensitivity of quartz, they grafted the filament on a blade of élinvar, a nonmagnetical iron-nickel alloy invented by Charles Edouard Guillaume at Sèvres having an elastic coefficient that varied little with temperature; the pendulum was silvered to guard against static electricity; the apparatus was placed inside a glass bottle at a hard vacuum and isolated in a Faraday cage. Ten minutes against a pocket watch gave a gravity reading. Maurain presented the new gravimeter to the Academy of Sciences in June 1930; Lejay did the same at the June meeting of the French Society of Physics and then in August at the Stockholm meeting of the International Geodesical and Geophysical Union.[94]

92. See ibid., pp. 50–53, for extracts from a letter by Lejay describing his recollections of the collaboration. See also SMB, Lejay to Maurice Burgaud, 16 Jan. 1928, outlining the pendulum-chronometer research.

93. PO, carton 3, II-B–1 c, Comité français de géodésie et de géophysique, minutes of general meeting of 13 Apr. 1921, to which is appended the report of the president of the geodesical section, relevant here. Ferrié proposed to carry out his research in collaboration with Georges Perrier and the astronomers at the Paris Observatory. Ferrié emphasized: "One can expect from it new facilities that are very important for determinations of gravity."

94. See Perrier, "Holweck," pp. 50–53; Lejay, *Notice*, p. 13, for Lejay's pendulum research; a brief remark on the gravimeter in SMB, Lejay to Father Superior, 27 July 1928, which along with Lejay to Burgaud, 16 Jan. 1928 (where mention is made only of the pendulum-chronometer), places bounds on the first meeting between Lejay and Holweck; and Louis Dumas, "Le nouveau pendule Holweck-Lejay," *BUA*, no. 26, 1932–33, pp. 26–33. See PIPG, Perrier to Maurain, 19 June 1930, for Perrier's regret at not having heard Maurain's academy presentation and his hope that Lejay would present the instrument at Stockholm. For the earliest communications, see Holweck and Lejay, "Etude préliminaire

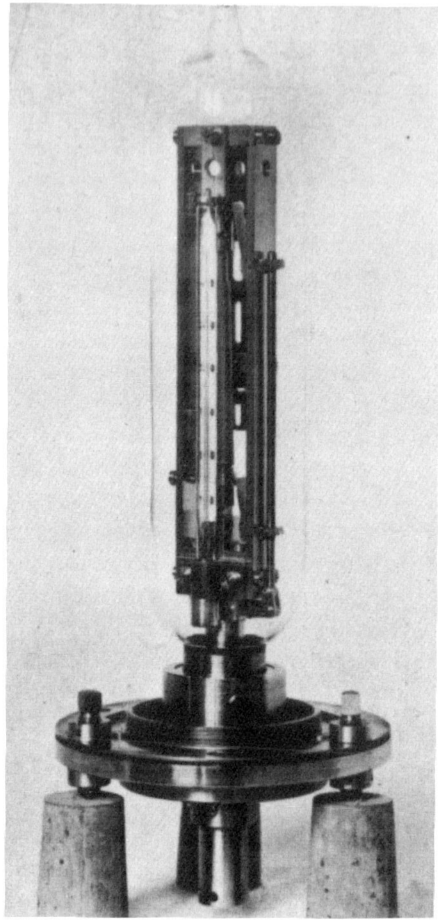

Holweck-Lejay gravity pendulum with its shields removed. The instrument revolutionized terrestrial gravimetrical surveying, doing for the land what Felix Andries Vening Meinesz's submarine pendulum gravimeter had done for the oceans. (From Fernand Holweck, "Description du nouveau gravimètre Holweck-Lejay: Résultats obtenus," *Bulletin géodésique*, no. 45 [1935], 295–303)

Lejay and Holweck continued to modify their device, and the central authorities received the innovation with sympathy. It was a key French

d'un diapason de quartz dans un vide élevé," *CRAS* 188 (1929), 1541–43; and idem, "Un instrument transportable pour la mesure rapide de la gravité," ibid. 190 (1930), 1387–88. A description of the instrument is in Lejay, *Développements modernes de la gravimétrie* (Paris, 1947), pp. 26–32. The Wiechert inverted-pendulum seismometer arrived at Zikawei around 1908 (*CIES*, p. 79).

invention, and for the preceding ten years, when French geophysicists dominated administration of the geodesical and geophysical union, French scientific contributions had been noticeably absent. In the field of gravity measurements, the scientific world had directed attention to the submersible pendulum invented by the Dutch geophysicist Felix Andries Vening Meinesz and used to measure gravity at many undersea locations. Vening Meinesz, furthermore, presented his results in French at the meetings of the union and carried out gravimetrical tests at the Paris Observatory.[95] During the middle 1920s, the French geophysicists had earmarked one hundred thousand francs for gravity measurements. They recognized that none of their apparatuses could be deployed outside a laboratory, and they preferred to develop a portable gravimeter of their own rather than to purchase one from Vening Meinesz.[96] Having enjoyed the collaboration of the Dutch navy, Vening Meinesz offered to lend the French navy one of his devices, to mount it on a French submarine, and to train French sailors to run it. After considering how Vening Meinesz's pendulum might be installed on a French boat, the general staff refused for a long time to deviate from a policy that excluded foreigners as submarine supercargo.[97]

Lejay's portable pendulum responded precisely to the needs outlined by French geophysicists during the 1920s. It was relatively inexpensive,

95. On Vening Meinesz, see *ER*, pp. 125–32. See also PO, carton 3, II-B–1, circular letter by Johan Jacob Aernoud Muller of 5 Feb. 1926, where Muller (president of the Dutch Geodesical Commission) provides the gravity values obtained by Vening Meinesz at Cambridge, Paris, and Uccle; and Georges Perrier, "La coopération internationale en géodésie et en géophysique," Bureau des longitudes, *Annuaire pour l'an 1928* (Paris, [1928]), pp. C.1–C.120, p. C.50 for Vening Meinesz's "sensational" results presented at the union's meeting of 1924 in Madrid, pp. C.81–C.86 for a glowing review of Vening Meinesz's communication at the union's meeting of 1927 in Prague.

96. PO, carton 3, II-B–2, Comité national de géodésie et géophysique, section de géodésie, draft of minutes for meeting of 6 June 1926.

97. Ibid., draft of minutes for meeting of 21 Apr. 1926, where use of Vening Meinesz's apparatus by French sailors was discussed in the presence of Rear Adm. Louis A. Pirot. According to the minutes, Georges Perrier paid close attention to the accounts of Vening Meinesz's early trials appearing in the British journal *Nature*, probably J. J. A. Muller, "Dutch Pendulum Observations in a Submarine with a New Apparatus," *Nature* 117 (1926), 531–33 (10 Apr. 1926). See also Muller, "Dutch Pendulum Observations in Submarines," ibid. 112 (1923), 393–94, 788; 113 (1924), 308–9, 641. The exclusionary policy is revealed in PIPG, Odend'hal (Chief of the Military Cabinet in the Ministry of Defense) to Maurain, 14 Mar. 1932. In 1933 the French military did finally agree to borrow an apparatus from Vening Meinesz, and they used it to carry out submarine gravity measurements (Georges Perrier, "Congrès de la recherche scientifique dans les territoires d'outre-mer, Première section, géodésie . . . , rapport préliminaire," *ACRACS* 13 [1937], 141–49, on p. 149). A second loan took place in 1936. As late as 1938 the ministry of education resisted authorizing Fr 160,000 to purchase a Vening Meinesz pendulum ("Congrès de la recherche scientifique dans les territoires d'outre-mer," ibid. 14 [1938], 69–79, on p. 72).

reliable, and independent of radio technology.[98] It would do for the land what Vening Meinesz's had done for the seas. Georges Perrier, France's leading military geodesical scientist, followed Maurain's presentation of Lejay's research with great interest. He threw his weight behind a gravimetrical survey of France using the new device along with a traditional one designed in the nineteenth century by Robert von Sterneck and acquired by Maurain's geophysical institute as a prize of war.[99] Maurain urged that the old Sterneck apparatus be abandoned in favor of Lejay's.[100]

The new gravimetrical survey of France received the sanction of the geodesical section of the French National Committee for Geodesy and Geophysics. Lejay found himself on a commission to this end along with Perrier, Maurain, Ferrié, General Jean-Baptiste Dominique-Pierre Viviez (head of the army's geodesical service), and two astronomers, André Danjon and Ernest Esclangon. As president of the commission, Perrier had budgetary authority. Maurain made sure that Holweck was eligible for the commission's largesse.[101] By the early 1930s the pattern was established: Holweck concentrated on improving and manufacturing the instruments, while Lejay—who in 1930 had risen to become director of the Zikawei Observatory—traveled to and from China taking measurements. Gravimetry was ideal for the Chinese milieu, Lejay exclaimed in 1928: "I believe that these short exercises, which do not require expenses and can be carried out during spare moments without continual supervision, are completely suited for Zikawei."[102] Lejay's hunch was amply confirmed. The Academia Sinica asked him to undertake a gravimetrical survey of China, which he accomplished admirably.[103]

Lejay did more than gravimetry. When in residence at Zikawei, he was a creative and strong-willed observatory director who insisted on research as a way of life. Holding the scientific work of his older colleagues—Louis Froc and Maurice Burgaud—in low regard, he urged

98. See PO, carton 3, II-B–2, Comité national, géodésie, 6 June 1926, for Marcel Brillouin's query whether wireless registration of a gravity pendulum would be more costly and more complicated than Vening Meinesz's photographical registration and for Georges Perrier's insistence on developing a portable apparatus.

99. See PIPG, Perrier to Maurain, 15 June 1930, for the survey; and ibid., Perrier, "Note pour la Commission de l'heure, des latitudes et de l'intensité de la pesanteur," 1931, for the origin of France's only Sterneck apparatus.

100. Dumas, "Le nouveau pendule Holweck-Lejay," *BUA*.

101. PIPG, Maurain to Director of Higher Education, 23 Nov. 1931, where the mechanics of the commission are specified and Holweck is mentioned.

102. SMB, Lejay to Father Superior, 27 July 1928.

103. Louis Dumas, "Le pendule Holweck-Lejay," *BUA*, no. 27, 1933–34, pp. 31–34.

Pierre Lejay (known also by his Chinese name, Ye Yue Fei) with his portable gravimeter. During the 1930s and 1940s Lejay supervised gravimetrical surveys around the world. (Courtesy of Shanghai Meteorological Bureau)

his staff to pursue timely questions of general interest.[104] He encouraged Louis Dumas to study for a licence and then take a doctorate in Paris for work done at Zikawei.[105] He bragged to his Jesuit superiors from a temporary posting at Versailles in 1931:

> It is rather stinging to note that when I pushed the observatory, against everyone, in the direction of research in pure science, I was reproached for drifting too far from the Chinese perspective to focus too exclusively on the international perspective; now the Chinese scientific services follow my lead, and they want to have me return quickly to benefit from my expertise in these areas.[106]

Among his pure scientific research projects he paid particular attention to the atmospheric ozone layer; he proposed that ozone transport be-

104. See SMB, Lejay to unknown, 26 Nov. 1928, for the judgments of Froc and Burgaud.
105. Ibid., Lejay to Father Provincial, 21 Feb. 1927.
106. Ibid., 18 Apr. 1931.

tween pole and equator correlated with anticyclonic progressions, and he offered radio sounding of the atmosphere as a way of anticipating typhoons.[107]

A corresponding member of the Academy of Sciences since 1935, Lejay was mobilized while on leave in France during the late 1930s. He spent the war in France, technically assigned to Francheville but in fact in Paris, supervising gravimetrical surveys. By war's end he had become a fixture of geophysics in the hexagone, and there he remained to the end of his life. Lejay advanced to become a nonresident member of the Paris Academy in 1946 (by virtue of his directorship *in partibus* at Zikawei); he followed Charles Maurain as president of the French National Committee for Geodesy and Geophysics in 1950. He never succeeded in verifying continental drift by means of longitude determinations.[108]

Lights of French Learning

In July 1898 one of the main supporters of the modernist movement in China, Liang Ki-tchao, asked the French ambassador in Beijing to contact the Catholic hierarchy in China about allowing Joseph Ma Siang-pei (Ma Xiangbo), a reform-minded writer, to direct a translators' college in the capital. The religious authorities were delighted at the prospect, as Ma was a former Jesuit with strong connections to the society, but they insisted that the college take shape at Shanghai under the surveillance of the Zikawei mission. The project went dormant, however, when the reformist Emperor Kouang-siu (Guangxu) was deposed by his mother, the dowager empress Ts'eu-hi (Cixi). Ma retired to the Jesuit orphanage of Tou-sè-wé, near Shanghai. There, in 1902, he began tutoring students from the nearby Nanya'ng Polytechical Col-

107. P. Lejay, "Travaux exécutés à l'Observatoire de Zi Ka Wei de 1933 à 1939," in Pacific Science Congress, Sixth, Berkeley, *Proceedings*, 6 vols. (Berkeley, 1940–43), 1:211–15; on the upper-atmospheric research, see Gherzi, "Scientific Work of the Catholic Church," p. 55. See also, generally, Lejay, *Notice*.

108. Tardi, "Lejay"; Lejay, Maurice Burgaud, and E. de la Villemarqué, *Coopération de l'Observatoire de Zi-ka-wei à la révision international des longitudes* (Shanghai, 1934) [Zô-Sè, Observatoire, Annales, 20]. Lejay's situation is from information supplied by Robert Brunet. In his autobiography, Ernesto Gherzi notes Director Lejay's decade-long sojourn in France (SJQ, Gherzi, "Autobiographie"). See also SMB, Lejay, "Administration de l'observatoire en l'absence du P. Lejay" [1935], where Lejay organizes the observatory for his extended absences. In retaining his title of Zikawei director, Lejay joined a long Vatican tradition. "By a bishop *in partibus infidelium* was signified not an evangelist, but a bishop taking his title from a people so insignificant, so remote, or so faithless that he might stay at his ease in Rome" (T. Clifford Allbutt, *Notes on the Composition of Scientific Papers* [London, 1925], p. 144).

lege, an imperial institution with a Chinese rector. When a strike paralyzed Nanyang in 1903, Ma's students urged him to organize a formal college.

Ma's college, the Aurora, opened that year in rooms on the Zikawei Observatory complex. He personally provided scholarships for indigent students. It was a democratic regime, with students running the administration under Ma's principalship. From the outset, Ma involved the Jesuits in the curriculum. Their vision, however, was not his: the Jesuits wanted French to be the dominant language of instruction, whereas Ma and the students preferred English; the Jesuits opposed student government and political involvement, which Ma encouraged. In 1905 Ma and his students broke with the Jesuits and used provincial funds to refound their school as Fudan University. The Zikawei Jesuits retained the name of the college (Aurora, or Zhendan), and they reopened it later that year. This time they shared power with no one, but not out of religious sectarianism: in the nascent university's program, religion would be something for the students to engage on their own time. Such were the origins of the most important French institution of higher learning in Asia.[109]

After an inauspicious beginning, the institution grew to 172 students in 1906–7. The campus relocated on seventeen acres at Lou-ka-wè, closer to the French Concession. There, with the observatory nearby, the Jesuits conducted a three-year preparatory course on the level of the French baccalauréat and two senior divisions, science and letters, which prepared students for a diploma corresponding to a French licence. In 1914 the curriculum expanded to include three senior divisions: letters and law, mathematics and civil engineering, and medicine. By 1917 the university had 27 instructors (among whom were one medical doctor and two civil engineers) and 203 students. Beginning in 1913, the French ministry of foreign affairs gave the Aurora one thousand francs yearly; in 1916 the amount rose to fifteen thousand francs and in 1917 to twenty-five thousand, most or all of this money going to the medical faculty. The French Concession of Shanghai granted some three hundred thousand francs in 1915 to relocate to the university the Zô-Sè Observatory and a natural history museum then at Zikawei.

109. "Une université française en Chine," *RC* 6 (1918–21), 69–83; Joseph Dehergne, "Notes sur la brève histoire de l'Aurore," *Etudes* 350 (1979), 613–17. According to Dehergne, the name derived from the comment of a Chinese supporter. See Ruth Hayhoe, "Towards the Forging of a Chinese University Ethos: Zhendan and Fudan, 1903–1919," *China Quarterly*, no. 94, 1983, pp. 323–41, which departs substantially from the two preceding accounts. I follow Hayhoe's description, which draws on interviews with Joseph Ma Siang-pei's descendants, but what precisely happened in 1905 is a matter for future research.

The naturalists moved in the early 1930s; the astronomers stayed where they were.[110]

The engineering section produced practical men (the first *brevet d'ingénieur civil* was awarded in 1916), but the French tradition of certifying technical adepts entailed a strong dose of pure learning. After seven years when the exact sciences were taught by Jesuits with little training in the subject, physics became the province of Louis de Jenlis. He had graduated from the Ecole centrale des arts et manufactures in 1895, attended the Ecole supérieure d'électricité, and worked as an engineer before joining the Society of Jesus in 1903 at age twenty-eight. De Jenlis spent the next twenty-seven years teaching his specialty at the Aurora. He realized his *oeuvre* as the architect of a number of the university's buildings. He sought "to educate men who, thanks to a general scientific culture that was as broad as possible, would know how to adapt to any situation." He deplored compartmentalized education in science. Timid and quiet, he never learned Chinese.[111]

Physics and chemistry acquired a scientific teacher in 1916, when Michel Vittrant arrived. He quickly organized separate laboratories for the two disciplines and then devoted himself entirely to physics.[112] In contrast to de Jenlis, Vittrant was an exponent of Chinese culture. He proposed that all scientific calculations be carried out in a modified Chinese system of weights and measures. "Progress must not be impeded by the pretext of unification or 'standardization,'" Vittrant wrote. "The law of real progress is the development of the past."[113]

Separate faculties for the exact sciences came into being around 1920, when the ministry of foreign affairs granted Fr 30,000 for labo-

110. P. Lefebvre, "L'Aurore," *Il pensiero missionario* 3 (1931), 251–60; Samuel Couling, ed., *Encyclopedia Sinica* (Shanghai, 1917), pp. 584–85; Michael Loy Sinclair, *The French Settlement of Shanghai on the Eve of the Revolution of 1911* (diss., Stanford University, 1973), pp. 335–39; W. Y. Chyne [Chuang, Wen-ya], *Handbook of Cultural Institutions in China* (Shanghai, 1936), p. 34. A summary appears in Ruth Hayhoe, "Catholics and Socialists: The Paradox of French Educational Interaction with China," in *China's Education and the Industrialized World*, ed. Hayhoe and Marianne Bastid (Armonk, N.Y., 1987), pp. 97–119, on pp. 105–6. On the natural history museum, see O. Piel, "Le 70e anniversaire du Musée Heude," *Revue nationale chinoise* 11 (1939), pp. 29–65, the move on pp. 50–51. Cf. *CIES*, p. 273.

111. Salvini, *Cent ans*, pp. 194–201; G. G., "Le père Louis de Jenlis, 1875–1938," *BUA*, no. 38, 1938–39, pp. 1–8, for citation; "Une université française," p. 80, for the brevet.

112. Henri Belval and P. Liang, "Histoire des laboratoires de chimie à l'Aurore," *BUA*, no. 20, 1929–30, pp. 58–63.

113. Vittrant, "Système chinois d'unités," ibid., no. 3, 1920–21, pp. 46–51, on pp. 50–51.

ratories, Fr 90,000 to endow three chairs, and Fr 450,000 for capital developments. In this way the Quai d'Orsay expanded its earlier commitment and purchased control of the university. It specified that the Aurora would have to be directed by a French citizen, that its professors would have to be mostly French, that the lay professors be graduates of French universities or grandes écoles, and that all these appointments existed at the pleasure of the civil authorities in Paris. The ministry of public instruction in Paris was glad to let the ministry of foreign affairs handle the Jesuits; public instruction contributed only Fr 5,000 for new construction and recognized the Aurora's secondary-school certificate as equivalent to a French baccalauréat.[114] Mathematician-cum-politician Paul Painlevé justified French investment as a bulwark against the Americans, who threatened to dominate Chinese education through the Rockefeller Foundation. "We cannot do too much to encourage and materially support this truly national enterprise," he emphasized in 1921.[115] As the French consul general in Shanghai wrote to the governor general of Indochina in 1917: "The Aurora University is the only organization on which we may rely to compete with the efforts of foreign powers to establish their moral influence over the Chinese; it is the only one that is capable of fighting effectively against German competition, if that should continue after the war, or of taking its place if it were to disappear."[116]

Professors under the new regime, like those under the old one, expressed little interest in research, but they did prepare students for scientific careers. Graduates of the exact-science faculty found employment as electrical or mining engineers. Tsiang Ping-jan came to direct the magnetical service at the Tsingtao Observatory, and another graduate worked as an astronomer there. (The institution had been founded by the Germans and then mismanaged by Japanese conquerors.)[117] Another graduate, Lieou Lin-chou, taught physics at Ki-nan University at Chengfu, and Wang Tsien-ming became an astronomer at the Bureau

114. See "L'Université 'l'Aurore,'" *Lettres de Jersey* 1 (1920), 132–34, for the funding from the ministry of foreign affairs and the equivalency from the ministry of public instruction; and *BUA*, no. 19, 1918–19, p. 2.

115. Zikawei, Jesuit Mission, *Nouvelles de la mission*, no. 745 (24 Jan. 1921).

116. Aix, Indochine 18375, French Consul in Shanghai to Governor General of Indochina, 25 July 1917.

117. Lefebvre, "L'Aurore," p. 259; *BUA*, no. 26, 1932–33, pp. 72–74, for Tsiang. The Tsingtao astronomer may have been Sié Tong-i, according to ibid., no. 35, 1936–37, pp. 36–37. Joseph de Moidrey reported in 1924, shortly after the Tsingtao Observatory returned to Chinese hands, that the observatory director "inundates me with questions. His first observer spent nearly a year here" (HJA, "Zikawei, 1901–1928," de Moidrey to E. F. Pigot, 19 Mar. 1924).

for Scientific Research in Nanking.[118] By the late 1920s, six Aurora graduates were teaching at their alma mater.[119]

The rector of the Aurora noted in 1933 that physics and mathematics drew few students because of their difficulty; in his view, mathematics there was on the level of *mathématiques générales* in the French licence.[120] Testimony from Aurora graduates suggests, however, that the Aurora functioned as a French secondary school. One graduate, Wang Pong, took courses at a lycée in the hexagone to prepare for the Ecole centrale des arts et manufactures. At the lycée he had been weakest in geometry and strongest in physics and chemistry, but in the entrance examination he succeeded in geometry and failed in physics and chemistry. He targeted the Ecole centrale, instead of the Ecole polytechnique as some of his schoolmates had, because "the program of my studies was not broad enough to allow me to go to other schools."[121] At the Ecole centrale, Wang found his progress impeded by his poor command of written French.[122]

During the wars that enveloped China in the 1930s, the Aurora beaconed stability and continuity. By 1939 it counted 452 students in its various faculties.[123] Furthermore, to judge from Aurora graduates who did manage to become certified researchers in Europe, the 1930s saw a steady improvement in the curriculum at the Jesuit university. Pan Tcheng-tao, who had projected four years for study in France, concluded that the time was insufficient. He had to spend two years preparing for the examinations to enter the grandes écoles and three years at one of these schools; Pan decided to study for the Institut de chimie and also the Ecole centrale.[124] He then acquired a doctorat ès sciences from Paris.[125] Tchang Yong-li, upon graduating from the Aurora in 1936, continued in physics at the University of Louvain. He took courses from Charles-Jean-Gustave-Nicolas de la Vallée Poussin on analytical functions, Gustave Verriest on group theory, and Georges Lemaître on

118. "Les fêtes de 25 ans," *BUA*, no. 17, 1925, pp. 41–43. The astronomer was possibly Hou Kang-fou, a member of the Academia Sinica (ibid., no. 26, 1932–33, pp. 72–74). Wang's identification (a translator in the geodesical bureau at Nanking) is in ibid., no. 27, 1933–34, p. 63.

119. Ibid., no. 17, 1929, pp. 41–43.

120. G. Germain, "L'enseignement des mathématiques et des sciences dans les écoles secondaires," ibid., no. 28, 1933–34, pp. 75–77.

121. *BUA*, no. 22, 1929–30, pp. 94–95, 100–101.

122. Ibid., no. 27, 1933–34, p. 64.

123. Ibid., no. 40, 1939–40, p. 103.

124. Ibid., no. 24, 1931–32, pp. 74–75.

125. "Le rayonnement de l'Aurore." Pan Tcheng Kao, *Etude expérimentale de la vitesse de propagation des ultra-sons dans des gaz* (diss., University of Paris, 1931). Pan's letters detailing his progress surely preceded the doctorate by a number of years.

relativity (the last was "rather a course on tensor calculus," he wrote). He spent most of the day in a physics laboratory. In April 1937 Tchang presented a paper to the Scientific Society of Brussels in which he calculated frequencies and normal vibrational modes of monodeutero-ethylene, expanding a determinant equation of the ninth degree and finding its roots; then he compared the results with Raman spectra and infrared spectra. Tchang's work appeared in the *Annales* of the Brussels society.[126] In 1939 he received a doctorate under Charles Manneback for a dissertation in astronomy, with a thesis in mathematical physics; his jury, which included cosmologist Georges Lemaître, awarded the diploma with "the highest distinction."[127] Ou Kiun-houo left the Aurora in 1937 and went to work in organic chemistry under Charles Prévost, *professeur sans chaire* at the Sorbonne; he quickly published three notes in the *Comptes-Rendus* of the Paris Academy and then took a doctorat ès sciences from Paris. Another graduate in 1937, Lin Che-kin, also took a Paris doctorate.[128]

The accomplishments of the Aurora University may be measured against a parallel initiative on the part of the ministry of foreign affairs. The evolution of the institution in question, the Institut franco-chinois d'industrie et de commerce, suggests the unwillingness of metropolitan authorities to create any institution of higher learning in foreign lands.

126. Georges Lemaître, Charles Manneback, and Tchang Yong-li, "Calcul des fréquences et modes de vibration de la molécule de monodeuteroéthylène," *Annales de la Société scientifique de Bruxelles* 57 (1937), 120–28; J. M. Delfosse, J. C. Jungers, G. Lemaître, Y. L. Tchang, and C. Manneback, "Raman Spectrum of Monodeuteroethylene," *Nature* 139 (1937), 1111–12, where it is indicated that Lemaître and Tchang carried out the calculations.

127. *BUA*, no. 36, 1936–37, p. 97; no. 40, 1939–40, p. 101, for the doctorate. According to the science faculty archives at Louvain, Tchang's dissertation was titled, "Etude des trajectoires des particules électrisées dans le voisinage d'un dipôle magnétique," and his thesis (presumably the object of oral interrogation) was "La fonction potentielle de la molécule d'éthylène." Professor D. Laduron, dean of the Louvain Faculty of Sciences, has kindly provided me with an archival extract of Tchang's defense. It may be noted that the Zikawei Jesuits sent graduates to Louvain—a Catholic university—as early as the 1910s (*RC* 6 [1918–21], 81).

128. *BUA*, no. 39, 1938–39, p. 81; no. 40, 1939–40, p. 101. Ou indicates working under Prévost at the Ecole normale, but the *Annuaire générale de l'université et de l'enseignement français*, 1938–39, lists Prévost only in the Faculty of Sciences. See also Ou Kiun Houo, "Sur un alcool alpha-gamma-biéthylénique," *CRAS* 208 (1939), 40–42; idem, "Sur une préparation d'alcools beta-éthyléniques, ramifiés en alpha," ibid., 528–30; Prévost and Ou Kiun Houo, "Sur une scission des alcools primaires et secondaires beta-éthyléniques," ibid. 206 (1938), 1661–62; and Lin Che-Kin, *Contribution à l'étude spectrale et à la synthèse des phtaléines dissymétriques et de leurs dérivés* (diss., University of Paris, 1939). Jean Perrin presided over Lin's examining jury, but Lin dedicated his thesis in part to his organic-chemistry professor, Pauline Ramart-Lucas.

In 1917 French police raided the campus of a German university in the French Concession, evicted the professors and students, and confiscated the furnishings. The establishment, the Tung-Chi University, offered the finest engineering education in China at the time. The Germans obtained protection and support from the Chinese educational authorities and relocated in Woosung, a suburb of Shanghai.[129] When article 134 of the Versailles peace treaty gave France clear title to the German property, the Quai d'Orsay transformed it into the Institut franco-chinois.

Unlike their Jesuit and their German counterparts, who were flexible about beginning university-level instruction modestly and optimistic about upgrading standards gradually, French administrators sought to fix for all time the school's scholarly magnitude in the educational firmament. French policymakers intended the Institut as a school for secondary technical instruction; it would produce "technicians, foremen, supervisors, and clerks."[130]

The director chosen for the school, however, had other ideas. He was Charles B. Maybon, a veteran Sinologist who had been an educational administrator in Indochina and a teacher in the first Indochinese University.[131] Maybon had graduated from the Ecole centrale des arts et manufactures and had gone on to a *doctorat ès lettres.*[132] Like his Shanghai counterparts, he sought to run a real institution of higher learning—not an école d'arts et métiers (in the French system, a secondary school) but a university-level school like his alma mater. From the beginning, he set out to reorient the Institut toward training "theoretical engineers." He recruited students with secondary-school diplomas. The Chinese educational authorities, who licensed foreign schools, decided to place the Institut in the category of professional schools, first-class, something short of university rank. Maybon would

129. See *CIES*, pp. 281–92, for the Tung-Chi University. According to the French consul in Shanghai: "When the French authorities closed the German school, the professors and students emigrated to Woosung, where they found a home. The institution was continually funded by the provincial government of Kiangsu, and it seems today that the Chinese government includes it in its official establishments" (Aix, Indochine 44218, "Enquête du Ministère des affaires étrangères sur l'enseignement du français en Extrême-Orient," E. Koechlin to Minister of Foreign Affairs, 20 Apr. [1929]).

130. PAE, Asie 1918–1940, Chine 485, Report of commission of inquiry into the Institut franco-chinois, appended to A. de Fleuriau to Minister of Foreign Affairs, 6 Apr. 1923.

131. Aix, Indochine 48042, "Université indochinoise," announcement of courses for 1907–8, which provides Maybon's various charges.

132. See PAE, Asie 1918–1940, Chine 485, Note for the Sous-direction d'Asie, 10 Aug. 1922, for Maybon's education.

not have it, as the schools of this category were in his view far below the Institut. The French ambassador in Peking and the French consul in Shanghai would not have Maybon. He was sacked in 1924. He left to direct the Ecole supérieure de pédagogie at Hanoi.[133]

In defending his "theoretical" direction, Maybon emphasized that he was responding to Chinese realities. One could not reflect the French system on the China coast without suffering some distortion, and in his view one had to proceed optimistically:

> One must absolutely not be mistaken about the value of this word [*university*], which represents not only an institution of higher learning as in France but also an establishment that, founded in imitation of American or English universities or colleges, does not distinguish between secondary instruction and higher learning, [between] the rigorous limits set in France by these two orders of instruction. From the point of view of technical instruction, which is of special interest here, an establishment like a French école d'arts et métiers, belonging to French secondary technical instruction, would be classified as an establishment of higher learning in China.[134]

Maybon's vision was indeed a special one. He could not obtain (or did not want to obtain) local funding for his school, that which the Jesuits and the Germans had tapped, and his reluctance to knock on neighborhood doors found little sympathy in the ministry of foreign affairs in Paris.[135] To operate in the black, he willingly sold the fine German furnishings his school had inherited from the previous occupants.[136]

On the Quai d'Orsay, French mandarins made their decision by the middle of 1923. The Institut would become a trade school, Maybon

133. The early history of the Institut may be found in documents surrounding the student strike and Maybon's dismissal, kept in PAE, Asie 1918–1940, Chine 485, Report of commission of inquiry, appended to de Fleuriau to Minister of Foreign Affairs, 6 Apr. 1923; Maybon's reply, "Observations sur le rapport de la commission désignée pour étudier la réorganisation de l'Institut franco-chinois d'industrie et de commerce," n.d.; and a cabinet note of Oct. 1925. See Zikawei, Jesuit Mission, *Nouvelles de la mission*, no. 741 (29 Sept. 1920), for Maybon's arrival to head an école d'arts et métiers whose graduates would go on to Lyons for further training. Maybon's Indochinese appointment is given in *Min.*, 1930, s.v. "Hanoi."

134. Maybon, "Observations sur le rapport."

135. PAE, Asie 1918–1940, Chine 485, Minister of Foreign Affairs to Ambassador in Peking, 22 Aug. 1922.

136. Ibid., Maybon, "Institut technique franco-chinois: Rapport sur la marche des études pendant le 1er semestre de l'année scolaire 1923–24." The sale, to the Société franco-chinoise de construction métalliques et mécaniques, netted less than "$51,000" (probably a figure in Chinese dollars, although it should be recalled that figures in Indochinese piastres and Mexican silver dollars were also indicated at this time by a "dollar" sign).

would be recalled to France, and the new Institut would be administered by the Chinese.[137] According to Aimé-Joseph de Fleuriau, the French ambassador in Beijing, the Institut had to be given "a purely technical character." Maybon "had judged exactly neither the spirit of young Chinese nor the situation of the country. His error is excusable," de Fleuriau emphasized, "but his enterprise failed."[138]

De Fleuriau was one of France's most trusted envoys, and his judgment carried weight in Paris.[139] Realities of the French position in China informed the ambassador's perspective. In the first place, as de Fleuriau noted in 1922, the time was not yet ripe for a major French offensive in higher education. The Americans, for example, had used much of their revenue from the Boxer indemnity (the French part of which contributed to support the Institut franco-chinois) for fine colleges, while the Chinese were bit by bit pushing them out the door, discouraging good directors and teachers and keeping only those who danced to the Chinese tune. Furthermore, according to de Fleuriau, a French teacher in China "risks becoming alienated from his country and not carrying out useful work. He will find himself in a situation that is hardly intellectual at all, very poorly educated, perfectly disorganized. Europeans will offer him few resources, for China is not, at the moment, a very pleasant place to live."

The Chinese were filled with inertia, and it was impossible to push them. "It is necessary to treat them like children, like big, vain children, who believe themselves to know everything without having learned it and who do not tolerate intellectual superiority from foreigners."[140] In the ambassador's view, France would do well to wait, let the Chinese organize things, and move in later. At the same time, the Chinese authorities in Paris sought a real école d'arts et métiers, the graduates of which could go on for further studies in France or find posts directly in industry. A representative of the Chinese ministry of communications

137. Ibid., telegram, Emmanuel-Marie-Joseph de Peretti de la Rocca (Paris) to de Fleuriau (Peking), 15 June 1923; Auguste Wilden (consul in Shanghai) to de Fleuriau, 28 Mar. 1923. When he became ambassador to Teheran, Wilden vigorously combated "Germano-American" cultural initiatives, in which "America would provide the money, Germany the men" (ibid., Asie 1918–1929, Perse 66, Wilden to Minister of Foreign Affairs, 30 Nov. 1929). Wilden intrigued against the eminent archaeologist Ernst Emil Herzfeld, who was appointed to head a language and archaeology school financed by "German-Jewish bankers" (Wilden to Minister of Foreign Affairs, 20 Mar. 1928).

138. Ibid., Asie 1918–1940, Chine 485, de Fleuriau to Minister of Foreign Affairs, 6 Apr. 1923.

139. De Fleuriau left China after a short tour of duty to become France's ambassador in London (*DBF*, s.v. "de Fleuriau").

140. PAE, Asie 1918–1940, Chine 485, de Fleuriau to Marx (Foreign Affairs), 16 Aug. 1922.

specifically asked that Maybon be relieved. Maybon was "quite a pleasant man . . . *but not a technician.*" His post was to go, according to the Chinese messenger, to an "eminent engineer" from Paris, someone who would be able to affiliate the Institut with a metropolitan school. The proposed new direction would especially help in the "fight against competition from the English and the Germans."[141]

French contempt for Chinese intellect combined with fear of being eclipsed by American and German rivals to bring an end to Maybon's enlightened plans. The Americans, amateur operators, posed no serious problem, but the Germans were to be treated with respect. Given the provenance of the buildings and appliances of the Institut franco-chinois, the Tung-Chi University loomed as a constant reminder of unrealized aspirations. Ambassador de Fleuriau noted in 1923 that the Tung-Chi medical faculty worked well, while the engineering faculty "evoked disappointment." De Fleuriau continued: "After eighteen months, the Germans had to recognize that their combined program of theoretical instruction and laboratory exercises was not suited to the country and that the Chinese student refuses to deliver himself to the manual tasks of the worker." They had been forced the divide the engineering faculty in half: one part for "theoretical engineers," the other for practical studies. In ignorance or perhaps in disregard of the important positions achieved by Tung-Chi graduates, the ambassador asserted that the engineering faculty had produced only two men "of quite mediocre value," one employed by Siemens and one by the Shanghai tramways. It was important to discount imperialist rivals because graduates of the Institut franco-chinois could find positions only with the French and Belgian railways in China.[142]

In 1924 Maybon handed over his institute to H. Civet, a graduate of the Paris Ecole d'arts et métiers, but not without a parting shot at what he felt was a recipe for disaster. Casting an eye over "the three most painful years" of his life as an educator, he cited the opinion of Jesuit missionaries that "manual labor is too discredited among the educated to permit technical education, that which we call *arts et métiers*, to have a chance of success."[143] He must have wondered what would become of the staff that he had assembled: mathematician Ho Lou, *licencié ès sci-*

141. Ibid., Note of a conversation with Tsang-Ou, delegate of the ministry of communications of the Republic of China, 2 Aug. 1922, recorded by Clichant of the Service des oeuvres, emphasis in the original.

142. Ibid., de Fleuriau to Minister of Foreign Affairs, 6 Apr. 1923. For a survey of the positions occupied by Tung-Chi graduates, see *CIES*, p. 285.

143. PAE, Asie 1918–1940, Chine 485, Maybon, "Institut technique franco-chinois . . . 1923–24," citing from the *Revue des missions du Tchély* as reported in *L'Asie française*, 1923, pp. 374–75.

On 14 May 1926 an earthquake stopped the pendulums at the Zikawei Observatory; here the Jesuits are looking for the "lost second." (Cartoon by Sapajou from the *North China Daily News*, in Zikawei, Observatory, *Cinquante ans de travail scientifique* [Paris, (1928)])

ences mathématiques from Louvain; electrical engineer Shan Tiwan, a graduate of the Institut polytechnique of Grenoble.[144] Such pretensions to higher learning as they may have had would never be fulfilled at the Institut franco-chinois.

Decisions made in Shanghai reflected general French policy throughout China. The authorities supported establishing a Franco-Chinese University in Beijing, insofar as it could serve as a vehicle to prepare French teachers for positions in Chinese schools and Chinese students for "more advanced study first in France and later in Beijing."[145] The Quai d'Orsay directed Boxer indemnity funds toward financing a French naval medical school in Tientsin, which had been operating since 1901, when Beijing provided a campus, but the idea had

144. Ibid. The teaching staff are listed in a report from Maybon and Hou Wen-yao to the Consul General of Shanghai, 8 June 1923.

145. Ibid., de Fleuriau to Marx, 16 Aug. 1922.

always been to send the best graduates to France for a medical doctorate. The Tientsin operation survived Chinese cutbacks in the early 1920s only through the benevolence and influence of the Bank of Indochina.[146] There is no evidence to suggest that any of the French functionaries who came to teach exact sciences at Tientsin, Shanghai, and elsewhere had much interest in research.

For policymakers in Paris, the central problem in China concerned the enormous appeal of English — the foreign language that opened doors and guaranteed employment. In 1902 one Jesuit in Tientsin reported on the situation in detail to his superior in France. French schools in a number of prefectures had been replaced by English schools, he wrote, and one viceroy had adopted English for the entire province. "The thing is reasonable. The Chinese have enough difficulties learning a single language, and the science courses cannot be given in all languages. When it was decided to have only one language, there was no doubt about the choice: English took command."[147] A quarter-century later matters were unchanged. In 1929 Foreign Minister Aristide Briand received the opinion of his ambassador in China, Damien-Joseph-Alfred-Charles de Martel: "The study of our language will again be in favor when it allows finding a well-paid position, if not to the same extent as English, at least with sufficient opportunities."[148] Even Maybon had to respond to this reality at the Institut franco-chinois. He wrote in 1923: "Instruction in English, at least in technical matters, having seemed necessary to us, we instituted it beginning in the preparatory school."[149]

By the late 1930s, when French citizens constituted about 1 percent of the total number of foreigners in China, the director of instruction for the French Concession in Shanghai, Charles Grosbois, reflected on the future of the French language. The outlook was dismal. The two major civilian French institutions of higher learning in China — the Franco-Chinese University in Beijing, founded in 1920, and the Institut franco-chinois in Shanghai — had between them produced only around 150 graduates; this was in contrast to the 500 graduates of the Aurora University and the more than 550 diplomas awarded by the Tung-Chi University. To support the Aurora and the Institut franco-chinois, the French Concession contributed $127,000 annually, in Chinese dollars

146. Ibid., Jean-Emile Saussine (French consul in Tientsin) to Dr. A. King (Director of the Tientsin Naval Medical School), 28 Dec. 1923.

147. Cormont., GMC 21, Father Becker to Father Desmarquest, 18 Sept. 1902.

148. PAE, Asie 1918–1929, Chine 482, "Professeurs en Chine," de Martel to Briand, 3 July 1929.

149. PAE, Asie 1918–1940, Chine 485, Maybon and Hou Wen-yao to Consul General in Shanghai, 8 June 1923.

(six francs to the dollar); more than a million francs went to send Chinese students to France and to support Sinological institutes in Paris and Lyons. These monies came from the Boxer indemnity. The Aurora and the Institut also received more than $56,000 from the French Concession, and the Caisse des oeuvres in the ministry of foreign affairs contributed substantial amounts to the Jesuit university. After an exhaustive discussion of institutions and budgets, Grosbois concluded that the French language had become dangerously irrelevant in China. He urged metropolitan authorities to overcome "blind confidence in the destinies of our language" and to avoid "weak resignation in our inferiority." French institutions and French money had to look eastward.[150]

Over the century before the advent of the People's Republic of China, French Jesuits operated the largest foreign scientific complex in East Asia. They focused on physics and astronomy, becoming China's finest exponents of these fields. Adaptable and enterprising, they obtained funding from private and government sources. Their inventories of geophysical and astronomical phenomena were received with confidence and admiration by secular colleagues; their work received substantial support from the French military. Their labor is one of the clearest examples in modern times of science as an anodyne to political convulsion.

150. PIPG, Grosbois, "Le problème de la langue française en Chine," accompanying Grosbois to Maurain, 16 July 1937.

7

Lebanon and Madagascar: Peripheral Territories

Holy Land and Sacred Sky

After the French invaded Algeria, the Holy See called on the recently reconstituted Society of Jesus to establish a Greco-Melchite seminary in Beirut; the seminarians were joined soon by others who went to minister to the Maronites of Lebanon. By the late 1840s a Jesuit school had been established in Ghazir, and the entire missionary operation had been confided to the Jesuit Province of Lyons. In 1875 the Ghazir collège descended from its mountain to relocate in Beirut, which, having been named an administrative seat by the Ottoman rulers, burgeoned in size and prestige under informal French protection. In direct response to the Syrian Protestant College, founded by American missionaries in Beirut, Pope Leo XIII granted the collège a university charter in 1881, allowing it henceforth to award doctorates in philosophy and religion following the model and standards of Rome's Gregorian University. Two years later, the Third Republic authorized the Beirut Jesuits to open a medical school at their Université St-Joseph—again, to keep pace with the American Protestants. The medical school obtained the designation of *faculté* from the French ministry of public instruction in 1888, five years after Algiers had similarly been invested; the Society of Jesus hired medical professors in the new school, but appointments were approved by the state educational authorities in Paris.[1] Beirut was the eastern anchor of France's cultural empire in the Mediterranean.

For thiry years the French Jesuits dispensed theology and medicine

1. Beirut, Université St-Joseph, *Université Saint-Joseph de Beyrouth* (Beirut, 1948), pp. 17–23; *Les jésuites en Syrie, 1831–1931*, vol. 5, *Le collège et le cercle, Beyrouth* [Beirut, 1931], pp. 7–8; Henri Jalabert, *La Vice-Province du Proche-Orient de la Compagnie de Jésus (Egypte, Syrie, Liban)* (Beirut, 1960), p. 92.

in Beirut, with the support and encouragement of the Third Republic. In 1894 the medical school began awarding diplomas for a four-year course of study countercertified by the medical faculty at Lyons; in 1897 the St-Joseph medical diploma received independent recognition from the French state. France, however, shared responsibility with Turkey for the program of study: Turkish army doctors sat on juries at the medical school. The various confrontations and invasions leading up to the First World War impelled the metropolis to tether its Beirut missionaries on a shorter leash. As the St-Joseph University answered to spiritual superiors in Lyons, it seemed reasonable for the University of Lyons to infuse Beirut with an aliquot of temporal learning. In 1912 and 1913 Lyons sent out two missions to survey Asia Minor for possible new French universities. The missions, constituted by professors and local businessmen, concluded by recommending creation of law and engineering schools at the Jesuit university. The two professional schools opened late in 1913, at a ceremony attended by academic, military, and diplomatic representatives from France. In the engineering school at least, the course of study lasted two years and the level of instruction was elementary.[2]

The Jesuits at St-Joseph found their authority threatened by American Protestants and Ottoman administrators; the French saw their intellectual influence in Lebanon diluted by German schoolmasters and engineers to the north and British administrators and engineers to the south. Required was a major establishment for the prosecution of pure learning. By virtue of their experience in dealing with the celestial realm, the Jesuits decided that astronomy fit the bill. They debated ways and means for two years before, in 1906, the Province of Lyons called Bonaventure Berloty, a fifty-year-old astronomer at the Jesuits' Spanish observatory in Tortosa, to place the new institution on its feet.

Berloty was a native of Lyons who, after signing on with the Jesuits, completed a doctorate at Paris and then became a professor in the Catholic faculties of Angers and Lyons. He had trained at the Stonyhurst Observatory in England and was entirely familiar with solar astronomy and seismology.[3] Berloty landed in Beirut in 1907 with a Jesuit colleague, Joseph Brachet, and several tons of instruments. He surveyed the terrain around Beirut before deciding to locate at a vineyard in Ksara, an abandoned village located high on the slopes of the Beqa'a Valley fifty-two kilometers by rail from Beirut. Berloty returned to Eu-

2. Jalabert, *Vice-Province du Proche-Orient*, pp. 70–72, 93.

3. Georges Perrier, "Bonaventure Berloty (1856–1934)," *Bulletin géodésique*, no. 44, 1934, pp. 296–98.

rope for a conference at The Hague, where the International Seismological Association voted to lend him a seismograph. He and Brachet then installed themselves at Ksara. The observatory's founder arrived unable to speak Arabic.[4]

The observatory rose through the ministrations of Berloty, aided by various Jesuit brothers and younger members of his order. He constructed sheds and rooms for meteorological instruments and for a small meridian telescope, a plinth for fixing the observatory's location on the globe, a cellar for housing seismographs, and a darkroom for developing photographical film. These preliminaries accomplished, Berloty received Charles Combier, who had completed a licence ès sciences in 1906, as a Jesuit assistant. Combier was replaced in 1910 by Charles Coste, who sported a licence ès sciences mathématiques; Coste, Berloty's last trained coworker before the war, left in 1911.[5] At the urging of the Egyptian Survey Department, Berloty began issuing a monthly meteorological report. A Mainka seismograph arrived in 1909 and was promptly installed, although its pendulum clock proved to be a constant headache.

Berloty then decided to tour sister institutions. At Cairo early in 1910 he visited Bertram Francis Eardley Keeling of the Helwan Observatory; at Algiers he consulted with François Gonnessiat. Berloty's principal destination, however, was the new Jesuit observatory at Cartuja, near Granada, Spain, which had just installed a 32-cm refracting equatorial telescope in addition to the usual meteorological and seismological instruments.[6] From there he went to Toledo, Madrid, and Ebre before traveling to several meteorological observatories in France. He visited Strasbourg for discussions with Carl Mainka, whose double seismograph he had just received.

It was not long before Berloty's promiscuous attentions were brought to account by temporal powers in Paris. Emile-Victorien Piat, the French consul general at Damascus, got wind of Berloty's establishment and invited him in for discussions. Piat promptly wrote to Paris about the matter. Berloty deserved support, the consul emphasized, be-

4. Berloty, "Notice historique sur l'observatoire fondé en Syrie par les pères de la Compagnie de Jésus," *Annales de l'Observatoire de Ksara: Mémoires* 1 (1921), 1–68, provides an exhaustive discussion of the early years; the history is continued in "Notice historique sur l'Observatoire de Ksara (suite)," ibid., 69–120, which, however, is dated 1927; when, subsequently, no reference is given, documentation may be found in Berloty's two-part, consecutively paginated history.

5. Henri Jalabert, *Jésuites au Proche-Orient* (Beirut, 1987), pp. 177, 250–51, for Combier and Coste.

6. *Vregille*, pp. 562–68, also lists a Vicentini seismograph.

cause "the Germans at Haifa and also the Americans at Beirut several years ago founded observatories in these two cities with the liberal assistance of their government and the learned societies of their country. It would become us no more, then, to ignore the project of Father Berloty than to ignore his observations, which, while being of service to science, allow us an advantage in struggling against our rivals."[7] The diplomatic report found its way to the ministry of public instruction, with the result that Ksara became a recipient of all publications of the Bureau of Longitudes, the ongoing *carte du ciel* and a related lunar map, and the *Annales* of the Central Meteorological Bureau. The Paris Academy of Sciences invited Berloty to sit in on one of its sessions and then made his observatory the beneficiary of its *Comptes-Rendus*.

In his own account of this avalanche of encouragement, Berloty presents himself as a passive actor. In fact he submitted a detailed funding proposal to Parisian administrators through the office of the consul general in Damascus. He asked for a 1,000-kg inverted-pendulum seismometer of Emil Wiechert's design at Fr 4,000, a vertical Vicentini seismometer at Fr 1,000, and a Galitzin seismometer at Fr 4,500; with other items, the request came to between Fr 20,000 and Fr 25,000. Berloty emphasized that the Americans at Beirut had a big equatorial telescope and a Milne seismograph, while the objectives of his telescopes measured only several centimeters in diameter and his continued access to a seismograph came only by way of sufferance. In helping him, "good deeds are thus done for the country and also for French influence." France, "with the marvelously lucid intelligence of her children, has always been a country of intellectual initiative and diffusion." Alfred Angot, the head of the Central Meteorological Bureau in Paris, seconded the Damascus consul's plea for a grant to Berloty's operation.[8]

Metropolitan mandarins continued their efforts on Berloty's behalf. More instruments went to the Beqa'a Valley. Late in 1911 the director of the Paris Observatory, Benjamin Baillaud, identified an amateur astronomer in the Levant who was interested in divesting himself of a 40-cm equatorial telescope. Berloty traveled to meet the man and see the lenses at Smyrna. A subsequent trip to view the housing persuaded Berloty that he did not want to pay for the cost of moving it to Ksara. Berloty went to Europe late in 1912 to find a more suitable instrument. He found what he wanted in a 16-cm equatorial owned by the comte de Puyfontaine. Baillaud came to Berloty's aid, asking the ministry of pub-

7. PAN, F17/13588, Piat to Jean Cruppi (Minister of Foreign Affairs), 15 Apr. 1911; Berloty, "Notice," p. 29, reports that he met Piat on 9 Feb. 1911 and 11 Apr. 1911.

8. PAN, F17/13588, Berloty, "L'Observatoire de Ksara (Syrie)," [1911]; Angot to Minister of Public Instruction, 19 May 1911. Quotation from the former.

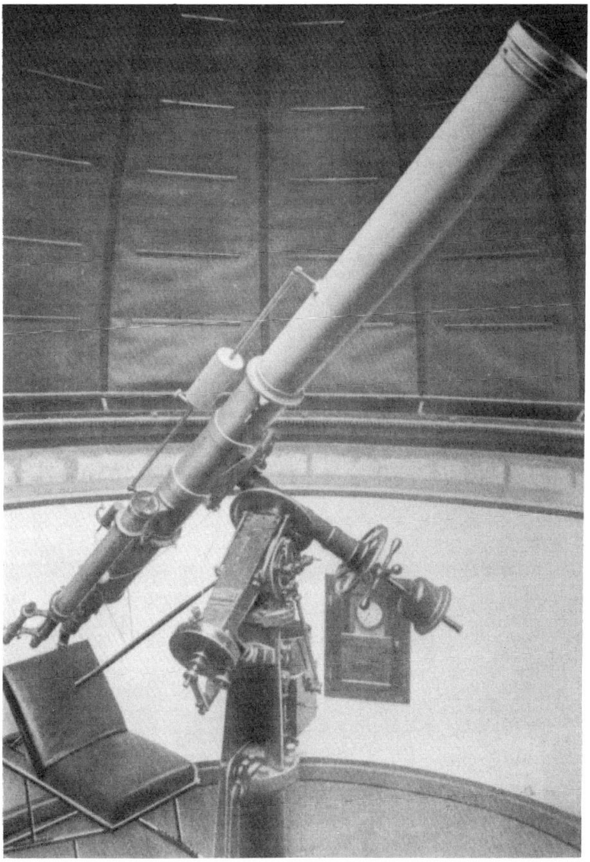

The 16-cm equatorial telescope at Ksara. The telescope was installed in 1914, before France and Turkey went to war. Vandalized during the war, it was reconstructed after the Jesuits returned to the observatory at war's end. (From Bonaventure Berloty, "Notice historique sur l'observatoire fondé en Syrie par les pères de la Compagnie de Jésus," *Annales de l'Observatoire de Ksara: Mémoires* 1 [1921–27], 1–120)

lic instruction to cover the cost of purchasing the telescope, some sixty thousand francs. Baillaud continued by dunning the ministry of foreign affairs. "There would be interest for French science in having this observatory provided with an equatorial of large dimensions, one capable of rivaling the big instrument that the Americans have set up at Beirut."[9]

Did the Third Republic pay for the Jesuits' telescope in Lebanon?

9. Ibid., Baillaud to Minister of Public Instruction, 30 Jan. 1913; Baillaud to Minister of Foreign Affairs, 20 Nov. 1913, for the quotation.

The archives are silent about an answer to Baillaud's proposal. Certainly Berloty would have been the first to credit such unusual benevolence, and in his exhaustive chronicle of the observatory's early days there is no hint of it. We are led to imagine that the Jesuits paid cash and perhaps persuaded the donor that a lower price would contribute to achieving a state of grace.[10] It is certain that by 1913 Berloty had arranged for purchasing de Puyfontaine's telescope and related equipment, including a heliograph. Refigured by Prin in Paris, the telescope parts arrived at Ksara early in 1914. A new cupola turned up in June, but the tower to shelter it was not yet complete when Europe went to war. Berloty worked through August and into September. When the hired masons left, the Jesuits continued on alone, preparing the cupola for mounting. The Ottoman empire declared war on France on 1 November. Berloty decided to install all the heavy parts of the telescope to make their theft more difficult; he hid the objectives and the clock movement. The Jesuits were escorted to Beirut and permitted to leave the country late in December 1914.

Berloty sat out most of the war in Cairo, where he spent a good deal of time at the Helwan Observatory, directed by Harold Knox-Shaw. He returned to Ksara four years after his eviction. Nothing remained of the observatory but its walls. All the portable parts of the instruments—and all the observatory's fixtures—were gone, as were the archives and library. In local villages Berloty recovered a portion of what had been pillaged during the four days between the exit of the German-Turkish detail that had occupied the observatory and the arrival of its British replacement.

Cordial Entente: Jesuits, Generals, and Professors

Berloty traveled to France in 1919 to repair some of his instruments and campaign for the refurbishing of his observatory. He succeeded in obtaining money from Georges Picot, high commissioner of France in Syria, and from the ministry of foreign affairs, as well as twenty thousand francs from the Loutreuil Fund of the Paris Academy of Sciences. The library was replenished from metropolitan and peripheral deposits, such as the Paris and Helwan observatories. Instruments arrived from all quarters. The big telescope and the heliograph were reconstructed.

10. Berloty, "Notice," pp. 34–35, mentions prizes of Fr 500 and Fr 2,000 given by the Lyons Academy of Sciences and Belles-Lettres in 1911 and 1913, grants of Fr 3,000 and Fr 5,000 by the Académie française in 1913 and 1914, two large sums from an anonymous donor, and a small grant from the ministry of foreign affairs in 1914.

Berloty handled everything by himself until April 1920, when he received two military meteorologists for six months. In their place came Charles Combier as the observatory's assistant director and Georges Horan, an English-Jesuit veteran of the Zikawei Observatory who had been unable to tolerate the climate of Shanghai.[11] By the middle of 1921 the observatory was again taking seismological and meteorological recordings; magnetical recordings began in 1924. Both the seismological and magnetical series were frequently interrupted by equipment failure. A 450-kg Mainka seismograph came in 1921, and a vertical Galitzin-Wilip (from the Paris Academy of Sciences) in 1935. The prewar 133-kg Mainkas began regular recording only in 1936.[12]

The army immediately asked Berloty to help survey its new acquisition, and Berloty responded with enthusiasm. Observation posts came to a number of military locations. In June 1921, Ksara became the central seat of the military meteorological services, with Berloty as its director.[13] Berloty's title confirmed a longstanding de jure role, since he had provided weather bulletins and preliminary geodesical surveys of the Beqa'a Valley in 1919. A military geodesical mission arrived in August 1920; at its head was Lt. Col. Georges Perrier, who had been part of the army's geodesical expedition to Ecuador early in the twentieth century. The army maintained a presence at the observatory through 1924. The most active of the resident military geodesists was Jean Govin, a polytechnicien and veteran of the European trenches who died in a plane crash in 1924.[14]

The Jesuits continued to work with the military authorities after the army decamped from Ksara. They established a network of meteorological stations throughout the Near East, a number of them at religious missions. The mandated states bore the cost, while the observatory itself received funding from the ministry of foreign affairs.[15] The Jesuits predicted weather in Syria until 1935, when the French National Meteorological Office relieved them of the charge.[16] Berloty's devotion

11. See Jalabert, *Jésuites au Proche-Orient*, p. 410, for Horan.

12. *Université Saint-Joseph de Beyrouth*, p. 288.

13. See Berloty, "Notice," pp. 63, 67–68, for the formal appointment.

14. Georges Perrier, "Jean Govin," *Bulletin géodésique*, nos. 1–4, 1924, pp. 227–33.

15. Berloty, "Notice"; Charles Combier, "La climatologie de la Syrie et du Liban," *Revue de géographie physique et de géologie dynamique* 6 (1933), 319–46. For the financing of the meteorological service, see *Les jésuites en Syrie, 1831–1931*, vol. 8, *L'Observatoire de Ksara et services de météorologie* (Paris, 1931), p. 25.

16. Emile Joulia, "Trois observatoires français d'outre-mer: Ksara, Ambohidempona, Zi-ka-wei," Toulouse, Société d'astronomie populaire, *Bulletin mensuel*, no. 197, 1939, pp. 122–39, on p. 126.

Commander and confessor: Captain de Volontat de Merveille (*center*) and Charles Combier surveying the French mandate in Lebanon, 1920. (From Bonaventure Berloty, "Notice historique sur l'observatoire fondé en Syrie par les pères de la Compagnie de Jésus," *Annales de l'Observatoire de Ksara: Mémoires* 1 [1921–27], 1–120)

to the cause of French imperialism brought him a corresponding membership in the Paris Academy of Sciences in 1925. He then handed over direction of the observatory and the meteorological service to Combier, while continuing on at Ksara until his death in 1934.

Meteorology and cartography occupied the Jesuit scientists for more than fifteen years. Late in the 1930s the occupying French army no longer required the Jesuits for either task. Ksara's seismographs oscillated in and out of repair.[17] Charles Combier, ever inventive, turned his telescopes to the earth's shadow across the moon.[18] He received assistance from Jean Delpeut, who arrived from the Jesuit observatory on Madagascar in 1936.[19] The Jesuits maintained the observatory until the 1970s, when the Beqa'a Valley succumbed once more to a hostile army.

The Jesuit operation at Ksara followed the fortunes of the St-Joseph University in Beirut. During the First World War, the medical school

17. A description appears in ibid., pp. 124–28.

18. Combier, "Sur quelques problèmes nouveaux posés par la photographie de l'ombre de la Terre," *CRAS* 208 (1939), 1520–21.

19. Jalabert, *Jésuites au Proche-Orient*, pp. 328–29.

was suppressed, and in its place the Turkish regime substituted the Ottoman School of Medicine from Damascus.[20] The university reopened in 1919. The first graduating class of engineers to follow the restoration, in 1922, had eleven students. Although it was a private institution, the school answered to metropolitan ministries as well as to the government of the French mandate. Professionals and academics from Lyons, who had breathed life into the engineering school, presided over examining committees; academical juries included university professors from France. The Beirut institution aimed to send graduates to the grandes écoles of France; seven went to the Ecole supérieure d'électricité in Paris, and one of these succeeded in obtaining a doctorate from the Sorbonne. In 1930 the school's diploma was recognized as representing adequate preparation for beginning a metropolitan doctorate of engineering.[21] Largely pedagogues (Berloty's former assistant Charles Coste taught higher mathematics in the engineering school until his death in 1930), several of the staff were infected with the research ethic. Among the latter, the professor of medical physics, François Dupré La Tour, focused attention on nuclear physics.

Born into a distinguished family in Ecully, Rhone, in 1900 and educated at the Marianist Collège Stanislas, Dupré La Tour entered the Society of Jesus in 1918. He passed his régence in Beirut under Charles Neyron de Méons, professor of biological chemistry at the St-Joseph medical school. Ordained a priest in 1928, Dupré La Tour found a secular vocation in physics.[22] In the early 1930s he worked with the two independently wealthy scions of Parisian physics, Maurice de Broglie and Louis Leprince-Ringuet, studying the effect of various radiations on fatty acids; de Broglie supervised his doctorat ès sciences in 1932. Affinities went beyond money: Leprince-Ringuet's Catholic faith would have been sympathetic to Dupré La Tour, Leprince-Ringuet had traveled widely in Africa, and the two were distantly related.[23] Dupré La Tour joined the faculty at Beirut in 1934, but he continued to study neutrons (then called beryllium rays), alpha particles, and positrons, subjecting organic matter to their mercies. By the late 1930s he was

20. *Cinquantenaire de l'Université St-Joseph, Beyrouth, 1875–1925* (Beirut, 1925), pp. 23, 33.

21. *Les jésuites en Syrie, 1831–1931*, vol. 4, *L'Ecole française d'ingénieurs* (Paris, 1931).

22. Jalabert, *Jésuites au Proche-Orient*, pp. 330–31; idem, *Le souvenir de nos morts* (Beirut, 1983), pp. 142–43.

23. See *Jean Puyo interroge Louis Leprince-Ringuet de l'Académie française* (Paris, 1976), p. 12, where Leprince-Ringuet indicates his favorite cousin Louise Dupré La Tour – one of his daughters married one of her sons; and p. 44 for the African travels. See Leprince-Ringuet, *Atoms and Men*, trans. Elaine P. Halperin (Chicago, 1961), pp. 34–35 for Maurice de Broglie's laboratory, pp. 11–12 for how de Broglie missed discovering the neutron.

experimenting with X-rays to determine the structure of complex organic molecules.[24] The war brought an end to Dupré La Tour's research career; in 1942 he became the university's chancellor.[25]

While the St-Joseph professors taught physicians and engineers and the Ksara geophysicists helped the French army survey newly conquered territory, one man in the Syrian mandate—Jean Chevrier—dedicated himself for the most part to pure research. His accomplishments derived from skillful use of all the French facilities in the Near East; his progress depended on the patronage of metropolitan savants.

Chevrier gained admission to both the Ecole polytechnique and the Ecole normale supérieure in 1920; he chose to attend the latter. Upon graduating in 1922, he worked in a university chemistry laboratory while studying for the certificates in physics and chemistry then required for a *licence homogène.* In 1925 he accepted a position at the lycée run by the French lay mission in Beirut. After three years Chevrier decided to write a doctoral thesis on atmospheric electricity in the Near East. He presented himself first to the director of higher education in the ministry of public instruction, then to Charles Maurain. To Maurain he emphasized: "So far from intellectual and material resources, which I would have found near you, I cannot perhaps think of works having an overly general nature that would be carried out more easily elsewhere under better conditions. The wisest thing would be then to draw on my 'geographical situation' for the special material that it can provide." Chevrier indicated that he had tried to take preliminary measurements of the atmospheric gradient but that the effort had not been successful. Combier offered him hospitality at Ksara. Chevrier asked Maurain whether he might look into atmospheric ozone or perhaps follow the lead of Pierre Lejay and study magnetical storms.[26] These questions, as we have seen, related to radio transmission and weather forecasting. They were of great interest to overseas military men during the interwar period.

Maurain welcomed Chevrier's inquiries, furnished instructions and

24. Dupré La Tour's publications are listed in Bernard A. Fiekers, "Bibliography of Jesuit Publications in Chemistry," *Jesuit Science Bulletin* 34 (1957), 71–94, on 73–74; the French titles are translated into English. See Bruce R. Wheaton, *The Tiger and the Shark: Empirical Roots of Wave-Particle Dualism* (Cambridge, 1983), pp. 274–75, for Maurice de Broglie's X-ray absorption experiments in the early 1920s and his laboratory collaborators.

25. Henri Jalabert, "François Dupré Latour," in *Dictionnaire du monde religieux dans la France contemporaine*, vol. 1, *Les jésuites*, ed. Jean-Marie Mayeur and Yves-Marie Hilaire (Paris, 1985), p. 106.

26. PIPG, Chevrier to Maurain, 29 Apr. 1928.

advice, and then saw to it that Chevrier's data appeared in the *Comptes-Rendus* of the Paris Academy as well as in the *Annales de l'Institut de physique du globe.*[27] Maurain recognized that the man in Beirut was no ordinary colonial expatriate. By the beginning of January 1930 Chevrier had plunged into quantum mechanics. He began to translate Arthur Erich Haas's *Materiewellen und Quantenmechanik*, with a view to having his second thesis topic (the one traditionally posed by the faculty) come from the information in it. He also worked his way through the latest Solvay Congress volume on electrons and photons. This interest in German-language theory was unusual for a French scientist, much less one located overseas. Chevrier's plans did not end there. He had recently contacted the local naval squadron about research in atmospheric electricity, and he hoped one day to take observations on the Red Sea aboard a French cruiser.[28]

Late in 1930 Chevrier took leave from his teaching post to devote himself entirely to magnetical and electrical research, for he had received support from the French army occupying Lebanon and Syria, as well as a scholarship from the French ministry of foreign affairs.[29] He persuaded the resident czar of culture, Gabriel Bounoure, to propose him as head of a meteorological service for the French mandate, which would be devoted to the needs of air traffic. (The two existing meteorological operations, run by the Jesuits at Ksara and by the army for the French National Meteorological Office, were ineligible for funding from the mandated states.)[30] Although Chevrier did not find his way to the mooted meteorological triptych, he drew on good will with Maurain and an introduction to General Gustave Ferrié to join the magnetical section of the French national geodesical and geophysical committee.[31]

27. Chevrier's first six notes to the Paris Academy (1929–36) are in the bibliography to his *Contribution à l'étude du champ électrique de l'atmosphère* (diss., University of Paris, 1938); Chevrier published seven articles in the *AIPG* between 1931 and 1938, the last one being his doctoral dissertation. All his work concerned electrical and magnetical observations in Syria and Lebanon.

28. PIPG, Chevrier to Maurain, 18 Jan. 1930. Haas, *Materiewellen und Quantenmechanik: Eine elementare Einführung auf Grund der Theorien De Broglies, Schrödingers und Heisenbergs* (Leipzig, 1928), into six editions by 1934. A French translation by A. Bogros and F. Esclangon appeared in 1937.

29. PIPG, Chevrier to Maurain, 22 May 1930 and 2 Jan. 1931.

30. Ibid., Bounoure (Inspecteur général des oeuvres françaises) to Maurain, 29 June 1931. The mandate of Syria and Lebanon occasioned diplomatic difficulties for the French authorities. Chevrier emphasized to Maurain that people in the French high commission were "very jealous" of the fact that they served in a country under mandate rather than in a colony or possession of France (Chevrier to Maurain, 9 Jan. 1935).

31. Ibid., Ferrié to Maurain, 18 Nov. 1931.

Early in the 1930s Chevrier had become a bright hope of French-language physics in the eastern Mediterranean. His rising star overshadowed the Ksara Observatory, which by this time claimed to oversee seismology, terrestrial magnetism, and meteorology. Chevrier complained to the director of higher education in Paris, Jacques Cavalier, about Father Berloty. Chevrier had to compare his magnetical theodolite with those at Ksara. Berloty had sent him absolute magnetical readings, but Chevrier believed that these were not sufficiently precise; Berloty would not allow him to be present when readings were taken. Because Chevrier could not calibrate his own instrument, two publications of his had systematic errors.[32]

Chevrier resolved matters with the Jesuits.[33] His research went toward a dissertation under Maurain's supervision; his livelihood came from the Beirut meteorological service and the American University, where he worked and taught.[34] He felt secure enough to decline Maurain's suggestion in 1936 that he apply for a vacant post in Tunisia (possibly the directorship of the meteorological service).[35] Chevrier defended his dissertation in 1938. It summarized observations on atmospheric electricity. His second thesis, suggested by jury member Francis Perrin, was "the quantum theory of chemical valence in diatomic molecules," although theory found no place in his printed dissertation.[36] Dr. Chevrier returned from his defense to become professor of mathematics in the engineering school at the St-Joseph University.[37]

32. Ibid., Chevrier to Cavalier, 3 Sept. 1932; Chevrier sent the same message to Maurain on 4 Oct. 1932.

33. Ibid., Chevrier to Maurain, 31 Dec. 1934; *Annales de l'Observatoire de Ksara, observations (Section magnétique), années 1930–1933*, where Berloty's successor Charles Combier thanks Chevrier for assisting Berloty and for taking charge of the magnetical measurements in 1933 and 1934, during Berloty's illness.

34. PIPG, Chevrier to Maurain, 31 Dec. 1934, where Chevrier signs his name as "attaché au Service météorologique"; 21 Jan. 1936, where Chevrier apologizes for not having made much progress on his thesis because of his teaching obligations in mathematics at the Ecole française and in chemistry at the American University. Chevrier wrote to Maurain on stationery of the American University's chemistry department from November 1937 to May 1938; he does not figure as a publishing member of the American University, however, in Denise Dominique d'Orsonnens, *La création et le développement de l'Université américaine de Beyrouth, de 1920 à 1940* (M.Sc. diss., University of Montreal, 1981), pp. 35–36.

35. PIPG, Chevrier to Maurain, 8 July 1936.

36. See ibid., 24 Nov. 1937, for the topic of the second thesis, which appeared only as "Propositions Given by the Faculty" on the title page of the dissertation, *Contribution à l'étude du champ électrique*.

37. *Min.*, 1938, s.v. "Beirut."

God's Astronomers Royal

A final instance of convergence between Jesuit scientists and French colonial ambitions is found on Madagascar. During the period leading up to the Franco-Madagascan War of 1883–85, when Britain and France competed for hegemony on the island, French missionaries, medical men, and navigators measured land and sky. The early Jesuit missionary Désiré Roblet took meteorological readings during the 1860s and 1870s. The earliest systematic recordings — daily temperatures and pressures — came from the hand and eye of Jean Laborde, a French adventurer and commercial exploiter, during the years 1864–78. The observations were then continued by the French Christian Brothers mission. The principal impetus for exact sciences, however, came from one of the early naturalists, Alfred Grandidier.[38]

Grandidier wandered across Madagascar from 1865 to 1870, returning home to find a chair in the Academy of Sciences. His voluminous Madagascan natural history began to appear at just the time that the Third Republic, flushed with success at having overrun Tunisia, decided to expand from several strategic enclaves to embrace all the peoples of the Grand Island of Madagascar. Having taken longitude measurements himself during his wanderings, and having edited reports compiled by Father Roblet (to whom he sent instruments and instructions), in 1880 Grandidier asked the Catholic hierarchy to establish a permanent meteorological station at Tananarive under the French Jesuit mission there. In the meantime, the Hova of Madagascar found French tutelage to be a mixed blessing; in 1882–83 they forcibly expelled the invaders. The French returned with more soldiers and bigger guns; the Jesuits regained their mission. In 1887 the Jesuits received a mandate from the French resident general, Charles Marie Le Myre de Vilers, to transform their meteorological station into an astronomical observatory. Edouard-Elie Colin arrived in 1889 to bring the project to fruition.[39]

38. Charles Poisson, "La physique du globe à Madagascar," *Bulletin de l'Académie malgache, numéro spécial du cinquantenaire*, 1954, pp. 117–34, on pp. 119–20; H. Dopp, "Un cinquantenaire: L'Observatoire d'Ambohidempona à Tananarive, 1889–1939," *Revue des questions scientifiques* 6 (1939), 408–21; *Hydrogr.*, pp. 230–31, for the naval surveyors; Paul Lesourd, *L'oeuvre civilisatrice et scientifique des missionnaires catholiques dans les colonies françaises* (Paris, 1931), pp. 199–200, on Roblet's early surveys. On Laborde's Madagascan manufacturing enterprise, beginning in the 1840s, see *GFOM*, p. 144.

39. *Vregille*, pp. 522–34, on Madagascar; Lesourd, *Oeuvre*, p. 200, for Grandidier's relationship to Roblet; Colin, *Travaux géographiques des PP. Roblet et Colin, S.J. à Madagascar* ([Paris?] 1895), the text of a lecture delivered to a group of alumni of the Ecole polytechnique on 27 Mar. 1895. For Grandidier's early cartography on Madagascar, see Alfred

Born in Graulhet, Tarn, in 1852, Colin received his earliest education in local Catholic schools, where he exhibited a precocious talent for music. As he matured he grew to appreciate the harmonies of mathematics. At age eighteen, during the Franco-Prussian War, he presented himself as a Jesuit candidate in Toulouse. He took his literary and philosophical studies in southern France and taught at various collèges. He was a theology student at Uclés in Spain when he was tapped by the superior of the Jesuit Province of Toulouse (who was designated as a recruiter by the vicar of Madagascar) to build the observatory. In 1887 he went off to study astronomy at Stonyhurst under John Perry and then meteorology at Montsouris. These techniques in exact sciences being child's play for a thirty-five-year-old Jesuit, Colin finished his apprenticeship, turned up at Tananarive, and settled on the granite heights of nearby Ambohidempona as the place for his telescopes.[40]

France was at this time a guest of the Hova monarchy, and Colin had to negotiate with the regent for cession of the land. In eighteen months he raised a magnificent brick and stone structure with three cupolas. The eastern tower housed a portable meridian circle (the survivor of a fire during the Paris Commune) lent by the Paris Observatory. To the central tower came a 22-cm Eichens equatorial, again courtesy of the Paris astronomers; it had served on the French Venus-passage expeditions and had since remained out of service. The ministry of foreign affairs paid the ministry of public instruction two thousand francs to refinish the Eichens and then covered the cost of sending it to Madagascar.[41] This was an abrupt change of heart from the preceding year, when foreign affairs had explicitly refused to purchase any of Colin's instruments.[42] We discover the reason for the change in a letter sent home by the French resident general in Tananarive: "At the English collège of Tananarive [there is] a lens of four inches, with which the

Lacroix, *Notice historique sur Alfred Grandidier, membre de la section de géographie et navigation, lue dans la séance publique annuelle du 18 décembre 1922* (Paris, 1922), pp. 16–17.

40. Charles Poisson, *Le père Elie Colin, directeur de l'Observatoire de Tananarive, 1852–1923: Un bon ouvrier de la science française* (Bourges, [1923]); Dopp, "Cinquantenaire."

41. PAN, F17/13588, Ministry of Public Instruction to Director of Paris Observatory, 22 Sept. 1890 (copy); PO, "Tananarive, 1880–1890," Ministry of Public Instruction to Admiral Mouchez, 28 Sept. 1888 and 9 Oct. 1888 for the offer of a 40-cm equatorial circle, and 23 May 1890, saying that two thousand francs will not be enough to render the large equatorial circle useful and proposing to send instead the 22-cm equatorial used for the passage of Venus.

42. PAN, F17/3827, Ministry of Foreign Affairs to Ministry of Public Instruction, 21 Sept. 1889.

The first observatory at Ambohidempona, Madagascar, destroyed in the unsuccessful Madagascan war of independence of 1895 and rebuilt in a less imposing style (From Charles Poisson, *Un cinquantenaire: L'Observatoire d'Ambohidempona à Tananarive, Madagascar, 1889–1939* [Paris, (1939)]

Madagascans are taught some astronomy; the one at the French observatory should then be more substantial and more finely constructed."[43]

Colin had the Eichens remounted and installed in a tube that resembled a cannon. The Madagascans charged with transporting it inland from the port of Tamatave, thinking it a monstrous weapon, pitched the harmless instrument in a thicket; Colin retrieved it and placed it in functioning order. From the Central Meteorological Bureau under Eleuthère Mascart came everything needed for observing weather and magnetical field; the instruments were installed in the south tower and generally along the observatory's balustrade, with the magnetical recorders relegated to a nearby shed. Colin and Roblet, who had since been assigned to the observatory, promptly began their astro-

43. Ibid., F17/13588, Resident General to Minister of Foreign Affairs, 16 Jan. 1890.

nomical careers by determining the observatory's place on the globe; in 1891 they concluded that they stood 1,402 meters above sea level, which put them among the world's highest astronomers. Colin also obtained a royal prefix for the institution from His Majesty Ranavalomanjaka III.[44]

Having established the observatory, Colin and Roblet left it in the hands of their Jesuit subordinates Father J-P. Combes and Brother Soula – neither of whom had training for or interest in scientific work – and proceeded to triangulate the island.[45] They walked across the country for four years, after which time the Hova again disabused themselves of French civilization: they threw out the invaders in 1895. This time the French would have no mistakes. The army enlisted Alfred Grandidier to produce maps of the island using Colin and Roblet's surveys. After the successful campaign of reconquest, the two Jesuit surveyors returned. They resumed their affiliation with the geographical bureau of the army's general staff, accompanying the troops between 1896 and 1904 to chart France's newest possession.[46] As a result of Colin's dedication to the cause of French imperialism, the scientist-priest received three prizes from the Academy of Sciences and two from the Société de géographie of Paris; in 1899 the Paris Academy honored him with a corresponding membership.[47]

Ravaged by fever, Colin had returned to France in 1894. The Hova war of independence, which saw all civilians and missionaries leave the island, occurred in his absence. The Madagascan vicar asked that the observatory be placed under royal protection, as it was a royal institution. When invading French troops threatened the capital of Tananarive, the Hova government ordered that the observatory, a fortified post commanding the capital, be demolished. The building was taken apart stone by stone, and the instruments were scattered across the terrain. The Hovas then set up an artillery battery to the north of the ruins. The French invaders quickly set the freedom fighters to rout. Following the occupation of Tananarive, the ruins of Ambohidempona offered

44. *Vregille*, pp. 523–31; PAN, F17/3827, Minister of Foreign Affairs to Minister of Public Instruction, 21 Sept. 1889, where Colin already has the title of royal astronomer; Eleuthère Mascart, "Sur l'Observatoire de Tananarive," *CRAS* 110 (1890), 608–9.

45. Toul., FMa 958, G. Vignoux to Provincial Superior, 25 Jan. 1895, reporting on the extreme disaffection of Brother Combes toward the observatory's mission and his relief at having been transferred away from it; but Combes, ordained as a priest, returned to take temporary charge of the observatory in 1899 during one of Colin's longitude missions for Governor General Joseph Simon Gallieni (Colin to Provincial Superior, 20 May 1899).

46. See Lesourd, *Oeuvre*, pp. 200–203, for Colin and Roblet's maps and Grandidier's patronage; see also Lacroix, *Grandidier*, pp. 21–22.

47. *Vregille*, pp. 531–34; Toul., FMa 958, Poisson to Provincial Superior, 20 May 1899, acknowledging congratulations on the corresponding membership.

shelter to a company of Senegalese soldiers; they destroyed what the Hovas and the French artillery had spared.

Eleuthère Mascart supported Colin's plea to the ministry of public instruction for help in reconstructing the observatory.[48] In 1896, while still in France, Colin met with the minister of colonies, who "reacted favorably to my [funding] proposal" and to his scientific program, especially its meteorological parts. The ministry of foreign affairs then referred the matter to the resident general at Tananarive.[49] Colin left Marseilles with instruments from Mascart and promises of substantial aid in the future.[50] Colin's ruined observatory gravely worried Maurice Loewy, director of the Paris Observatory, who still hoped to enlist the Jesuits for his photographical survey of the heavens. Loewy emphasized to the ministry of public instruction that France had to endure much shame because "while England possesses in its colonies eight observatories, of which four or five, at least, carry out very serious astronomical activity, France possesses none in the Southern Hemisphere."[51]

After several years, Colin decided to rebuild on the same site. With ten thousand francs from the local French authorities and with one hundred impressed Madagascans (furnished by the conqueror–turned–governor general Joseph Simon Gallieni), he reduced the ruins to their foundations and raised a new structure. It was a less imposing design. In 1899, with money from one of his prizes, Colin ordered a cupola and reinstalled his Eichens equatorial, repaired as best he could, adding two smaller meridian telescopes. His impoverished operation—which a decade previously had held such promise as France's Southern Hemisphere observatory—forced him to abstain from participating in the sky-chart survey. He focused on meteorology, using the telescopes for a time service. In 1902 Gallieni provided assistance for erecting a shed to monitor terrestrial magnetism; the results of Colin's subsequent magnetical campaigns across the islands were received with skepticism by metropolitan colleagues.[52] Colin nevertheless found an outlet for his observations in both metropolitan and colonial journals, especially the bulletin of the local learned academy.[53]

48. Toul., FMa 958, Colin to Provincial Superior, 22 Feb. 1896.

49. Ibid., 1 Apr. 1896.

50. Ibid., 23 May 1896.

51. PAN, F17/13588, Loewy to Minister of Public Instruction, 16 Dec. 1899; extract from minutes of the council of the Paris Observatory, 9 Dec. 1899.

52. *Vregille*, pp. 525–31; Aix, Madagascar 5 (13) D4, "Rapport sur l'Observatoire de Tananarive, 1889–1914"; PSG, colis 25, "Rapport sur les travaux scientifiques du P. Colin de la mission de Madagascar, 1897–1905." See Poisson, "La physique du globe à Madagascar," p. 124, for the skeptical response.

53. Gallieni founded the Académie malgache in 1902 to encourage literary and ethnological work; its scope extended to the natural sciences in 1912, the date when Colin

Even after Colin had proved himself as a patriot and friend of the army, the colonial government expressed strong reservations about funding Jesuit science. The problem, the colonial secretary general in Tananarive noted in 1898, turned on the distinction between secular and religious functions: "Until now we have not funded any religious minister, and we have not omitted to have noted on every occasion that the support of the colony was granted entirely for the pedagogical work of the missions, and that the administration did not expect to see diverted to the ends of religious propaganda, to which it remains completely estranged, the disbursements that it has been able to grant."[54] Writing to Gallieni in 1901, the administrator of the eastern region of Madagascar emphasized that, indeed, the Jesuits should expect to line up for government funding behind priests of secular orders. The "secular priests," he wrote, "counterbalance the hollow influence of the members of the Company of Jesus."[55]

It was something of a problem for the Third Republic to have depended on Jesuit missionaries for vital support in pacifying the island. By 1905 the government decided to relieve the Jesuits of responsibility for geodesy. "It is probable that I will no longer be charged with *official* scientific missions," Colin wrote. "The minister of colonies wrote to say that all these works will henceforth be confided to officers. . . . I am not complaining, because, in all sincerity, this trade of geodesist is very tiring (if one wants to do it conscientiously) and is in no way rewarded."[56] The ministry of colonies sent an inspector to size up Colin's operation. Following the inspection, Colin hoped that he would be left in peace: "My aim was above all to show him that the missionaries in Madagascar perform useful services for France and that, consequently, they do not merit being expelled from the colony." In the event, subsequent surveying of the island went to military men. Between 1896 and 1911 the army's geographical service in Madagascar spent more than Fr 1,200,000 on maps.[57]

What Colin could not have officially he received in confidence. In

---

started publishing in its *Bulletin* and the probable date of his appointment as a member. See Aix, Madagascar 6 (9) D23, "La science à Madagascar: Etablissements scientifiques de l'Ile, leur utilité, leur budget," 1935, for a discussion of the academy.

54. Ibid., Madagascar 6 (4) D2, Secretary General, "Note" [1898].

55. Ibid., J. Martin to Gallieni, 12 July 1901.

56. PSG, carton C, Colin to Secretary of the Société de géographie of Paris, 15 July 1905.

57. Ibid., 15 Aug. 1905, for the quotation. On the naval mapping, see *Hydrogr.*, pp. 238–45. PSG, colis 22, contains a detailed report of one coastal mapping campaign of 1908–9, led by the naval hydrographer Henri Roussilhe. Aix, Madagascar 5 (17) D1, "Rapport du général Riou commandant supérieur des troupes du groupe de l'Afrique

1901 Gallieni funded the first meteorological network for the colony, which he placed under his agricultural portfolio. In matters of acute concern, however, Gallieni continued to make demands on Jesuit expertise by asking Colin to analyze the course of tropical storms. Impressed by Colin's analysis, the governor general decided to institute a maritime meteorological service in 1903. He kept it under the general secretariat instead of relegating it to a colonial portfolio. With weather prediction – rather than general climatology – as an aim, Gallieni provided ten island stations with a telegraph link to a central office. Gallieni placed Father Colin at the nerve center of the telegraphical network in the unofficial capacity of director; Gallieni's successor in 1907 made the appointment official, giving the Jesuits technical supervision as well of all the posts previously controlled by the agricultural authorities.

Things did not work out as planned. The science of meteorology was centralized on the heights of Ambohidempona, but its administration remained with the governor general, its surveillance depended on the good will of the various provincial administrators, and its on-site staffing related to nothing more than chance. Incoming meteorological reports did not always find their way into Colin's hands; meteorological syntheses were strong on facts and weak on analysis.[58]

During the generation after the destruction of the first observatory, the scientific influence of the Madagascan Jesuits was limited to weather reports and inconstant magnetical readings, the former supervised by Colin and the latter first by the scientific neophyte Soula and then, in 1908, by Soula's successor Aurand, who sported a licence ès sciences. Roblet, who had continued to assist Colin in occasional cartographical projects, died in 1914. In 1915 Aurand went off into the bush on a mission of religious mercy, and magnetical registration at Ambohidempona became sporadic.[59] The observatory's printing press gave out in 1915, and in 1919 magnetical readings were no longer taken. The main building needed major repairs.[60] Colin kept charge of his meteorological re-

orientale sur le fonctionnement et l'organisation antérieurs et actuels du Service géographique."

58. See Charles Poisson, "Notes de météorologie malgache," *BEM* 23, no. 2 (1926), 1–18, pp. 8–14 for the prewar history; Colin, "Résumé météorologique de l'année 1914 à Tananarive," ibid. 15, no. 1 (1915), 141–43, for a typical yearly summary.

59. Dopp, "Cinquantenaire"; Poisson, *Colin*, p. 7. According to Charles Maurain, magnetical recording went on from 1902 to 1919, when funding evaporated (Maurain, "Les observations magnétiques en France jusqu'en 1921," *AIPG* 1 [1922], 9–19).

60. Toul., FMa 956, [Charles Poisson], "L'observatoire, novembre 1928"; PIPG, Colin to Maurain, [ca. 1921], where Colin refers to his having ended magnetical recordings in 1919 as a result of budgetary restrictions; Colin replied to Maurain's inquiry of 25 Nov. 1921.

Elie Colin and Désiré Roblet. Their careers illustrate the complementary interests of overseas Jesuits, the French military, and the colonial administration. (From Charles Poisson, *Un cinquantenaire: L'Observatoire d'Ambohidempona à Tananarive, Madagascar, 1889–1939* [Paris, (1939)])

cording and weather-prediction service through the war and into the 1920s.[61] He expired in 1923. His successor Charles Poisson shared Colin's affinity for professional killers.

Poisson's father owned a sawmill in Rion-des-Landes and was the longtime mayor of his commune. Charles was the eldest of nine children. Of his five brothers, four died in the First World War; one of these was a polytechnicien, two were Jesuits, and one was a naval

61. PMN, Angot to Colin, 18 June 1920, where Angot asks for recordings from the third trimester of 1919, specifying that he received reports for the preceding and following months; a note on Angot's draft indicates that Colin sent the missing data.

aviator-priest. Of his three sisters, two wore the habit of the Sacred Heart. After passing through collège at Tivoli, Poisson studied for the Ecole navale under the Jesuits at Jersey. He entered the naval school in 1900, graduated, and worked his way up the hierarchy. The exact sciences occupied his attentions, and his early observations appeared in the navy's *Annales hydrographiques*. He served in the Moroccan campaign of 1907, where he suffered from chronic enteritis. Three years later he resigned his commission to join a younger brother in the Society of Jesus. Mobilized in 1914, Poisson took his place as an officer in a regiment of naval fusiliers. He saw heavy fighting in Dixmude, Verdun, Champagne, Belgium, and the Somme; in 1916 he published an account of his first battles as a book. Demobilized, he studied theology at Enghien and, in 1920, received ordination as priest. When Colin died, forty-year-old Poisson was at Enghien taking his "third year" as a novice.[62]

If a Jesuit can be called to perform any task, his talents are generally put to good use by perceptive superiors. Poisson, having fought for France at home and abroad and having been schooled by the navy in the art of celestial navigation, was just the man to take over the attenuated Madagascan observatory. He promptly moved to Paris for schooling at Charles Maurain's geophysical institute and Emile Delcambre's National Meteorological Office. The Parisian geophysicists inspired him on to colonial glory. "Everywhere in the scientific institutions I am told that there is an opportunity to seize because of a political struggle between France and England, who contest for superiority in meteorology and physics. One must take advantage of the occasion." The Madagascan Jesuits would do well to emphasize that their work contributed to French meteorological hegemony in the South Indian Ocean.[63] Poisson reported an uncomfortable meeting with Delcambre, a man not favorably disposed to Jesuits. Delcambre thought that Poisson's predecessor Colin had dedicated himself too exclusively to astronomy, magnetism, and geodesy, his work in meteorology containing "only columns of figures lacking in real importance." Delcambre favored "a state service . . . organized by officers," and he seemed to have a candidate in mind.[64] Delcambre would have his way within a decade.

From the moment of his arrival at Ambohidempona, Poisson was a solitary man, with whom close cooperation was difficult. After three months, he drove the observatory's assistant Aurand back into the bush.

62. J[ean-Baptiste] C[oze], "Le père Charles Poisson, 1882–1965," *Maduré-Madagascar*, no. 116, July 1965, pp. 218–23.

63. Toul., FMa 981, Poisson to Provincial Superior, 31 July 1923.

64. Ibid., 24 Aug. 1923.

Then he began routine meteorological and geophysical observations. "He worked still more for himself and eventually his successors," Poisson's biographer notes,

> writing numerous notebooks about problems that he had to treat, and tables of the propagation of earthquakes adapted to Madagascar. But if everything can be found in them, their use is rather difficult, for he had an often confused mind – let us say, rather, a complicated one – and he placed on the same level the essential and the ancillary. He was able to find his way very well through his notebooks, written in a faint and spidery hand, but he would certainly have been a poor teacher, not knowing how to provide a clear explanation and not tolerating questions.

Poisson's magnetical recordings, taken between 1925 and 1938, were not of great value; his seismological recordings, made by subordinates on two 450-kg Mainka seismographs installed by the Paris Academy, found users among metropolitan savants. He received the usual salutes from colleagues at the seat of empire: affiliation with the Academy of Colonial Sciences and corresponding membership in the Academy of Sciences.[65]

To make anything of the observatory undoubtedly took the discipline of a Jesuit and a soldier. The situation facing Poisson in 1924, upon his arrival, was disastrous: "Seismology is in ruins with the hope of resurrection; magnetism is in ruins through weakness and discouragement; astronomy kept only a strict minimum of means and allowed lovely equipment to be ruined." The big telescope had to be replaced at a cost of Fr 120,000.[66] Poisson required, in addition, a good assistant.

> The man will have done enough mathematics so as not to mix up formulas and to know moderately well how to use them. He will know physics moderately well, without having taken a doctorate. I would ask him above all to be practical, rather a handyman, with regard to setting up the pendulums, and to release me from the responsibility of maintaining the instruments. In astronomy and meteorology he would be trained on site.[67]

Louis Cattala served as the seismological assistant for several years beginning in 1925, and Jean Delpeut, whom illness had forced out of the Ecole navale before the First World War, fit the bill for terrestrial magnetism beginning in 1928.[68]

65. Coze, "Poisson," pp. 222–23; Dopp, "Cinquantenaire," for the Mainkas.

66. Toul., FMa 956, Poisson to Toulouse Provincial, 20 July 1924. The discouraging state of seismology is revealed in Poisson's "Tremblements de terre de Tananarive, 1889–1924," *BEM* 21 (1925), 256–59, where measurements of earthquake intensity (Rossi-Forel) begin only in 1924.

67. Toul., FMa 956, Poisson to Provincial Superior, 31 Oct. 1924.

68. On Cattala, see Bernard Leurent, "Le père Louis Cattala, 1904–1963," *Maduré-*

Charles Poisson on the meridian circle at Ambohidempona, Madagascar. Despite its favorable location for observing the southern stars, the observatory never developed an active research program in astronomy. (From Charles Poisson, *Un cinquantenaire: L'Observatoire d'Ambohidempona à Tananarive, Madagascar, 1889–1939* [Paris, (1939)])

By the late 1920s Poisson had succeeded in obtaining Fr 20,000 per year from the colony to run the meteorological and time services. He installed electrical lighting, again provided by the governor general. The Paris Academy granted Fr 30,000 for his Mainka seismographs, and the Bureau of Longitudes, Fr 5,000 for terrestrial magnetism. His financial situation, however, remained precarious. It took Fr 2,500 to

Madagascar, no. 109, Oct. 1963, pp. 359–62; and Charles Poisson, "L'oeuvre scientifique du père Louis Cattala," ibid., pp. 363–67, on p. 363. On Delpeut, see Jalabert, *Jésuites au Proche-Orient,* pp. 328–29.

keep a Jesuit on site, and an additional Fr 960 went for basic amenities. He asked permission from his Jesuit superiors at Toulouse to spend some of the observatory's capital of Fr 83,000, a request that was declined. He found that lack of staff denied him the privilege of operating on the level of other Jesuit observatories, notably those at Zikawei and Ksara; recruitment of local assistants was difficult, given the low salaries (from Fr 1,200 to Fr 3,000 per year). As early as 1928 he found Delpeut unreliable as a seismologist; within three years Delpeut had lost his mind.[69]

Climate of Church and State on Madagascar

Even with modest support from government and metropolitan academics, the Ambohidempona Observatory was an operation that could be steered only by someone with Poisson's experience and discipline. Late in the 1920s, however, the metropolis sought to displace him by assuming control of Madagascan geophysics. It did not reckon with Poisson's tenacity and diplomatic skill.

In 1920 meteorology obtained the status of a service under the public works portfolio, and Colin collaborated with an engineer to maintain the island's observation posts. From 1924 to 1927 meteorology went to the colonial maritime service, which was uninterested in maintaining regularity or standards. At the same time, as Poisson obtained instruments from the metropolis, his patrons began to question his ability to provide regular geophysical reports. In 1928 he emphasized that he had agreed to accept the governor general's invitation to provide the "technical direction" of a new colonial meteorological service, which would be staffed by officers of the army's geographical corps at a nearby barracks. The army would be responsible for assuring that the observatory and the service remained in communication. Poisson suspected that local meteorological predictions might be taken over by military aviators, following the model in France, but he thought that the Jesuits would be able to negotiate with any new institutions of this kind from a position of strength.[70] Poisson knew that his observatory was a plum to be

69. See Toul., FMa 956, Poisson, note of 1928; [Poisson], "L'observatoire, novembre 1928"; Jean-Baptiste Coze to Provincial Superior, 28 Feb. 1931, and Poisson to Provincial Superior, 11 Apr. 1931, for Delpeut's insanity. Poisson writes in the latter: "I must tell you that if Father Delpeut is not truly crazy — and I believe that he is — he is in any case on the border of insanity."

70. Toul., FMa 956, "Observatoire de Tananarive," Dec. 1928; Poisson, "Notes de météorologie malgache," pp. 15–18 for the postwar developments; J. Laborde, "Service géographique: Rapport sur les travaux exécutés de 1924 à 1927," *BEM* 25, no. 2 (1928), 101–32, for a description of the attenuation and virtual extinction of the army's geographical service on Madagascar in the period 1914–24 and its subsequent regeneration.

plucked by either Emile Delcambre in the ministry of the air or Henry Hubert in the ministry of colonies.[71]

Early in 1929 Hubert advised Poisson of metropolitan privileges retained by the new colonial meteorological service.[72] Following a letter from Poisson, Hubert offered a clarification:

> It is in no way the intention of the department to direct, from Paris, the meteorological services of the colonies. This honor, and the responsibilities stemming from it, will reside entirely with the heads of the meteorological services. The role of the ministry will be to guarantee the necessary liaison between the services and between them and those of France and foreign colonies, to bring about standardization of common questions, and to coordinate efforts of a general kind.[73]

The ranks of senior colonial meteorologists were to be filled following discussions between metropolis and colony, in which both savants and administrators took part. Poisson wrote to the Madagascan governor general, asking for clarification about how he—a Jesuit and naval reserve captain—might be involved in this new colonial civil service.[74]

Poisson's provincial superior in Toulouse judged the situation to be impossible. Poisson would have to cede much of meteorology to retain the independence from the tentacles of the Third Republic that was essential for the Jesuits.[75] Poisson did not abandon hope of an *entente cordiale* between church and state. The new, civilian Madagascan meteorological service, he wrote to his superiors, was independent of the army and would include an *ingénieur-en-chef* as director, an ingénieur-adjoint, and an ingénieur in charge of an ancillary station. There would be state meteorologists to run a network of stations and a full complement of technicians, accountants, secretaries, and service personnel at the central office in Tananarive. Five of the stations would be equipped better than Ambohidempona; furthermore, Poisson, whose help figured in the proposal, would not be provided with funds for an assistant. A separation of functions was the logical outcome.

> The rival installation does not need in any way to be an astronomical, magnetical, and seismological observatory, which would be costly and difficult to provide with personnel. No one here needs astronomy, and our instruments

71. Toul., FMa 956, Hubert to Poisson, 12 Mar. 1929 (copy), where Hubert asserts his interest in Madagascan meteorology and geophysics. The army's geographical service had no interest in absorbing the observatory at Ambohidempona (Laborde, "Rapport").

72. Toul., FMa 956, Hubert to Poisson, 12 Mar. 1929 (copy).

73. Ibid., 2 June 1929.

74. Ibid., Poisson to Governor General Marcel Olivier, 3 Aug. 1929 (copy).

75. Ibid., Provincial Superior to Poisson, 11 Sept. 1929.

> are hardly used. No one is concerned with terrestrial magnetism, though perhaps the geodesists of the geographical service will be led one day or another to undertake explorations denied us by lack of personnel. No one, finally, will think of founding a seismological station to rival ours, which is an official one, for it would be necessary to justify it by better results, and to do that it would have to obtain extremely costly instruments.

The new meteorological service, in taking over some of the observatory's functions, would not erode the Jesuits' prestige. In Poisson's view, the latter had already faded through lack of men and means.[76] Poisson still hoped for a place in the civil lists. He emphasized that directors of other Jesuit observatories, notably that at Ksara, were state employees. He looked forward to the day when the colony would send him to international meetings, as the French Concession had sent Zikawei Jesuits to the Pan-Pacific Science Congress on Java.[77]

In anticipation of the new colonial scientific regime, the heads of the aeronautical, geographical, and meteorological services met to establish jurisdiction over various parts of meteorological Madagascar. The threesome agreed to give Poisson an ingénieur-météorologiste trained in the metropolis and a météorologiste-adjoint drawn from local talent. The central office of the new colonial weather service would be separated physically from the geographical service; it would have sections for climatology and prediction. Poisson, however, would retain control of observations in Tananarive, and he would be in charge of "meteorological research" on Madagascar. The three directors estimated that it would all cost between one and two million francs, plus an indeterminate amount for construction and equipment.[78] Poisson thought that Madagascan weather would furnish good research material for his new senior assistant:

> Meteorological science is young. In America, in England, and in Norway, researchers have found new approaches full of promise. They are justly proud of what they have done. But their theorems or their laws have something too absolute about them; to adapt them to tropical conditions it will be necessary to work with them practically for a long time, to soften the living bones, in short to reduce them to the essential. There is something excellent that promises to our future coworker that material for study will not be lacking.[79]

76. Ibid., Poisson to Provincial Superior, 5 Sept. 1929.
77. Ibid., 15 Sept. 1929.
78. Ibid., Poisson, "Note sur le Service météorologique," 6 Sept. 1929.
79. Ibid., Poisson to Provincial Superior, 4 Feb. 1930.

Governor General Marcel Olivier, for his part, was glad to have Poisson on a contractual basis.[80] The men in the colonies did not take account of metropolitan principles.

"Concerning Father Poisson, I should like first of all to pay my compliments to his high learning and great experience," the minister of colonies wrote to the Madagascan governor general in 1930. "But it is not possible for me to authorize you to invest him by contract in the position of head of the meteorological service of Madagascar." If the governor general wanted to have the Jesuits work for the colony, then he had only to increase the grant for work in terrestrial magnetism and seismology. After all, the minister noted, Poisson really only wanted to carry out his own research, and this could easily be accommodated by providing him access to the observations taken by the colonial meteorological service.[81]

Poisson knew that his hand was strong. In the first place, the ministry was not yet in the position to send a trained senior man to Tananarive; in the second place, while the colony published and circulated the Jesuits' seismological results, it did nothing about terrestrial magnetism. In the worst case, Poisson would receive some financing for his magnetical work.[82] Joannès Thomas, the new head of the meteorological service, arrived in 1931. He had, Poisson observed, "recently graduated, was still last year a science teacher *(agrégé)* at the lycée in Réunion, and . . . he had spent a study leave at Ambohidempona (January–February 1930) to familiarize himself with cyclones." As a condition of his appointment, Thomas had insisted that Poisson be made technical head of Madagascan meteorology.[83]

Thomas's meteorological service began with five European staff: Thomas was licencié ès science in chemistry; one of his assistants was a bachelier, and three had diplomas from commercial school. None had formal training in meteorology. The crew of neophytes received Fr 69,000 for matériel. By 1932 Thomas was pulling in Fr 529,000 for his meteorological network, Fr 100,000 of this going to European staff.[84] The total budget rose to Fr 612,500 in 1935, all of it provided by the colonial government. In the budgets of "scientific" enterprises on the island, only the antimalarial campaign and the geographical service sur-

80. Ibid., Governor General Olivier to Poisson, 2 Dec. 1929 (copy).

81. Ibid., Colonial Minister François Piétri to Governor General Olivier, 26 July 1930.

82. Ibid., Poisson to Provincial Superior, 14 Sept. 1930.

83. Ibid., 14 Feb. 1931; PIPG, Poisson to Maurain, 1 Mar. 1931, reporting the appointment of Thomas in January. Thomas's first name appears in *APGFOM*, 1937, p. 128.

84. Aix, Madagascar 5 (13) D5, "Rapport sur la situation du Service météorologique de Madagascar," 1931.

passed meteorology.[85] The ambitious program came to very little. Thomas attended an African meteorological congress in Lusaka in 1936, where he invited the delegates to meet at Tananarive in 1939, but he worried that the gathering would result in embarrassment unless his service were upgraded. "Unfortunately we strongly believe that nothing will be done before then and that this conference at Tananarive will put us at a disadvantage in the eyes of foreign visitors."[86] The secular meteorologist nonetheless served his colonial masters well. He rose to become ingénieur-en-chef third-class in 1937.[87]

As in Algeria twenty years previously, in Madagascar meteorology fell into the hands of an autodidact secondary-school teacher. The experienced Jesuit geophysicists were glad to receive only a modest grant from the colony. The Jesuit observatory and the new governor general, Léon Cayla, concluded a formal entente in 1932. The observatory received what Poisson had asked for several years earlier. It provided local weather reports and carried out research with government equipment. It determined the official time of day. It produced a monthly earthquake bulletin, published and circulated at government expense. It was to register the components of terrestrial magnetism and report them to Charles Maurain in Paris. As a temporary measure, it also provided weather prediction for air navigation. In exchange for these services, the colony paid Poisson sixty thousand francs yearly.[88] The marriage was made in heaven.

At just this time heaven's vault beaconed to metropolitan savants. Late in 1932 Jules Baillaud, the head of the sky-chart project at the Paris Observatory, lectured on the problems facing French astronomy and French observatories. He emphasized that "in France we have nothing more than obsolete tools that can absolutely not allow us to cultivate fruitfully the fields where astronomers of other nations are finding abundant harvests." French observatories were fewer than those of other nations, and what was worse, "their instruments are practically without value for the work before us." Astronomers in other countries were photographing the Milky Way, measuring stellar parallax, and observing stellar spectra. No French observatory had equipment for such studies. As a result of such backwardness, "all the most fecund fields are

85. "La science à Madagascar." It is well to emphasize that the Académie malgache limited its modest enterprises to local natural history and ethnography, even after 1912, when it admitted members from the exact sciences. Its total budget in 1935 was Fr 176,800.

86. Aix, Madagascar 5 (13) D7, "Rapport annuel sur le Service météorologique de Madagascar pendant l'année 1936," 15 Mar. 1937.

87. *APGFOM*, 1937, p. 128.

88. Toul., FMa 956, "Convention," 10 May 1932.

forbidden to us." France needed new telescopes and new observatories, especially one in the Southern Hemisphere. The Americans, the British, the Dutch, the Argentines (with American and German help), and the Russians (in New Zealand) were either operating or preparing to set up major telescopes. France had to do the same on Madagascar, where it could rely on the cooperation of the Jesuits at Ambohidempona.[89]

Enter Louis Cattala, who had spent several years as Poisson's assistant and was then studying at Lille for a licence ès sciences. The director of the Toulouse Observatory, Emile Paloque, was a friend of Cattala's brother-in-law. Paloque alerted Cattala to Baillaud's plans for Madagascar. Baillaud then elaborated on his ideas to Cattala. Baillaud emphasized that any Madagascan installation would be set up in concert with the Jesuits, of whose cooperation he had been assured by the director of the Jesuit observatory at Zikawei, Pierre Lejay. He proposed a modest operation:

> The basis of my ideas is that we cannot create in the Southern Hemisphere a complete French astrophysical observatory, not, that is, a self-sufficient one with the complement of astronomers and assistants that would be necessary. We lack personnel in our observatories in France. We will not find enough of them to export. We will not be given, furthermore, the money for creating new, attractive, and durable positions for the French of France. But it does not seem necessary to me for a useful undertaking to have a staff of *astronomers* assigned to the station. All interesting observations are done through photography and are in reality routine work. Once the apparatus is set up and the observational program fixed, observers repeat the same actions for one or two years. The essential point for pursuing my proposal is to know if there would be a staff of observers on Madagascar inclined to be trained and conscientious enough to carry out the work.

Baillaud continued that he wanted to place the observatory staff under the watchful eyes of the Jesuits. He had spoken about atmospheric transparency with Delpeut (whom he had known during the war) before Delpeut left for Madagascar. Delpeut thought that an equatorial of 15–20 cm might be used for photographical work.[90] Cattala replied that from the astronomical side, the Ambohidempona Observatory had been used only for latitude and longitude observations and for determination of local time. Madagascan climate, he added, was not propitious for star-gazing.[91]

The possibility of an astronomical observatory on Madagascar con-

89. Ibid., extracts from a lecture by Baillaud before the Society of Popular Astronomy of Toulouse, 11 Dec. 1932, copied by Louis Cattala.

90. Ibid., Baillaud to Cattala, 15 Oct. 1933 (copy).

91. Ibid., Cattala to Baillaud, 26 Oct. 1933.

tinued to be discussed for several years.[92] In 1938 the astrophysicists considered setting up a branch operation there. Poisson wrote to Cattala, who was spending the summer at the Brussels Observatory, about using an old apparatus for photographing the sun. Cattala spoke to Fernand Moreau at Brussels, and Moreau contacted solar physicist Lucien d'Azambuja at the Meudon astrophysical observatory. Cattala traveled to Meudon and discussed the matter. D'Azambuja sought to enlist the Madagascan Jesuits for the international program in solar observations, and he promised them that Meudon would be able to provide the necessary equipment if only they agreed to devote two hours a day to the sun. Poisson balked at the conditions. "We cannot begin any new commitment before your return," he wrote to Cattala. "And even then we would not be able to apply ourselves to observations requiring two hours per day if a definitive and firm decision from the *supérieurs majeurs* did not come through to assure us against any reduction of Jesuit personnel at Ambohidempona." If either Poisson or his assistant Jean-Baptiste Coze were removed from the observatory as a consequence of Cattala's appointment there, "we would become completely incapable of carrying out anything regular in astronomy." He continued that the greatest danger to the observatory did not come from external threats, even though the colonial government could always reduce its financial contribution; rather, everything depended on having sufficient Jesuit staff.[93]

It was not to be. While Cattala studied science and theology in France, Poisson struggled with temporary assistants; the colony's promised meteorological assistant never arrived. Delpeut slowly recovered his wits at the local seminary, beyond reach of Poisson.[94] Coze, who had studied engineering for several years, assisted Poisson from 1930 to 1933. For the next two years Poisson received Henri Gruyelle, a Jesuit from Champagne who had obtained a licence ès sciences before entering the Society of Jesus. Gruyelle returned to Europe for theological studies and died prematurely. In his place came André Gauvin, a Jesuit brother and former student of Jean Cabannes's at Montpellier; Gauvin was Poisson's single European assistant from 1935 to 1938.[95] Poisson

92. Thomas, "Choix dans l'océan Indien d'une région favorable aux observations astronomiques," *APGFOM*, 1937, pp. 62–63.

93. Toul., FMa 1001, Cattala to Father Confessor, 17 Dec. 1938.

94. Ibid., FMa 956, Cattala to Baillaud, 26 Oct. 1933 (copy), where Delpeut is indicated as having left the observatory to teach science at a seminary in Tananarive, and Poisson to Cattala, 17 Feb. 1937 (copy), where Delpeut is said to have left Madagascar for Ksara in Lebanon. Delpeut died as a Jesuit in 1977 (Jalabert, *Jésuites au Proche-Orient*, pp. 328–29).

95. Maurice Burgaud, "Le père Jean-Baptiste Coze," *Maduré-Madagascar*, no. 158, Jan.

begged his superior in Toulouse for a permanent replacement. "No pure mathematician," he cautioned early in 1937. "There is no place here for him. We do geophysics and thus remain closely in contact with reality. Astronomy itself is entirely secondary. We don't do it, and there is no interest in doing it with the equipment at our disposal."[96] He looked forward to the day when Louis Cattala would arrive to prepare for taking over direction of the observatory; in 1938 he received Coze, who had completed his theological preparation in Europe. What God had provided Poisson accepted with philosophical resignation. Two Jesuits could keep the observatory running.[97]

In 1937, at a conference devoted to overseas science, Poisson underscored the principle that "pure scientific research, undertaken without thought of immediate, practical applications," regularly leads to important technical progress. He hoped that the colony would be generous in its support of the geophysical work at Ambohidempona; with a larger staff, his observatory would join the company of Zikawei and Ksara.[98] The Jesuits knew better than to rely on inconstant support from temporal lords. Their observatories ran on manpower estimates projected more than a dozen years into the future. In 1938 Louis Cattala, preparing at Enghien for his assignment to Tananarive, recalled to his father confessor an observation that he had made two years previously, namely, that around 1940 thought would have to be given to recruiting a young Jesuit for Tananarive. The man tapped for Madagascan geophysics would arrive at the observatory "a dozen years later, at the time when Father Poisson would be seventy to seventy-five years old." Cattala hoped that the Jesuit Province of Champagne would give him a young scholastic to replace Gruyelle.[99]

What the Jesuits could only imagine, the colonial bureau of mines achieved. André Savornin, at the head of the bureau, and Henri Besaire,

---

1976, pp. 30–31. See Toul., FMa 956, Cattala to Provincial Superior, 30 Dec. 1936, for scientific studies by Coze and Gruyelle. See Adrien Boudou, *Madagascar: La mission de Tananarive* (Antanimena-Tananarive, 1941), p. 240 for Gruyelle and Gauvin. See also PIPG, Gauvin to Maurain, 29 Jan. 1938, where Gauvin writes as a teacher at the Tananarive lycée, on leave in Angoulême, France. Gauvin, then working for diplomas or a doctorate on the subject of solar radiation, indicates that he studied for the agrégation under Cabannes.

96. Toul., FMa 956, Poisson to Provincial Superior, 15 Jan. 1936; Poisson to Cattala, 17 Feb. 1937 (copy), for the quotation.

97. Burgaud, "Coze"; Toul., FMa 1001, Cattala to Provincial Superior, 17 Dec. 1938, citing a letter of Poisson's.

98. Charles Poisson, *L'Observatoire dans le plan de développement économique et social de Madagascar* [Tananarive, 1937], two-page text in Toul., FMa 956.

99. Toul., FMa 1001, Cattala to Father Confessor, 13 Feb. 1938.

Henri Gruyelle taking magnetical measurements on Madagascar. Magnetical surveying served to indicate mineralogical deposits (From Charles Poisson, *Un cinquantenaire: L'Observatoire d'Ambohidempona à Tananarive, Madagascar, 1889–1939* [Paris, (1939)]

from the middle 1920s chief of the geological laboratory in the service, produced detailed magnetical maps of the island. Through the 1930s they searched for magnetical anomalies that might indicate deposits of useful minerals. The focus on geological prospecting, while not unusual in the French colonies, assumed special significance on Madagascar. Grandidier had stimulated much interest in mineralogy there. In his wake came Alfred Lacroix, professor at the Natural History Museum in Paris and, in 1914, perpetual secretary of the Academy of Sciences. Although Lacroix saw the island only once on a short visit, his scientific reputation rested on analyses of Madagascan minerals; his tireless activity depended on local suppliers, whose collaboration was assured by the anticipation of academic favors.[100] Charles Poisson, indeed, found it convenient to work closely with the government's magnetical surveyors.[101]

100. *GFOM*, pp. 145–48.

101. Poisson and André Savornin, "Les anomalies magnétiques au sommet du Rantoandro," *CRAS* 202 (1937), 806–9. For Poisson's interest in prospecting, see Poisson, "Le magnétisme terrestre à Madagascar," *APGFOM* 13 (1935), 40–42.

To the republic that had meted out its support for Madagascan geophysics and astronomy in grams, Poisson had the honor of delivering a eulogy. In November 1940 he urged his followers in Madagascar to remain loyal to the Vichy government of Henri-Philippe Pétain:

> The chief, do not look for him. We have him. The highest authority of France, the National Assembly, has legally invested him in power, expressly giving him the right to modify, as he sees fit, the laws and the rules that coordinated, before the catastrophe, the rights and the civil duties of all our fellow citizens.
>
> Is it necessary to remind you of the extent to which this chief deserves our confidence and our complete devotion?

Poisson described Pétain's courage and intelligence during the First World War. God had sent him back to rebuild France, which had been defiled by perversion and corruption. Referring to Pétain's policy of "national revolution" for Vichy France, Poisson continued:

> It is sometimes said that the restoration of France is a peaceful and legal revolution. You have just seen that nothing less is required than a profound change in our minds and in our hearts to guarantee, among us, indispensable union, fealty to the chief, and obedience for the rights of God. It is through faith in God, through confidence in the chief, and through loyalty to the country, that we will carry the victory.[102]

When Poisson delivered this charge, French forces on Madagascar, like those in Lebanon, pledged allegiance to the government at Vichy. The Jesuit scientists, always resourceful, simply sought to remain on good terms with the temporal lords of the hour. Each regime in the overseas dependencies had a distinct character, but in practice the Jesuits collaborated with two colonial masters: the local military and civilian authorities, on the one hand, and various ministries in Paris, on the other hand. Operating in concert, the masters sheared the Jesuit observatories and university of scientific independence. The Jesuit operations revived in a new, postwar order, only to decline with the middle twentieth-century crisis of faith.

102. Poisson, *Allocution du R. P. Poisson à la cathédrale d'Andohalo le 11 novembre 1940* [Tananarive, 1940], in Toul., FMa 991.

IV The New World

In 1830 the new Latin American republics looked to France for leadership in scientific and intellectual matters. The authoritarian, scientistic spirit of Claude-Henri de Rouvroy, comte de Saint-Simon and Auguste Comte, in which engineers would reconstruct society, found an especially favorable reception. French administrators capitalized on this interest by sending savants to organize Latin American observatories and schools. The army and the navy also sent scientific missions to the New World. During the years between the twilight of the Second Empire and the First World War, France played a major role in astronomy and physics in Chile, Argentina, Brazil, Mexico, and Ecuador. Cultural imperialism was the motor of this costly involvement. France hoped to attract attention away from the political and commercial designs of its imperialist rivals, notably Germany and the United States.

As positivism declined, so did French influence generally. In Paris, attention shifted away from grandiose international projects in the exact sciences to settle on French overseas environments. Two institutions in the New World received special consideration: a geophysical observatory on Martinique and a scientific institute in Montreal. At other locations—in Argentina and in Mexico—interest in French norms persisted until 1940.

The institutions, projects, and missions discussed in part 4 all engaged in pure science to promote the prestige of France. Even the initiatives with a practical component—the Mexican scientific commission, the Ouro Preto mining school, the Martinique observatory—emphasized abstract knowledge. The example of French science was designed to stimulate French trade and arouse sympathy for France in international controversies, such as the First World War. The ambassadors of French science, however, were less concerned with discovery or invention than with representing French culture. Some men assimilated into their new environment, but most riveted their attention on Paris. Their comportment was not essentially different from that of colleagues in French overseas dependencies.

A number of physicists and astronomers who were active in the New World also served France in other settings. Whether he lectured in Montreal or supervised examinations in Beirut, Georges Déjardin incarnated French science; François Gonnessiat had comparable responsibilities as director of observatories at Quito and Algiers; Milan Ratislav Štefánik viewed Ecuadorans much as he viewed Tahitians. This reservoir of international expertise contributed to placing France at the head table in the salon of international protocol. Practical advantages also accrued to scientific institutions in the French empire. By the 1930s scientists in North Africa and Indochina had benefited from direct experience with research in diverse settings.

French activity in physics and astronomy in the New World projected French colonial ambitions. This was cultural imperialism French style, where Paris judged overseas researchers primarily as administrators and functionaries. New results mattered less than representation of France's political, and hence cultural, interests. In Chile and Argentina, French cultural imperialists stand out self-consciously against a background of German and American competitors, who generated a more impressive corpus of knowledge. The thin record of French research in Mexico and Brazil is especially instructive in this regard. This modus operandi remained essentially unchanged between 1830 and 1940. French impact attenuated because client scientific communities turned to imperialist rivals imbued with the research ethic that overseas French scientists suppressed.

8

Imperialist Rivalries in the Southern Cone

Inventories in Chile

French and German men of learning dominated the exact sciences during the four generations following Chilean independence from Spain. Administrators in Santiago assembled a broad spectrum of foreigners in an attempt to accelerate their integration in Chilean society. Chile assimilated the immigrants more easily than their learned discourse. The nation produced no center of scientific excellence.

In the first decades of nationhood, Chilean authorities sought to inventory their country, and a topographical map ranked first on the list of priorities. French geographers, following in the wake of revolutionary armies, had set the standard for cartography, and so Chilean eyes turned to Paris. Charles Ambroise Lozier, a thirty-eight-year-old French geographer, was engaged in 1823 as a teacher for the city of Concepción (Lozier had been trained in the pedagogical methods of Joseph Lancaster) and charged to produce a map of Chile. He rose to direct the Instituto nacional, Chile's premier educational institution, in 1826, but the appointment proved unhappy. Students revolted against his pedagogical regime, and the government sacked him. He abandoned his geographical work, labored as an engineer, and ended as an anchorite. The Chileans tried again to recruit a French savant. In 1828 they hired Claude Gay as teacher of physics and natural history in the French-staffed *colegio* founded at Santiago by Pierre Chapuis. In 1830 Gay received a contract to study the natural history of Chile. He spent the next nine years collecting information, toured Peru, and in 1842 returned to Paris, where he died in 1873. His labors resulted in a general survey of the country, published in 1855.[1]

1. Ernesto Greve, "Don Amado Pissis y sus trabajos geográficos y geológicos en Chile," *Revista chilena de historia y geografía*, no. 107, Jan.–June 1946, pp. 71–170, on pp. 72–79. Greve's sensitive study provides an excellent introduction to the history of exact

The last of the heroic French geographers, Aimé Pissis, arrived in 1848. Born in the Haute Loire in 1812, he studied at the Ecole polytechnique and the Ecole des mines. In 1837 he took part in a geological expedition to Brazil; he returned to France five years later. A second South American voyage began in 1847. This time he measured the heights of various mountains in Bolivia's interior. He was planning to return to France in 1848 when he received a call from Chile to direct a geological and mineralogical survey of the country. He produced detailed maps and geological studies, in the process garnering membership in the French Legion of Honor. In his case the honor seems superfluous, for he was no agent of French political designs. He attracted the best European assistants he could find, both French and German.[2]

As if to balance Gallic geology, Chilean authorities took advantage of the revolutions of 1848 to call in a German astronomer. External events propelled them to do so. In 1849 the U.S. navy sent an expedition to Chile to observe the retrograde motion of the planets Mars and Venus and, in conjunction with Northern Hemisphere measurements, thereby obtain a precise measure of the solar parallax. The naval officer in charge, James Melville Gilliss, had previously been delegated the task of planning the Naval Observatory in Washington. Gilliss based his Chilean research program on the calculations of Marburg astronomer and polymath Christian Ludwig Gerling. The Americans spent three years at Santiago de Chile. Their activity attracted the attention of Chilean president Manuel Montt and Montt's educational adviser, Ignacio Domeyko. The Chilean government offered to purchase the expedition's equipment for eight thousand pesos and use it to establish a permanent observatory. Beginning in 1850, three Chilean assistants worked with Gilliss; one of them, Gabriel Izquierdo, was detached from Pissis's geographical survey.[3]

---

sciences in nineteenth-century Chile. On Lozier's brief experience with higher education in Chile, see Amanda Labarca H., *Historia de la enseñanza en Chile* (Santiago, 1939), pp. 81–82.

2. Greve, "Pissis," pp. 82–83, 107–8. Among Pissis's assistants were Friedrich Wilhelm Döll, a surveyor educated at the polytechnical school of Cassel (p. 137) and Martin Drouilly Laurent, a French engineer (pp. 142–43).

3. Friedrich W. Ristenpart, 'Astrónomos alemanes en Chile,“ pp. 177–93 in *Los Alemanos en Chile*, the first volume of the festschrift of the German Scientific Society of Santiago, published for Chile's independence in 1910; R. Grandón M., "El Observatorio astronómico nacional: Renseña de su desarrollo y actividades en sus primeros cien años de existencia,“ *Anuario del Observatorio astronómico nacional de la Universidad de Chile*, 1952, pp. 7–35; Philip C. Keenan, Sonia Pinto, and Héctor Alvarez, *The Chilean National Astronomical Observatory (1852–1965)* (Santiago, 1985). Pages 17–52 of the Keenan-Pinto-Alvarez volume, which contains identical texts in English and in Spanish, concern the period

President Montt and Domeyko then turned to their principal scientific confidants, the brothers Philippi. Bernhard Eunom Philippi was a German naturalist-turned-entrepreneur who, through the middle of the nineteenth century, actively supported German colonization in southern Chile. Rudolf Amandus Philippi directed the polytechnical school of Cassel; when political changes forced him out in 1850, he went to Chile, the promised land of his younger brother.[4] Seeing the advantage of signing on a student of Gerling's to continue Gilliss's work, Bernhard Philippi enthusiastically recommended one to his friend Salvador Sanfuentes Torres, the administrator of the southern province of Valdiva who had since gone on to become secretary general of the University of Santiago.[5] In 1850 twenty-five-year-old Karl Wilhelm Moesta arrived as Gilliss's heir apparent. Until Gilliss left, Moesta received an appointment, ironically, as assistant to Pissis.

Moesta assumed direction of the observatory with Gabriel Izquierdo as an assistant, although Izquierdo soon left for the Chilean corps of engineers. Moesta asked the Chilean consul general in Hamburg, who was then acting as an agent for colonization, to find a German coworker. An offer went out to another of Gerling's students, who declined in favor of a German post. In 1855 the consul succeeded in hiring Hermann Volckmann, at a salary of a bit more than half that given to Moesta. Volckmann was such a prize that Pissis offered him a raise and captured him for the geographical survey. Moesta again looked to Germans. In 1859 he received, on the recommendation of Gotha Observatory director Peter Andreas Hansen, Richard Schumacher, the son of the director of the Altona Observatory who had been Hansen's teacher and patron. Once more Pissis wooed the new astronomer to the staff of the geographical survey. For a time around 1859 Moesta employed Volckmann's brother Friedrich in an honorary capacity; Friedrich

---

before 1923; the Gilliss expedition is considered in all three treatments. For the internal, political dimension, see Wayne D. Rassmussen, "The United States Astronomical Expedition to Chile, 1849–1852," *Hispanic American Historical Review* 34 (1954), 103–13; and John P. Harrison, "Science and Politics: Origins and Objectives of Mid-Nineteenth Century Government Expeditions to Latin America," ibid. 35 (1955), 175–202.

4. George F. W. Young, *Germans in Chile: Immigration and Colonization, 1849–1914* (New York, 1974); Jean-Pierre Blancpain, *Les Allemands au Chili (1816–1945)* (Cologne, 1974) [Lateinamerikanische Forschungen: Beihefte zum Jahrbuch für Geschichte von Staat, Wirtschaft und Gesellschaft Lateinamerikas, 6]; a summary account where sciences make no proper appearance is available in Gerardo Jorge Ojeda-Ebert, *Deutsche Einwanderung und Herausbildung der chilenischen Nation (1846–1920)* (Munich, 1984) [Beiträge zur Soziologie und Sozialkunde Lateinamerikas, 30].

5. Greve, "Pissis," p. 94.

Volckmann went on to become an associate professor at the University of Marburg.[6]

Notwithstanding the perennial problem of finding trained assistants, Moesta pursued a vigorous and innovative program in observational astronomy. He relocated the observatory to a more suitable site and oversaw construction of new buildings. He mounted expeditions to observe solar eclipses. He observed the declination of Mars at its 1862 opposition, in cooperation with his predecessor Gilliss, who had risen to direct the Naval Observatory in Washington; the result provided a new value for the solar parallax and hence for the distance of the earth from the sun.[7] In the middle 1860s Moesta grew weary. His teaching responsibilities at the university in calculus and astronomy (with some twenty students) sapped his energy. In 1865 he applied for leave to attend an astronomical congress in Leipzig. He never returned. At issue may have been the terms of his contract, as well as the destruction by shipwreck of a new 24-cm refractor that he had just ordered. His departure was amicable, for beginning in 1869 he acted as Chilean consul general in Dresden. His second Chilean star catalogue appeared in 1875.[8]

Moesta's successor was José Ignacio Vergara, an assistant whom he had trained. Vergara's ambitions, however, lay in the domain of politics, and he rose to occupy important ministerial portfolios. Vergara presided over the steady deterioration of Moesta's old operation. Finally, in 1883, he asked Moesta to find two astronomers for Santiago. Moesta complied. Adolf Marcuse and Wilhelm Wickmann arrived as the first and second astronomers. Marcuse had obtained a doctorate at Berlin under Adolf Hirsch; he was a veteran of the 1882 German expedition to observe the transits of Venus, and he had spent some time observing at Pulkovo in Russia. Wickmann seems never to have published a scientific paper. The two Germans did not get along well together.[9] Marcuse provided information about the observatory's poor condition, which appeared in a book criticizing Vergara's management of the institution; Vergara promptly fired him. Marcuse returned to Germany,

6. Ibid., pp. 92–95ff., 139, 143–44; see also *DSB*, s.v. "Hansen" and "Schumacher," and *Pogg.*, s.v. "Volckmann."

7. See Keenan, Pinto, and Alvarez, *Observatory*, pp. 19–21, for the international project involving nine observatories.

8. C. G. Moesta, *Observaciones astronómicas hechas en el Observatorio nacional de Santiago de Chile en los años de 1856 hasta 1860* (Santiago, 1875). This and a companion volume by Moesta, *Observaciones astronómicas hechas en el Observatorio nacional de Santiago de Chile en los años de 1853, 1854 i 1855* (Santiago, 1859), contain historical introductions; the books may be found in the United States Naval Observatory Library. A Dresden imprint of the second volume is indicated in Chile, Ministerio de educación pública, *Exposición retrospectiva de la enseñanza: Obras de los profesores, catálogo* (Santiago, 1941), p. 143.

9. Ristenpart, "Astrónomos alemanes."

where in 1907 he finally received an associate professorship at the University of Berlin. Wickmann resigned soon thereafter to direct the Quito Observatory from 1887 to 1895. The only German to remain at the observatory, from 1883 until his death in 1902, was the mechanic Luis Grosch, who devoted himself first to meteorology and then to seismology.[10]

While the Germans handled academic astronomy, a French presence in exact sciences had been assured by the navy. In 1856 the Chilean ambassador in Paris was instructed to hire a director and an assistant director for the naval school at Valparaiso. Forty-seven-year-old capitaine de frégate Jules-Jean Feillet, having supervised practical astronomy courses at the Brest naval school for the preceding five years, accepted the call. He signed on for five years, retiring from the French navy but keeping his rank in the Chilean navy.[11] As Feillet's assistant director came Ensign Sylvain Louis Simonnot, whose posting to Réunion had been replaced by the Valparaiso assignment. Twenty-three-year-old Simonnot died from lung inflammation scarcely six months after arriving in Chile.[12]

In 1882, just at the time that Vergara had commissioned Moesta to hire assistants, the French navy intervened again in Chilean astronomy.[13] The occasion was a major expedition (one of four such under naval command) sent out by the Academy of Sciences to observe the transits of Venus. At its head was naval lieutenant Octave Marie Gabriel Joachim de Bernardières, a veteran of Maximilian's Mexican wars who was accomplished in terrestrial magnetism and geodesy. His second was Lt. Léon Barnaud, exactly his age (thirty-seven) but with little scientific experience.[14] They set up two telescopes, an 8-inch equatorial and a 6-inch equatorial, at Cerro Negro, a short distance south of Santiago, and established their geographical position telegraphically by exchanging observations with stations at Valparaiso, Buenos Aires, Lima, and Panama. De Bernardières took his measurements over the course of a year. His progress was sped by Vergara, who in November 1881 had stepped down as minister of public instruction to devote all his attention to the

10. Keenan, Pinto, and Alvarez, *Observatory*, pp. 27–28; Ristenpart, "Astrónomos alemanes," pp. 11–12.

11. VM, s.v. "Feillet," José Francisco Gana to Admiral Blanco Encalada (Ambassador of Chile in France), 15 Oct. 1856, French translation; note of 28 Jan. 1857; Chilean Embassy to Ministry of Navy and Colonies, 2 Mar. 1857.

12. VM, s.v. "Simonnot," Chilean Embassy to Ministry of Navy and Colonies, 19 Mar. 1857; Commandant General of the Chilean Navy to Minister of the Navy, 18 Nov. 1857.

13. See Luis L. Zegers, *Transito de Venus por el sol* (Santiago, 1883), pp. 252–68, for Vergara's commission to Moesta and the poor condition of the observatory in 1882.

14. VM, s.v. "de Bernardières" and "Barnaud"; ibid., BB4 1342, "Dossier du passage de Vénus sur le soleil en 1882."

Santiago Observatory (whose directorship Vergara had kept throughout his years in politics). Vergara treated the French as he would treat all comers. He installed a Belgian team, for example, near the grounds of his observatory.[15]

The French navy provided additional direction to Vergara's search for men to replace the unhappy Germans. Late in 1886 Admiral Mouchez, on the lookout for observatories that might be persuaded to participate in his sky-chart survey, invited Vergara to Paris for an international conference to discuss the matter. Vergara replied that he was too busy to attend, especially as he was trying to issue the observatory's observations, "unpublished since 1865." He nevertheless expressed interest in the project.[16] Through novelist Alberto Blest Gana, the Chilean ambassador in Paris who had helped arrange the Venus-passage expeditions of 1882, Vergara asked Mouchez to provide Chile with the same Gautier refractor that had been sent to La Plata in Argentina and to find observers for the project.[17] Sought were an *adjudant-astronome* at the equivalent of Fr 11,250 yearly; a second adjudant at Fr 9,375; and two third-class adjudants at Fr 7,500.[18]

Mouchez quickly found a man for one of the third-class positions, Alphonse Boinot, and Blest Gana agreed to sign him on as soon as Mouchez had trained him; Boinot, an assistant at the Paris Observatory, promptly went to work with the brothers Henry in constructing the sky-chart telescopes. For the senior post, Mouchez proposed Edouard Jean Pierre Marie Sylvain Perrin, a thirty-five-year-old naval lieutenant who, after traveling to Argentina and Chile in 1882 on one of France's Venus-transit expeditions, had gone to the Montsouris Observatory in Paris to polish his astronomical and meteorological techniques.[19] But Perrin (who rose to become vice admiral in 1913) would not be moved,

15. See Zegers, *Transito*, pp. 43–163 on the French expedition, pp. 167–81 on the Belgian one, p. 252 on Vergara's forsaking politics for astronomy.

16. PO, "Santiago, 1888–1889," Vergara to Mouchez, 9 Feb. 1887.

17. Ibid., Blest Gana to Mouchez, 2 May 1887 and 23 May 1887. See Zegers, *Transito*, pp. 231–38, for Moesta's correspondence with Blest Gana concerning the Venus transit of 1882.

18. PO, "Santiago, 1888–1889," Blest Gana to Mouchez, 20 July 1887.

19. VM, s.v. "Perrin"; PO, "Santiago, 1888–1889," Blest Gana to Mouchez, 31 July 1887: "I will wait the result of the inquiries that you have kindly undertaken with regard to Mr. Perrin. If this man requires a salary of more than Fr. 11,250 to fill the post of first adjudant at the Santiago Observatory, I would have to refer the matter to my government, because I am not authorized to offer more. I would beg you also to kindly note on this subject that it is not the post of director of the observatory, but that of adjudant of this establishment, that is offered with the salary mentioned above." On Boinot, see *Astron.*, s.v. "Boinot."

and Mouchez quickly tapped Jean Albert Olbrecht, a twenty-nine-year-old astronome-adjoint at the Paris Observatory, who in 1884 had completed a doctorate at the Sorbonne on eclipses of Jupiter's satellites.[20] Filling the other adjudant positions required a bit of negotiation. Mouchez first asked Emmanuel Esmiol, a former aide-astronome of his at Paris who recently had been transferred to Marseilles; Esmiol declined, citing family responsibilities.[21] For the position of junior assistants Mouchez finally settled on Irénée Lagarde, a twenty-year-old assistant of Guillaume Bigourdan's at Paris, and Xavier Devaux, on leave from a post at the Algiers Observatory.[22] Florencio Maturana, an assistant at the Santiago Observatory, went to Paris to join Boinot at the Henrys' shop.[23]

The three French astronomers were happy with their situation in Chile. After a year at Santiago, Devaux expressed his satisfaction to Mouchez. Things were going well, "the country being pleasant and our director rather accommodating." The only difficulty was to "fight against an unbelievable inertia" that afflicted everything in Chile. Their principal work concerned longitude measurements for a new map of the country.[24] Early in 1889 Mouchez indicated that the Chilean equatorial was proceeding on schedule.[25] Then Vergara died. Mouchez wrote to Ambassador Blest Gana to ask for Chile's intentions. Obrecht had been designated as Vergara's successor, but Mouchez avowed that he did not know whether Obrecht would continue Vergara's commitments. The equatorial had been completed and could be sent to Santiago; Boinot, engaged verbally by Santiago, continued with the Henry brothers; Maturana had been delegated to represent Chile at an upcoming conference devoted to the sky chart.[26] Chile offered itself to France as an astronomical satrapy.

20. PO, "Santiago, 1888–1889," Blest Gana to Mouchez, 4 Aug. 1887; *Astron.*, s.v. "Obrecht."

21. PO, "Santiago, 1888–1889," Edouard Stephan to Mouchez, 17 July 1887; *Astron.*, s.v. "Esmiol."

22. *Astron.*, s.v. "Lagarde" and "Devaux." Mr. Philippe Véron informs me that Devaux resigned from the army to accept the Santiago post. To keep his seniority in the French civil service, however, he was named aide-astronome at Algiers, a position he held until his appointment as *élève-ingénieur des télégraphes* in 1891. He never observed in Algiers.

23. *Astron.*, s.v. "Maturana."

24. PO, "Santiago, 1888–1889," Devaux to Auguste Fraissinet (Secretary of the Paris Observatory), 20 Oct. 1888. For an early result of the surveying, see Obrecht, Devaux, and Lagarde, *Coordenadas jeográficas de algunas ciudades de Chile* (Santiago, 1890), reprinted from the *Anales* of the observatory.

25. PO, "Santiago, 1888–1889," Mouchez to Vergara, 4 Feb. 1889.

26. Ibid., Mouchez to Blest Gana, 30 July 1889.

American Environment, European Visions

Obrecht took charge of the Santiago Observatory as Chile careened toward civil war. Devaux returned to the Paris Observatory in 1890; Lagarde set off for the north of Chile to undertake cartographical work and returned to a Parisian career four years later.[27] Boinot and Maturana continued on with the Henrys, Maturana dying at his post there in 1893. Later in 1893 the Henrys' 33-cm telescope arrived in Santiago. Obrecht mounted it, but lack of instruments for measuring the plates prevented him from beginning the sky-chart survey.[28] He withdrew from the project of his native land. In place of positional astronomy, he focused attention on positional geography; he filled his spare hours with minor elaborations of celestial mechanics and pedagogical expositions of higher mathematics.[29] His finest initiative was to employ Alfred Krahnass as a meteorologist.

Krahnass, a native of Poitiers, studied at the Ecole centrale des arts et manufactures; in 1865, at age twenty-five, he relocated as a civil engineer in Chile. He worked in a number of technical capacities until, in 1891, Obrecht named him to replace Grosch as the observatory's meteorologist. When the Chilean legislature suppressed the meteorological service in 1892, Obrecht appointed Krahnass as the observatory's second astronomer and specified that his realm would be physical astronomy; it was a way of continuing practical meteorology under the cover of pure science. In 1897 Krahnass added responsibility for optics to his charge. He constructed instruments of all kinds and measured gravity at Santiago with a Kater pendulum. In 1909, the same year as his retirement, he received an appointment as head of the observatory's meteorological section.[30]

At just the time that Obrecht was thinking about the sky-chart work, in 1891, he helped organize a learned academy to radiate French influence – the Scientific Society of Chile. The society grew out of monthly meetings held by the Union of French Schools in Santiago. Initially, its officers came from French professors in Santiago (with Obrecht presiding for the first year), and the society's council drew on the talent of commercial and professional émigrés. The founders – and only French

27. *Astron.*, s.v. "Devaux" and "Lagarde"; Grandón M., "Observatorio," p. 16, has Devaux and Lagarde both resigning from the Santiago Observatory in April 1890.

28. Juan Taulis, "Fotografía celeste," in J. A. Obrecht, *Anuario del Observatorio astronómico* (Santiago, 1897), pp. 215–44, esp. 216–18 for not beginning the sky-chart work.

29. Grandón M., "Observatorio," pp. 18–19.

30. F. W. Ristenpart, "Prefacio," in Observatorio astronómico de Santiago de Chile, *Publicaciones*, no. 4, Meteorology Section, *Observaciones meteorológicas de 1909* (Santiago, 1910), pp. v – vii, for a description of Krahnass's career.

citizens could qualify for the designation of "founding member"—hoped that the society would bring French ideas to Chile by holding monthly meetings, establishing a library, and publishing learned proceedings and memoirs.[31] The task in Chile was great, Alphonse Noguès emphasized in his presidential address of 1892. Only a single real laboratory, that in chemistry run by a German professor, existed in the country; the physics collection at the university was "a beautiful museum containing numerous and beautiful apparatuses carefully kept in glass cabinets, tightly closed off and spotlessly clean." French academicians sent encouragement. Pierre Puiseux, an astronomer at the Paris Observatory, wrote to Obrecht in 1892 about the honor that would accrue from having Chileans "consider France as their natural instructor." Such "moral influence" was especially important, in Puiseux's view.[32]

Obrecht and his colleagues devoted years of their lives to the society, but their plans failed. Despite the fact that the proceedings appeared bilingually beginning in 1892, by 1896 membership had stalled at around 200, and attendance at the society's sessions never passed ten. Ambitious, not to mention pretentious, monographs appeared in the society's *Notes et mémoires*, but the level of discourse descended to the ridiculous. Obrecht published a text on celestial mechanics, and the associate director of the Santiago natural history museum, Fernand Lataste, launched an extended polemic arguing against Darwinian evolution, but these exercises in high science rubbed elbows with a long-running debate on solutions to the problem of the trisection of an angle.[33] The society lost two prominent supporters when Lataste returned to France in 1892 after the death of his wife and two of his children and when Noguès died in 1895.[34] We can only speculate whether Lataste, who had explored North Africa and Senegal, would have been able to create a new school of natural history had he stayed in Chile.[35]

31. Emilio Rodríguez Cerda, "Renseña histórica de la Société scientifique du Chili," Société scientifique du Chili, *Actes* 24 (1914–16), 86–89.

32. See Société scientifique du Chili, *Actes* 1 (1891), xx, for the society's French charter and character; 2 (1892), xxvi, for Noguès's remarks; and 2 (1892), xcviii, for extracts from Puiseux's letter. Noguès, an *ingénieur des mines*, was a professor of industrial physics and technology at the University of Santiago (M. Vega E., *Album de la colonie française au Chili* [Santiago, 1904], p. 90).

33. J. A. Obrecht, "Sur une nouvelle méthode de détermination des orbites des planètes et des comètes," Société scientifique du Chile, *Notes et mémoires* 3 (1893), 65–104, 171–205. In a number of articles from this period, Lataste challenged the "transformist" observations of the French naturalist Remy St-Loup. The trisection debate began in *Actes* 3 (1893), lxxxviii, xcvi – xcviii, cxviii – cxix.

34. Vega E., *Album*, pp. 82–83, 90.

35. See *Publications scientifiques de Fernand Lataste, candidat à la chaire de biologie de l'enseignement populaire supérieur de la Ville de Paris* (Bordeaux, 1889), where the title page

Certainly he and Obrecht were the strongest researchers in the scientific society, and where two men with complementary interests might have carried the association, one man with substantial administrative commitments could not do so alone. In any event, by 1896 funds no longer existed to continue the ambitious publications schedule and to set in order the growing collection of books and periodicals that had been arriving as gifts and as a result of corporate exchanges.

His energies diluted by involvement with the French scientific society and the mapping of Chile, Obrecht saw his observatory eclipsed by the initiative of an imperialist rival. In 1893 Americans from Lick Observatory mounted an expedition to Chile to observe a solar eclipse. They found observing conditions there so promising that in 1903 they returned to set up a major observatory on San Cristóbal, a mountain 250 meters above Santiago. The aim was to measure radial stellar velocities (and hence determine if there were a preferred frame of reference for the universe), a consuming interest of Lick astronomers. They set up a 92-cm parabolic mirror, a close copy of one that had been used to good effect on Mt. Hamilton, in a Cassegrain system. Financed by one of Lick's patrons, the "D. O. Mills Expedition" quickly began to take spectroscopical data.[36]

The leader of the Lick expedition, William Hammond Wright, found Chilean astronomy lacking. He saw Gilliss's old telescope, recently remounted at the naval school by a naval instructor, Weinroth, and used for time signals; when the government declined to finance Weinroth's work, he abandoned the telescope in favor of a Spanish sextant. Weinroth, according to Wright in charge of rescuing Chilean meteorology from "a hopeless condition," would make a suitable acquisition for the Californians.[37] As to Obrecht's observatory, "a few public improvements have been added since Gilliss's time, but the people are pretty much the same as then." Obrecht's principal astronomer, Ernesto Greve, provided Wright with much assistance; he was "a bright fellow,

---

provides a list of Lataste's exploits. He was funded in 1884 by the scientific mission to explore Tunisia (Eusèbe Vassel, "L'auteur de la découverte des phosphates africains et la mission d'exploration de la Tunisie," Institut de Carthage, *Revue tunisienne* 5 [1898], 401–22, on p. 414).

36. Erich P. Heilmaier K., "El Observatorio del San Cristóbal y su historia," Universidad católica de Chile, *Revista universitaria* 25 (1940), 5–11. When the Lick astronomers decided to end their work on San Cristóbal in 1927, Manuel Foster Recabarren purchased the observatory and donated it to the Catholic University of Chile (Donald E. Osterbrock, John R. Gustafson, and W. J. Shiloh Unruh, *Eye on the Sky: Lick Observatory's First Century* [Berkeley and Los Angeles, 1988], pp. 134–36).

37. LO, "Chile," Wright to William Wallace Campbell, 22 Apr. 1903.

young and full of energy. . . . I am sorry he is tied up here where of course he will never amount to anything."[38]

The French scientific society attempted to counter the rising profile of Germany in Chilean affairs. During the civil war of the early 1890s, a German officer who had been reorganizing the Chilean general staff played a dominant role. The officer, Emil Körner, recast the army in a German mold and over the next fifteen years sent dozens of Chileans to study the art of war in his native land.[39] In the 1880s, Chilean educator Valentín Letelier argued for a national normal school, along the lines of German pedagogical seminaries. He finally persuaded the ministry of education to hire six German professors as the core of his Pedagogical Institute. Their contracts were set initially for six years, and their annual salaries ranged from 1,000 to 2,800 pesos. Along with the senior professors came dozens of smaller fry for the secondary schools. Among the leading men over the period before the First World War were a number of mathematicians and physicists, such as Richard Poenisch and August Tafelmacher. They had no resources for research. In their pedagogical institutions, physics was a descriptive science in which mathematics had no part; in the early 1890s and through at least 1903 the Pedagogical Institute lacked a physics laboratory.[40]

Germans were by far the most numerous among foreigners in Chilean educational institutions.[41] In view of the rising importance of things Teutonic, and given Obrecht's mediocre performance as observatory director, it is not surprising that he lost his position when Pedro Montt became president of Chile in 1906. Like his father Manuel Montt, the new president was partial to German culture. He called a German astronomer to replace Obrecht (who nevertheless continued teaching at the local university) as director of the Santiago Observatory.

The new man, Friedrich Wilhelm Ristenpart, arrived in 1908. He had obtained a doctorate from Strasbourg in 1891 and for the next nine

38. Ibid., 26 Apr. 1904. Wright suggested to Campbell that he make use of Greve's services for the Spanish eclipse of 1905. Greve was "not only an efficient observer, but an exceedingly high-toned gentleman" (ibid., 3 Apr. 1905).

39. Frederick M. Nunn, *Yesterday's Soldiers: European Military Professionalism in South America, 1890–1940* (Lincoln, Nebr., 1983), pp. 102–6; Blancpain, *Allemands au Chili*, pp. 702–50.

40. Blancpain, *Allemands au Chili*, considers the invasion of the German pedagogues in detail on pp. 666–701. On pp. 772–73 he provides a list of the first thirty teachers hired in 1889–90 for the Pedagogical Institute and for the secondary schools; contracts are considered on p. 675, the situation of physics laboratories on p. 693.

41. The entire teaching corps of Chile, their nationality, assignment, and salary, are provided in Chile, Ministerio de instrucción pública, *Personal de instrucción pública: Instrucción superior, secundaria i especial* (Santiago, 1913). Listed are 49 Germans, 23 French, 21 English, 16 Italians, 12 Spaniards, and 10 Swiss, among other nationalities.

years had moved from one assistantship to another, finally landing a position with the Berlin Academy of Sciences in 1900. He advanced to become privatdocent at the University of Berlin in 1904. When he left Berlin, he had the foresight to obtain a two-year leave of absence from the Prussian Kultusministerium, his employer, although shortly after arriving in Santiago he felt optimistic about spending the rest of his life there. Ristenpart brought in two younger men to serve on his staff, Walther Zurhellen and Richard Prager. Zurhellen had a doctorate from Bonn, and Prager had one from Berlin; like Ristenpart, both Zurhellen and Prager were experienced and published astronomers.[42]

When Obrecht failed to begin work on the French sky chart, Maurice Loewy and his advisers at the Paris Observatory reassigned the Chilean celestial zone to Montevideo, where a spirited, self-taught astronomer had been busy making observations. The astronomer, Enrique Legrand, had been educated at the Ecoles chrétiennes de Passy and at the Ecole central des arts et manufactures in Paris. In 1900 he was thirty-nine years old and still without an affiliation at an institution of higher learning. Loewy imagined that Legrand would be able to organize an observatory. *Fin-de-siècle* Uruguay, however, had nothing approaching a stable government; insurrections and civil war extended through the first decade of the new century. Legrand's observatory failed to materialize.[43]

Legrand, like Obrecht before him, defaulted on his promise, and in 1910 Ristenpart received a portion of Obrecht's former mandate. He set Zurhellen to work on the sky chart while Prager supervised the calculations. Ristenpart had a broader vision. He wanted to relocate the observatory to a better site (electrical street illumination, smoke, and dust were destroying his optical window), obtain more modern instruments, and secure better salaries. He found the site and supervised the transfer of the observatory; he ordered the equipment, especially a 60-cm refractor from Grubb, Parsons and Company. And he continued to publish.

President Pedro Montt died in 1910, soon after arriving in Germany for medical treatment. Montt's successor was not merely ambivalent about astronomy: in 1912 he froze appropriations for the new observatory site and initiated an audit into alleged financial improprieties. Ristenpart found opposition from his large staff. His style was perceived as autocratic and inflexible. After being publicly humiliated by Ristenpart,

42. In addition to the general histories of astronomy in Chile, see Prager's obituary of Ristenpart in *Vierteljahrschrift der astronomischen Gesellschaft* 49 (1914), 14–22. Keenan, Pinto, and Alvarez, *Observatory*, is particularly good on Ristenpart's regime.

43. See Luis Morandi, *La labor científica de E. Legrand* (Montevideo, 1938), pp. 13–14, for Legrand's involvement with the *carte du ciel.*

Zurhellen resigned in 1912; he found a position at the Berlin-Babelsberg Observatory, and he died fighting in France four years later. After Ristenpart offered Zurhellen's position to a succession of young German astronomers, an assistant of Johannes Hartmann's at Göttingen, Franz Pingsdorff, accepted.[44] The furor surrounding Ristenpart's administration reached the press, and it led the government to cancel his contract and have him replaced by the lawyer and editor who had negotiated his hiring, Enrique Matta Vial. Word reached William Joseph Hussey, the American astronomer who directed the La Plata Observatory in Argentina. "There seems to be some defect of character," Hussey noted about Ristenpart, "that does not run true, and which prevents his making good in the best sense."[45] Ristenpart, who in only a few years had restored confidence in Chilean astronomy, was mortified. Several days after Matta Vial took charge of the observatory, Ristenpart committed suicide. He left a son by his first marriage and a wife and three daughters by his second. He did not live to see the birth of his third daughter.[46]

It took the government less than two months to call back Professor Obrecht as scientific director of the observatory. The curatorial administration went to an assistant-director confidant of the government's, Ismael Gajardo Reyes. It was a power-sharing arrangement that Ristenpart had explicitly rejected in his last months, but Obrecht did not mind returning to the observatory in a diminished capacity. The French notion of *cumul de fonctions*, according to which a scientist or scholar held positions at a number of institutions, had become entrenched in Chile. The position in astronomy carried a large salary, and with it Obrecht became one of the highest-paid academics in the country.[47] That he was

44. AIP, Karl Schwarzschild papers, Schwarzschild to his parents, 8 Oct. 1912, where he reports Ristenpart's request for an assistant. Schwarzschild tapped Kurt Felix Bottlinger, a student of Hugo von Seeliger's whom he had also advised, and Kasimir Graff at the Hamburg Observatory, but both declined in the end to leave Germany (Bottlinger to Schwarzschild, 22 May 1912, 26 Nov. 1912, 7 Dec. 1912; Bottlinger, *Die Gravitationstheorie und die Bewegung des Mondes* [diss., University of Munich, 1912], where Bottlinger investigates the possibility of classical modifications of the inverse-square gravitational law). Keenan, Pinto, and Alvarez, *Observatory*, p. 47, also indicates offers to Heidelberg privatdocent August Kopff and Vienna Observatory *Adjunkt* Adolf Hnátek. LO, "Chile," Ralph E. Wilson to Campbell, 21 Sept. 1914, where Pingsdorff is cited as having been at the Santiago Observatory for two years; Pingsdorff, *Der Sternhaufen in der Cassiopeia Messier 52* (diss., University of Bonn, 1909), under Karl Friedrich Küstner.

45. Ann Arbor, University of Michigan, Bentley Historical Library, Hussey Family papers, box 3, diary entry by Hussey, 4 Mar. 1913.

46. LO, "Chile," Joseph Haines Moore to Campbell, 15 Apr. 1913: "It would seem that his troubles unbalanced his mind and in a moment of 'mental perturbation' the terrible deed was done."

47. Chile, Ministerio de instrucción pública, *Personal de instrucción pública*, p. 167, re-

selected to succeed Ristenpart is testimony to strong support from his years of lecturing at the university's School of Engineering and Higher Mathematics.

Obrecht had arrived at a point in life when astronomical research meant little. Certainly he was uninterested in continuing commitments made by his predecessor. He argued in favor of abandoning the new site for the observatory, which had been chosen by Ristenpart and already partially occupied. He discounted Ristenpart's plans for preparing charts of the southern skies down to the tenth magnitude, and he was unenthusiastic about continuing Santiago's role in the sky-chart project, which he eventually gave up for a second time. His staff eroded. Prager returned to Germany. Pingsdorff, having arrived in Santiago only the day before Ristenpart's suicide, found life difficult under Obrecht—especially after the beginning of the war. "The Chileans have broken their contract with him in the manner of paying him his salary, paying him paper at 90% premium on gold when gold is at a premium of 180%," the American observer at San Cristóbal noted in September 1914. "With the present intense feeling of bitterness against the Germans all over the country, there is every prospect that he may lose his position altogether."[48] Obrecht forced Pingsdorff to move out of his house and then gave it to Gajardo, the administrative chief; Pingsdorff resigned in 1915. Richard Wüst, Ristenpart's mechanic, alone among the Germans remained under Obrecht's regime. Obrecht suffered a stroke in 1922. Gajardo, who had risen in 1916 to become associate "scientific" director, succeeded him. There were no funds for new staff, and until Gajardo left in 1929, the observatory had no real scientific program.[49]

In the game of musical chairs played out by French and German scientists, one of Ristenpart's senior appointments endured through the next administration. In 1910 Ristenpart replaced Alfred Krahnass in the meteorological service with Walter Alfred Knoche at the head of a new and independent Central Meteorological and Geophysical Institute. Knoche had recently completed a doctorate at Berlin under meteorologist Wilhelm von Bezold and had served as assistant at the Prussian meteorological service and the weather office in Berlin. Like Risten-

---

cords his salary as being paid in sterling, not (as with most teachers and professors) in pesos or (as with several savants of foreign origin) in gold; the figure amounted to 25,150 pesos. The rector of the university, Solar Domingo Amunátegui, received a salary of 27,855.35 pesos (p. 92).

48. LO, "Chile," Wilson to Campbell, 21 Sept. 1914.

49. Keenan, Pinto, and Alvarez, *Observatory*, pp. 47–54; LO, "Chile," Moore to Campbell, 24 Jan. 1911, for "Würst"'s experience working for Zeiss in Jena.

part, he was an accomplished and dedicated researcher. Knoche trained local assistants and published a portion of his results in Spanish. In one undertaking, he worked with Jakob Laub, Albert Einstein's collaborator, who was then his counterpart at the La Plata Observatory in Argentina, on an extensive magnetical and meteorological study of the Brazilian eclipse of 1912.[50] Knoche's embryonic weather service accomplished little in the way of scientific analysis. After a decade in Chile, his interests turned toward climatology and the connection between environment and civilization. In 1919 he became director of an institute for medicine, technology, and public health in Santiago. He moved to Madrid in 1932, only to return to Chile in 1936. He finally went to work for the Argentine national meteorological service, and in 1945 he died in Buenos Aires.[51]

The French approach to exact sciences in Chile resulted in geographical chronicles rather than theoretical or observational undertakings. Obrecht was uncomfortable with the exact sciences that had emerged late in the nineteenth century, which interrelated observation, experiment, and theory. Ristenpart flourished under such a regime. Obrecht was content to run the observatory as an institution for sinecured *érudits* and an office for addressing the practical concerns of the nation; Ristenpart took charge of a factory for producing new knowledge. The counterpoint of French and German styles of scientific activity is seen here with striking clarity. The work of another French scientist in Chile, Fernand, comte de Montessus de Ballore, provides a variation on the theme.

Fernand Jean-Baptiste Marie Bernard, comte de Montessus de Ballore, received his training at the Ecole polytechnique, where he graduated in 1873; among his classmates was the future marshal of France, Ferdinand Foch. When he was thirty years old, Artillery Captain Montessus headed a military mission to San Salvador. There he trained Salvadorans and led them against Guatemalan forces. The local environment awakened an interest in seismology. He returned to teach at his alma mater and in 1888 published a major statement on earthquakes in Central America. As he climbed upward in the military hierarchy of France, he continued to assemble material about seismology. At his retirement from the army in 1906, he had a card catalogue of more than

50. See *CIES*, p. 201, for Laub.

51. Knoche, "Resumen de la organización del Servicio meteorológico," in *Proceedings of the Second Pan American Scientific Congress*, 11 vols. (Washington, D.C., 1917), 2:730–35; "Walter Knoche," *Geofisica pura et applicada* 8 (1946), 190–91; *Pogg.*, 5, appendix, s.v. "Knoche." The *Anuario meteorológico de Chile* for 1914, published in 1918, has Carlos Henriquez listed as director of the meteorological and geophysical institute.

100,000 historical earthquakes, and he was the author of scores of seismological studies, among them several books identifying global earthquake zones.

The great Valparaiso earthquake of 1906 led the Chilean government to establish an independent seismological service (seismology until then had been the province of Luis Grosch at Obrecht's observatory). The authorities called Montessus de Ballore as director. At the time, German and British geophysicists were taking enormous strides in the science of seismology, the analytical and observational foundations of which were laid in the years before 1914; the Chileans saddled themselves with a fifty-six-year-old amateur who had strengths in neither instrumentation nor mathematical analysis. Montessus, a natural-history fact collector, lectured at the University of Santiago on construction in earthquake-prone areas. He extracted broad generalizations from his catalogue until the very end of his life; between 1906 and 1921 he published twenty-six notes in the *Comptes-Rendus* of the Paris Academy of Sciences. His last book, a treatise in French on seismology, was edited by his mathematician brother Robert. No disciple followed the path that he took.[52]

Early twentieth-century Santiago offers a natural setting for contrasting styles of French and German scientific activity. The substantial French colony there—sixteen thousand to eighteen thousand—opposed waves of German immigrants.[53] The distinction between French and Germans attracted the comments of André Bellessort, a French traveler. Following a visit to Bolivia in 1895, Bellessort wrote that French immigrants represented "unintelligent individualism" and desired immediate personal gain; the Germans, however, sought to consolidate their colony and thought of the superior interests of their "race." Bellessort emphasized: "One group [the Germans] say, let us conquer a new province; the other group say, let us amass enough to return quickly to our own land."[54] Given these competing styles, and the metropolitan energies spent to mobilize them, it is a triumph of

52. Charles Davison, *The Founders of Seismology* (Cambridge, 1927), pp. 160–76; A. Renier, "Fernand-Jean-Baptiste-Marie Bernard comte de Montessus de Ballore," in Montessus de Ballore, *La géologie sismologique* (Paris, 1924), pp. 453–58; Jean Pierre Rothé, "La vie et l'oeuvre de Montessus de Ballore (1851–1921)," *Acta geographica*, no. 24, Dec. 1975, pp. 31–45.

53. Henri Lorin, "Impressions du Chili: Les Chiliens et la France," *Revue des deux mondes* 7 (1912), 192–215, on p. 210; J-P. Blancpain, "Francisation et francomanie en Amérique latine: Le cas du Chili au XIXe siècle," *Revue historique*, no. 544, Oct.–Dec. 1982, pp. 365–407.

54. Bellessort, *Le jeune Amérique: Chili et Bolivie* (Paris, 1923), p. 325.

Chilean statesmanship for the country to have remained neutral during the First World War.

Reflections on Silver Shores

Under the enlightened presidency of Jorge Sarmiento, Argentina established educational traditions that would guarantee its place as Latin America's leading center of scientific research for the next century. Sarmiento was the architect of Córdoba's Academy of Sciences, which he populated with a corps of naturalist *sabios* imported from Germany. The academy, while suffering internal dissension, promoted scientific explorations of the country.[55] Sarmiento's most astute appointment came in the form of Benjamin Apthorp Gould, whom he brought in from the United States to direct Córdoba's new national observatory. During his tenure at Córdoba, from 1870 to 1885, Gould and his assistants surveyed the southern skies, pioneered stellar photography, and produced various catalogues, some of which are still consulted by astronomers.[56]

Given Gould's energy and accomplishment, why should nineteenth-century Argentina have been the site for a second major observatory? The answer lies in Argentine politics. Centrifugal forces dominated the early Argentine confederation to the point of periodical military struggles between provincial and federal chieftains. The outcome of one such struggle in 1880 displaced the government of the Province of Buenos Aires to a new city, La Plata, located fifty kilometers south of Buenos Aires on the River Plate. The province was Argentina's richest one, enjoying revenues on the order of those at the disposal of the national government. The provincial capital, accordingly, was invested with institutions of higher learning to rival those of the federal capital. The most impressive of these institutions became the astronomical observatory.[57]

The idea of founding an observatory at La Plata came in conjunction

55. *CIES*, p. 151, and the references cited there; Susan Sheets-Pyenson, *Cathedrals of Science: The Development of Colonial Natural History Museums during the Late Nineteenth Century* (Montreal, 1988), p. 61.

56. The best general account of Gould at Córdoba is Enrique Chaudet, *La evolución de la astronomía durante los últimos cincuenta años (1872–1922)* (Buenos Aires, 1926), pp. 20–65. See also Marcelo Montserrat, "La introducción de la ciencia moderna en Argentina: El caso Gould," *Criterio*, no. 1632, 25 Nov. 1971; John E. Hodge, "Benjamin Apthorp Gould and the Founding of the Argentine National Observatory," *Americas* 38 (1972), 152–75; and David S. Evans, *Under Capricorn: A History of Southern Hemisphere Astronomy* (Bristol, 1988), p. 134 for present utility, although in general Evans's account is useful only for English-language institutions.

57. See *CIES*, pp. 152–53, for the founding and aspect of La Plata.

with the transit of Venus in 1882. The provincial government ordered a 20-cm refractor from Gautier in Paris, through which French naval lieutenant Edouard Perrin observed the event at the town of Bragado. The arrangements were unusual, even for the time and place. Perrin was assigned to a ship at La Plata, and he transmitted his observations to the Paris Academy of Sciences, but Argentina bore his travel costs.[58] Perrin suggested the endowment of a permanent facility. With the encouragement of the governor, Dr. Dardo Rocha, the province agreed to purchase a small meridian circle, two astronomical clocks, a chronograph, and meteorological instruments for an observatory. Governor Rocha went to the French navy for a director. He chose Perrin's friend and collaborator, Victor François César Beuf, for the job.[59]

At the time he was tapped, in 1881, Beuf was a forty-seven-year-old lieutenant. He had contracted dysentery during the Mexican invasion and returned to France in 1863, where he asked for a port assignment. He served as treasurer at Toulon for a number of years before rising to direct the observatory there in 1874. Fluent in Italian and modern Greek, he had no trouble learning Spanish.[60] He arrived in the new city of La Plata in time to supervise capitaine de frégate Georges Ernest Fleuriais's passage through Buenos Aires for the transit of Venus.[61] He set about to build the observatory.

Beuf traveled to France to acquire zenith telescopes, chronometers, and chronographs for mapping the province.[62] By 1886 he had acquired a staff and several buildings, as well as authorization to purchase an 80-cm refractor, a 6-inch photographical refractor, a Thollon solar spec-

58. VM. s.v. "Perrin," Perrin to Head of the Etat-Major of the Navy, 31 Aug. 1882.

59. William J. Hussey, "A General Account of the Observatory," Universidad nacional de La Plata, Observatorio astronómico, *Publicaciones* 1 (1914), 1–64; Chaudet, *Evolución*, p. 11. See VM, s.v. "Beuf," Hervé Faye to Minister of the Navy, 16 June 1881, for Faye's encouragement of Beuf and Perrin's interest in observing lunar occultations on the high seas.

60. See VM, s.v. "Beuf," for the vita. The dossier makes clear that Beuf retired from the French navy to take the La Plata position in 1881 and that he was marked for the position early in the year (Maritime Prefect of the 5e arrondissement to Minister of the Navy, 14 Jan. 1881).

61. PO, "La Plata, 1886–1907," Beuf to Mouchez, 28 Oct. 1886: "I remember very well the passage of Fleuriais here in '82. I had to show him myself the transits of the Bouquet [de la Grye] apparatus at the naval school. And later during the longitude with Valparaiso, he never would have been able to work things out if I had not been there." See also Jean-Philippe-Ernest de Fauque de Jonquières, "Sur quelques essais, faits, à la mer, avec le gyroscope-collimateur de M. le capitaine de vaisseau Fleuriais," *CRAS* 104 (1887), 1039–41; and *DBF*, s.v. "Fleuriais."

62. PO, "La Plata, 1886–1907," Decree of the Province of Buenos Aires for constructing a provincial map, 21 Apr. 1886, French translation, referring to Beuf's initial mandate in 1883 of determining fifty points relative to La Plata.

troscope, a large altazimuth, and a master clock for the provincial time service. He anticipated having buildings for the equipment, as well as a palatial director's residence. Beuf planned to set up a provincial meteorological network.[63] "I am rather proud of having given this country (and it is a singular instance in all South America) a daily meteorological service following the model of the Bureau central in France," Beuf bragged to Mouchez early in 1886. Beuf saw his operation as a de facto branch of the Paris Observatory, and in the late 1880s he wrote to Paris on a monthly or semimonthly basis. "Without you," he emphasized to Mouchez in 1885, "I would not perhaps succeed in creating an observatory here." Beuf apologized for taking so much of Mouchez's time, but Mouchez had sired La Plata. "Since you are the father of one [the La Plata Observatory], it is appropriate that you occupy yourself with the education of the child."[64] In these lines we read more than the deference of a former naval officer addressing a French admiral.

Throughout the financial crisis of the middle 1880s, Beuf's observatory continued to receive appropriations. The country was bankrupt, Beuf noted to Mouchez in 1885, and the value of Argentina's paper money had plummeted. The province, however, rode out the panic "without much trouble." By the middle of 1886 Beuf had received well over Fr 100,000 for astronomical and meteorological instruments from the young, provincial minister of public works, Manuel B. Gonnet. The largest portion of the money went on deposit with the bank of Baring Brothers in London, and a smaller sum went to the Banque maritime in Paris. Mouchez could draw on both accounts.[65] Beuf wanted the finest equipment available. "Don't worry about the price," he admonished Mouchez about a precision astronomical clock.[66] Having obtained funding from the provincial government (money that had to be spent by the end of the year), Beuf fretted about receiving his instruments. "How shall I reply to the people who say to me here: Well, when are you beginning? Are you going to set up the meteorological stations soon? or When will you measure the longitude of your observatory?"[67] The barometers and thermometers arrived late in October 1886: "Here

63. Hussey, "General Account," p. 5.

64. PO, "La Plata, 1886–1907," Beuf to Mouchez, 26 Apr. 1886 and 27 Aug. 1885.

65. Ibid., 17 Oct. 1885; Beuf to Auguste Fraissinet, 28 May 1886, indicating that Fr 95,000 went to Baring Brothers and Fr 20,000 to the Banque maritime; Gonnet to Beuf, 9 June 1886, indicating that Fr 40,600 was sent to Mouchez for meteorological instruments; Beuf to Mouchez, 16 June 1886, where Beuf describes Gonnet as "quite a young man, scarcely twenty-eight years, lawyer by profession but full of zeal and ardor for everything that can contribute to raising the intellectual value of his country."

66. Ibid., Beuf to Mouchez, 18 June 1886.

67. Ibid., 16 July 1886.

is a beginning. We are at least going to be able to install the meteorological stations; the absence of chronographs has required us to postpone other really serious work."[68]

With the promise of commanding the finest observatory in the Southern Hemisphere, Beuf volunteered for Mouchez's sky-chart project, at the admiral's request. In 1886 Mouchez canceled Beuf's order for a 6-inch photographical telescope and persuaded La Plata's provincial government to substitute one of the Henry brothers' new astrographic lenses, which were to be standard issue for the sky chart. While in Paris the following year (when Beuf accepted a seat on the sky chart's permanent committee), Beuf ordered a 16-inch refractor at Fr 102,500 for observing comets and faint stars outside the meridian and an 8-inch Gautier meridian circle for cataloguing faint circumpolar stars. In addition, Beuf ordered an 80-cm reflecting telescope from Gautier and the Henrys at a cost of Fr 56,000.[69]

Care had to be taken, however, to marginalize the national observatory at Córdoba, where Gould's American successor, John Macon Thome, had just begun to establish his own research priorities. From the beginning Beuf had worried about the Americans, who were supported by the federal government. Córdoba had obtained privileged exchanges with European observatories, Beuf complained in 1885, and as a result there were virtually no astronomical publications at La Plata; and Gould had tried to interfere with Beuf's mandate to map the province.[70] There was no contact at all between the two astronomical operations. "The Córdoba Observatory has not breathed a word," Beuf wrote to Mouchez about preliminary plans for the sky chart. "I believe that this has completely escaped them. Furthermore, even in the time of Gould this establishment hardly shined in its activity, all resources being directed to the mechanical work of a catalogue. But it seems that today it is in complete decay." The explanation was obvious to Beuf: the Province of Buenos Aires was "much richer than the nation."[71]

Late in 1886 Mouchez decided not to involve Thome with the sky chart. Beuf worried about the consequences of the snub: "Allow me . . . to express the regret that it was not done because of the extreme jealousy that the Córdoba Observatory has for La Plata, a situation that now and again makes enemies for us. It should not be forgotten that Gould, who certainly has some personal value, is a charlatan of the

68. Ibid., 12 Dec. 1886.

69. Hussey, "General Account," pp. 10–11, 14, 45–46, 52.

70. PO, "La Plata, 1886–1907." Beuf to Mouchez, 28 Aug. 1885, for the origin of Córdoba's astronomical library; fragment, n.d., probably Beuf to Mouchez, where allusion is made to Gould's interference with the provincial mapping scheme.

71. Ibid., Beuf to Mouchez, 10 May 1886.

highest order, who, although gone from the observatory, does not cease directing it from afar." Beuf thought that he would say something when the time was right about the nest of Anglo-Saxons at Córdoba. "They are almost all Germans or North Americans."[72]

The arrival of the sky-chart instruments coincided with a period of financial stringency, and in the middle of the 1890s, when all his big telescopes had been mounted, Beuf's health began to fail. In 1890, after a trip to Paris, he wrote to Auguste Fraissinet, the secretary of the Paris Observatory, that he had "never felt the problems of this country so acutely as since my return." In 1891 he believed that though things had taken a turn for the worse in the economy (it was the time of the collapse and reorganization of Baring Brothers), the observatory would continue to receive support. But later that year the provincial legislature canceled funds that had previously been authorized for operating expenses. The uncertainties drove Beuf to distraction. He scrawled a letter to Fraissinet early in 1892: "The present situation . . . is still too sad to allow me a state of mind sufficiently free and disengaged from worry to discuss freely and pointedly with a friend." In the middle of 1892, after Mouchez died, the money placed in his trust went to pay for Beuf's large equatorial.[73] An assistant then broke the objective of the sky-chart Gautier by applying too much pressure on its screw mountings. By 1895, his health having deteriorated and his funding having evaporated, Beuf withdrew from the sky-chart project. "Political events seem to be opposed to the systematic undertaking of the work," was how the men at the Paris Observatory put it in 1895.[74]

The Argentine-educated engineer Virgilio Raffinetti, who succeeded Beuf, was more direct: lack of funding had brought the observatory to ruin.[75] As the century ended, Beuf made no progress toward another of his mandates — mapping the province — beyond erecting piers at various points for taking longitude measurements, although he did set into operation a provincial meteorological network.[76] Beuf was a loyal agent of French scientific interests — he sent young Argentine talent to the Paris Observatory — but the enterprise of a fawning subordinate has little intrinsic value to a superior.[77] It is not surprising that immediately

72. Ibid., 12 Nov. 1886. Beuf regularly referred to Gould as a "charlatan" (see, e.g., 10 Apr. 1886).

73. Ibid., Beuf to Fraissinet, 21 July 1890; Beuf to Mouchez, 21 July 1891; Beuf to Fraissinet, 21 Oct. 1891 and 4 Jan. 1892; Fraissinet to Beuf, 19 Apr. 1893 (draft).

74. *Bulletin du Comité international permanent pour l'exécution photographique de la carte du ciel* 2 (1895), 453; Hussey, "General Account," p. 53, for the inexpert mechanic.

75. Raffinetti, 1901, cited in Chaudet, *Evolución*, pp. 14–15.

76. Hussey, "General Account," pp. 21, 60, 62. Beuf's son Victor ran the meteorological network from 1885 to 1899.

77. See PO, "La Plata, 1886–1907," Beuf to Fraissinet, 25 Oct. 1889, where Beuf sends

following Beuf's death in 1899, La Plata's celestial zone went to her rival Córdoba.[78]

Director Loewy had been angling for several years to hook Córdoba. In 1898 he expressed strong support for Córdoba when the astronomers there faced retrenchment.[79] When Thome agreed to take on Beuf's celestial zone, Loewy arranged for all the necessary instruments.[80] Thome's funding came courtesy of the experienced federal minister of state Senator Carlos Pellegrini. Loewy had received Pellegrini at the Paris Observatory, and Thome highly valued his patronage. "A man of great influence" who "continues to be a good friend of the observatory," Thome described him to Loewy in 1901.[81]

The sky-chart equipment arrived late in 1901, and Thome anticipated welcoming an assistant from Paris to help set it up. Nothing, however, was certain in Argentina, and especially so because Beuf's death had placed Raffinetti in charge of the La Plata Observatory. "Our rascally neighbors are again making trouble, and there is great excitement and confusion in Buenos Aires, and it is doubtful whether the Congress will give me an adequate budget." Thome estimated that he needed sixteen hundred to two thousand francs per month for the project. While only half that amount was in hand, Thome felt that his recent receipt of the Paris Academy's Lalande prize would eventually secure him what he needed.[82] Early in 1902 the telescope was mounted and the money obtained: "*In consequence* of that event [the Lalonde prize], our Congress voted the sum of Fr 1,600 monthly for the photographical work; whereas *without* it, they would probably have voted only Fr 600. You have, therefore, *again* rendered me and the Observatory a great service, and we are all very much in your debt."[83]

---

along a letter for Octavio Pico, "one of the rare friends I have in this country," who had been charged with surveying the boundary between Argentina and Chile; and 25 July 1894, where Beuf recommends Luis J. Dellepiane, a captain of engineering on a geodesical buying trip to France. Both Pico and Dellepiane later became professors at the University of Buenos Aires.

78. "Conferenza internationale per la carta fotografica del cielo," *Memorie della Società degli spettroscopisti italiani* 29 (1900), p. 3 of a ten-page monographic reprint.

79. PO, "Cordoba, 1898–1910," Loewy to Thome, 10 June 1898; undated testimonial letter of Loewy's in Chaudet, *Evolucíon*, p. 12.

80. See PO, "Cordoba, 1898–1910," Thome to Loewy, 6 Apr. 1901, where Thome looks forward to receiving all the instruments; and Thome to Loewy, [1901?], where Thome specifies disbursement of Fr 34,150 left in Loewy's charge when Thome visited Paris.

81. See ibid., 12 June 1901, for Pellegrini's visit to the Paris Observatory and the evaluation.

82. Ibid., 26 Dec. 1901.

83. Ibid., 5 Mar. 1902.

Thome's most difficult technical problem concerned photographical plates, which were shipped from the Lyons firm of Auguste and Louis Lumière. Frequently these arrived fogged. In 1903, for example, the culprits were Argentine customs agents, who insisted on opening the Lumière box and so spoiled half the shipment.[84] Thome's most annoying political problem concerned La Plata. In 1905 the university there was nationalized, and the observatory came along with it as one of the federal government's crown jewels. Raffinetti ceded early in 1906 to the professor of astronomy from Genoa, Francisco Porro di Somenzi, who heralded a new era of astronomical research.[85]

Porro reported enthusiastically to his Italian astronomer colleague, Giovanni Virginio Schiaparelli, about his call to La Plata. Both he and his family would be treated well, he emphasized. He looked forward to observing variable stars in the Southern Hemisphere. He would bring a mechanic from Italy, and he also hoped to persuade Luigi Carnera to follow him once Carnera's work with the International Geodesical Association had been completed; his older son would remain in Italy for military service and university studies.[86] Porro retained his enthusiasm upon arriving in La Plata. He affirmed to Schiaparelli that, all things considered, he would not exchange places with any other astronomer in the world, even though Beuf's old instruments were in a deplorable state of disrepair. "I do not dream of *records* in the North American fashion, but I will be able to do the best that I desire."[87] While Porro steered the telescopes, he also served as the dean of La Plata's science faculty.[88] Loewy quickly took advantage of the new breeze. In 1906 he congratulated Porro upon his appointment and asked if he would take up Beuf's sky-chart zones.[89]

Thome cast a jaundiced eye toward his new federal rival. Porro, Thome noted to Loewy, had been a professor of astronomy without an

84. Ibid., 23 Nov. 1903; see ibid., 29 Apr. 1904, for another instance of destruction at the hands of customs.

85. Not to be confused with the geodesist Ignazio Porro. See PO, "La Plata, 1886–1907," Beuf to Mouchez, 18 June 1886, asking Mouchez to order geodesical instruments from either the Società filotècnica founded by Ignazio Porro at Milan in 1865 or from Porro's sister workshop in Paris; *DSB*, s.v. "Porro"; and Paolo Brenni, "Italian Scientific Instrument Makers of the Nineteenth Century and Their Instruments," in *Nineteenth-Century Scientific Instruments and Their Makers*, ed. P. R. de Clerq (Amsterdam, 1985) [Nieuwe Nederlandse bijdragen tot de geschiedenis der geneeskunde en der natuurwetenschappen, 17], pp. 183–203, on pp. 196–97.

86. Milan, Osservatòrio astronòmico di Brera, Archives, 445, Porro to Schiaparelli, 5 Dec. 1905. I am grateful to Dr. Pascuale Tucci for making available this letter and other material from the Brera archives.

87. Ibid., 9 Jan. 1906, emphasis in the original.

88. *CIES*, p. 186.

89. PO, "La Plata, 1886–1907," Loewy to Porro, 20 June 1906.

observatory. Lacking competent observers at La Plata, his task was to "train young Argentines as assistants." In anticipation of great things to come, and with the death of Thome's patron Pellegrini, Porro had come to command the lion's share of federal astronomy. His budget was more than twice Córdoba's, "even though it has no proper instrument for making an observation and they have already spent Fr 100,000." Porro, in fact, received "Fr 3,000 per month, almost three times what they give me." Raffinetti, "who collected all the instruments when he was interim director, an Argentine and a useful man to the rector [Joaquín V.] González, receives two times what I make." Thome concluded: "I do not believe that they will succeed in making an observation or determining the instrumental errors, but they are all good Argentines."[90]

Porro followed up a longstanding geodesical interest by arranging to measure the latitude of Oncativo in association with geodesist Friedrich Robert Helmert; his was a southern station for studying the periodical variation of latitudes – a question that had absorbed the International Geodesical Association for many years.[91] For the rest, he accomplished little before returning to Italy in 1910. Porro was followed at La Plata by the American William Joseph Hussey and then, after a temporary Argentine appointment, by the German Johannes Hartmann; Hussey and Hartmann fared somewhat better than Porro.[92]

As for the sky chart, it went to La Plata's northern rival at Córdoba. Thome, admirably prepared by tradition and temperament to carry out a portion of Mouchez's dream, set to the task at hand. He died in 1908, and his American successor, Charles Dillon Perrine, completed the zone.[93] By the eve of the First World War, however, interest in the sky

90. Ibid., "Cordoba, 1898–1910," Thome to Loewy, 29 July 1906. According to Fortunato J. Devoto, "Noticia descriptiva," in La Plata, Universidad nacional, Facultad de ciencias físico-matemáticas, *Anuario*, 1910, pp. 65–74, on p. 69, Porro had directed the Turin Observatory from 1897 to 1901.

91. Devoto, "Noticia," p. 71; see Osservatòrio di Brera, Porro to Schiaparelli, 5 Dec. 1905, for the geodesical interest. On the variation of latitudes (i.e., the migration of the poles), see Ernest Lebon, *Histoire abrégée de l'astronomie* (Paris, 1899), pp. 173–75; Georges Perrier, "La coopération internationale en géodésie et en géophysique," Bureau des longitudes, *Annuaire pour l'an 1928* (Paris, [1928]), pp. C.1–C.120, on pp. C.14, C.43.

92. *CIES*, pp. 187–205, 220–26. Hussey held Porro in low regard. He explained to a colleague that Porro had visited the Zeiss works in Jena and "placed orders for 50,000 marks' worth of instruments, in an hour or two, Porro saying, 'You make them as you think they should be, you know more about it than I do'" (Ann Arbor, University of Michigan, Bentley Historical Library, Hussey Family papers, box 1, Hussey to Ralph Hamilton Curtiss, 26 July 1911).

93. John E. Hodge, "Charles Dillon Perrine and the Transformation of the Argentine National Observatory," *Journal of the History of Astronomy* 8 (1977), 12–25.

chart had waned. The two astronomical communities with the most experience in cataloguing stars — the Americans and the Germans — had never been enthusiastic about the French scheme, and with the new century they turned to radial velocities, double stars, astrophysics, and cosmology.

The image of France, cracked and faded, continued to attract Argentine physicists, some of whom drifted to the bistros and laboratories of Paris. Astronomer-priest Fortunato Javier Devoto seems to have stayed the longest — from 1914 to 1917; Luis J. Dellepiane, sent to the Paris Observatory by François Beuf, lectured on the "theory of progress" at the Sorbonne; Argentine physicist Adolfo T. Williams claimed to have completed a doctorat ès sciences around 1920 and favored the *Comptes-Rendus* with some of his publications.[94] France's greatest impact on the exact sciences in Argentina came in the way of persuading an Argentine physicist to become a writer.

The disenchanted physicist was Ernesto Sábato, who late in 1936 received the ninth doctorate in the subject awarded by the University of La Plata.[95] Sábato, the tenth of eleven children born to a mill owner and his wife who had immigrated from Italy, received elementary education in his native town of Rojas. He went to La Plata at age twelve for secondary school in one of Argentina's finest colegios. Sábato recalled: "For a year I was terribly alone. I felt isolated in class and I felt ridiculous, a country boy; I felt that the world was hostile and horrible, imperfect." Geometry opened a new world to the adolescent. When he saw a geometrical proof for the first time, "I experienced a sort of ecstasy: a perfect and exact world, beautiful and incorruptible, opened up to me." While at the La Plata school, in the late 1920s, Sábato and sev-

94. A ranking member of the Argentine clergy, Devoto took a licence in France and, upon his return, founded an observatory for cosmic physics in San Miguel, near Buenos Aires; for his labor he was admitted to the French Legion of Honor. Eduardo L. Ortíz, Antoni Roca i Rosell, and José M. Sánchez Ron, "Ciencia y técnica en Argentina y España (1941–1949), a través de la correspondencia de Julio Rey Pastor y Esteban Terradas," *Llull* 12 (1989), 33–150, p. 86. Hussey, "General Account," p. 9; Ernest Martinenche, "L'action du 'Groupement' depuis sa fondation jusqu'à la fin de l'année 1913," in Groupement des universités et grandes écoles de France pour les relations avec l'Amérique latine, *Bulletin de la Bibliothèque américaine (Amérique latine)* 4 (1913–14), 225–31, on p. 228 for Dellepiane; *CIES*, p. 236 for Williams, who claimed the doctorate in *Pogg.* On the Groupement see Patrick Petitjean, "Le Groupement des universités et grandes écoles de France pour les relations avec l'Amérique latine, et la création d'instituts à Rio, São Paulo et Buenos Aires (1907/1940)," in *Anais do Segundo Congresso latino-americano de história da ciência e da tecnologia, 30 de junho a 4 de julho, 1988*, ed. Ubiratan D'Ambrosio (São Paulo, 1989), pp. 428–42, where one sees no long-term French visitors in the exact sciences.

95. José Federico Westerkamp, *Evolución de las ciencias en la República Argentina, 1923–1972*, vol. 2, *Física* (Buenos Aires, 1975), p. 97.

eral friends (one of whom was the son of La Plata physicist and anarchist Rafael Grinfeld) discussed left-wing politics.[96]

Sábato took his *bachillerato* in 1928 and went on to study physics and mathematics at the University of La Plata. In 1930, as the Germanophile general Félix Uriburu removed the senile president Hipólito Yrigoyen from power, Sábato abandoned school and family to join the Argentine Communist party. In 1934 the party sent him to Brussels for a communist congress against war and fascism, presided over by Henri Barbusse. There he revolted against the Stalinist line and, instead of continuing on for indoctrination in Moscow, fled to Paris, where he lived from hand to mouth. A communist janitor at the Ecole normale supérieure offered him a place to sleep until he was taken in by sympathetic Latin American students. He contemplated suicide, but the harmonies of pure mathematics gave him a reason to live. He stole a book on analysis by Emile Borel and abandoned himself to "the fascination of this universe of eternal and incorruptible objects." He resolved to return to Argentina and continue studying for a doctorate in mathematics.

Late in 1936, when he completed his doctorate in physics on the excitation and ionization potentials of the krypton atom, Sábato felt "that the world of mathematics was foreign to the world of men. This universe of symbols seemed to me like a Platonic opium." He won a postdoctoral scholarship from the Argentine Association for the Advancement of Science to the Radium Institute in Paris. The choice in favor of France would have been encouraged by the short-term presence, during the 1930s, of distinguished French mathematicians – Emile Borel, Jacques Hadamard, and Georges Darmois.[97] In 1938 in Paris, Sábato "worked among electrometers during the day," while at night he stole away "to find the surrealists at the Dôme, just like an honorable housewife who, at night, became a prostitute." Sábato believed that science "was in the process of unleashing the deepest crisis known to man, that of his technological alienation." He witnessed Frédéric Joliot-Curie and his colleagues preparing for a nuclear apocalypse.[98]

Following the Munich crisis, Sábato left Paris to work with Manuel

96. Sábato, *Mes fantômes*, trans. Lucien Mercier and Jean-Marie Saint-Lu (Paris, 1988), pp. 19–22; Joaquín Neyra, *Ernesto Sábato* (Buenos Aires, 1973); Harley Dean Oberhelman, *Ernesto Sábato* (New York, 1970), pp. 17–21; *CIES*, p. 239, for police raiding Grinfeld's house in 1930.

97. Elías A. De Césare, "La matemática en Buenos Aires y La Plata," in *Evolución de las ciencias en la República Argentina, 1923–1972*, vol. 1, *Matemática*, ed. Luis A. Santaló (Buenos Aires, 1972), pp. 12–54, on pp. 47–51.

98. Sábato, *Fantômes*, pp. 22–23.

Sandoval Vallarta on cosmic rays at the Massachusetts Institute of Technology. He published a short paper in the *Physical Review* in which he argued against Hannes Alfvén's suggestion that cosmic rays might be produced by particles accelerated in a double-star system, analogous to the way that particles were accelerated in a cyclotron.[99] Upon demolishing Alfvén's cosmical cyclotron, Sábato returned to Argentina, where he married Matilde Kusminsky-Richter, taught at La Plata in the university and the normal school, and began to publish in experimental physics.[100] His literary passion grew. For a time in 1943 he moved to a modest ranch in Córdoba, courtesy of Federico Valle, a wealthy pioneer of Argentine cinema. While there, Sábato spoke frequently with two resident sabios, Enrique Gaviola and Guido Beck. Gaviola, Argentina's finest native physicist, had returned from a Guggenheim Fellowship at the California Institute of Technology to direct the Córdoba Observatory; Beck, an Austrian physicist, had fled to France (where he was interned) and then to Portugal before joining Gaviola at Córdoba.[101] Gaviola and Beck failed to persuade Sábato to remain a scientist. They heard him deliver a farewell address to physics.

The valedictory had been brewing since the late 1930s, when Sábato and Enrique Loedel Palumbo, a Uruguayan-Argentine physicist at La Plata, began scrutinizing the concept of temperature.[102] Sábato recalled that he grew interested in the topic at the time of his doctoral defense. He sketched his ideas during an examination with two La Plata physicists, Ramón G. Loyarte and Teófilo Isnardi. Isnardi expressed the view that thermodynamics was a harmonious construction incapable of refinement. Later, in Paris, Sábato explained his ideas to Paul Langevin, who was entirely skeptical.[103] Sábato believed that the traditional concept of temperature was inexact; it depended on the transitivity of thermal equilibrium. He preferred to derive temperature from the second law of thermodynamics, and in the process avoid introducing ideal gases and Carnot cycles. It was, in his words, a "phenomenological" ap-

99. Sábato, "On Alfvén's Hypothesis of a 'Cosmic Cyclotron,'" *Physical Review* 55 (1939), 1272–73.

100. Sábato and Rogelio McLoughlin, "Medida de pequeñas variaciones de grandes tensiones e intensidades de corriente," *Anales de la Sociedad científica Argentina* 126 (1938), 251–68; Sábato, "Equipo para estabilización de voltaje," La Plata, Universidad nacional, Facultad de ciencias físico-matemáticas, *Publicationes*, 2d ser., *Revista*, no. 139, 1940, pp. 425–32.

101. See *CIES*, pp. 240–46, for Gaviola; and Westerkamp, *Evolución*, p. 23, for Beck.

102. Sábato and Loedel Palumbo, "Una nueva forma de introducir la temperatura," *Anales de la Sociedad científica Argentina* 127 (1939), 271–81; idem, "Contribución a la fundamentación de la termodinámica," ibid. 133 (1942), 222–38.

103. See Neyra, *Sábato*, pp. 105–7, for Sábato's commentary.

proach. The final form of Sábato's text, an impressionistic derivation of standard results, prepared the way for its author to enter the world of literature.[104]

Sábato read his statement on thermodynamical phenomenology at the third meeting of the so-called Nucleus of Theoretical Physics, held in La Plata in 1944. These meetings had been organized by Beck, and they paved the way for creating the Argentine Physical Association later that year. At the 1944 meeting, Sábato obtained support from Loedel Palumbo, who sparred with Isnardi about the best way to represent temperature.[105] The young writer's involvement with physics, however, had come to an end. He abstained from joining the new physical association, although he apparently continued teaching at La Plata until he fell victim to Juan Domingo Perón's restructuring of the academic world in 1945.[106] Sábato placed the idol of physics aside without destroying it. He joined the company of writers who in their novels infused science with human sensibility.

Parisian physics repelled Ernesto Sábato. In his research publications, he wandered from experimentation to theory to instrumentation, finally to end in philosophical speculation. Scientists in relative isolation may reasonably be drawn to explore the human meaning of natural knowledge. The significance of French involvement in Argentine and Chilean scientific life, however, goes beyond academic philosophy. It provides a graphic illustration of nineteenth-century French cultural imperialism and its slow attenuation.

The decline in French prestige must not be seen as an erosion in the quality of French learning. It reflects the rise to scientific prominence of Germany and the United States, and it signals a reorientation of French overseas *oeuvres* away from autonomous regions and toward areas where France exercised clear authority. Such a reorientation was coherent with France's desire to tie her colonies more closely to the hexagone and to exploit them for material gain. Mercantile America – North and South – had relatively little to offer French investors of financial and intellectual capital.

104. Sábato, "El concepto de temperatura en la termodinámica fenomenológica" (1945), in ibid., pp. 109–23.

105. Westerkamp, *Evolución*, pp. 25–26.

106. Oberhelman, *Sábato*, pp. 21–22, gives Sábato teaching until 1945. Perón became president early in 1946. The military coup of 1943 resulted in dismissal of more than one hundred Argentine educators for subversive activity; many were reinstated in 1945, only to be dismissed once more in 1946. The repressions especially affected physiologists (José Babini, *La evolución del pensamiento científico en la Argentina* [Buenos Aires, 1954], p. 210).

9

Mexico and Brazil: Reign of Positive Thinking

France's Man in Havana

In 1850 Andrés Poey, the son and grandson of distinguished Cuban naturalists, established a small, private meteorological observatory in Havana. He traveled to Paris in 1855 and returned to have himself named interim director of a larger, state-sanctioned institution in 1856.[1] Through his family connections Poey attracted the attention of Havana's premier learned corporation, the Royal Economics Society.[2] The society, which had sent him to the Paris Exhibition of 1855, granted him a thousand pesos for the purchase of French instruments.[3] Late in 1858 Director Poey began to plan for an office and tower to house his "physical-meteorological observatory of Havana." Meteorology, he reminded the members of his patron society, had applications in public health, agriculture, navigation, and industry; in addition to a real building, Poey needed guarantees that the society would publish and circulate his observations. These observations, he continued in another request for funding, involved thermoelectricity, electromagnetism, atmospheric optics, celestial photometry, and the theory of materials. Early in 1859 he petitioned for more than nine thousand pesos to enlarge his room at the society's building into a real observatory.[4]

The decree regulating Poey's physico-meteorological observatory

1. The early history is recounted in PMN, Poey, draft of resignation letter of 2 Apr. 1869; see also Director General of the Ministry of War and Overseas Territory to Poey, 8 Dec. 1860, referring to Poey's temporary appointment as director of the observatory in 1856.

2. Salvador Vilaseca Forne, "Matemáticas y astronomía en la historia de Cuba," *Quipu* 2 (1985), 185–212, pp. 200–202 for the Economics Society, founded at the end of the eighteenth century.

3. PMN, Serafín Massana to Poey, 8 Oct. 1855 and 2 Aug. 1857.

4. Ibid., Poey to Director and Members of the Royal Economics Society, 14 Sept. 1858 and 22 Dec. 1858; Poey to Director of the Royal Economics Society, 1 Feb. 1859.

appeared late in 1860. The observatory would "acquire all the scientific data that can enrich the sciences in the Spanish possessions of America." It was to take daily meteorological observations at different hours and also observe special events, such as hurricanes and shooting stars. The observatory's director was responsible to the captain general of Cuba "in everything that does not relate to the material applications of science." Although the observatory was technically private, the civil authorities could specify its tasks. The director would be responsible for making geophysical measurements, publishing the results, corresponding with foreign meteorologists, and charting the migratory patterns of Cuba's birds. The decree also specified the tasks of the observatory's assistants and its janitor.[5]

Poey received an official appointment with a salary of nine thousand pesos.[6] He obtained small grants for instruments, translated some of his meteorological works into French, began teaching a course in his specialty (as one of its tasks, the Economics Society surveyed Cuban instruction), and waited for his new observatory to materialize.[7] Poey's chronicle of his activity in the early 1860s is the record of a maverick institution-builder who would have served his interests better by publishing fewer speculations and inventories based on secondary sources. After 1861 he failed to circulate systematic observations, and this inattention—not unusual at the beginning of a new observatory—later proved disastrous.[8] Just as he was setting his house in order, Poey fell in with the imperial designs of France, the country whose culture he most admired.

Soon after the French invasion of Mexico in 1864, the innovative minister of public instruction in Paris, Victor Duruy, assembled a collection of his ministry's savants for a scientific mission to complement the Mexican expeditionary force. He allocated Fr 200,000 for a thorough exploration of the newly acquired sphere of influence. The enter-

5. "Reglamento para el régimen del Observatorio físico-meteorológico de la Habana," *Gaceta de Madrid*, 19 Dec. 1860, the decree dated 8 Dec. 1860.

6. PMN, Director General of the Ministry of War and Overseas Territory to Poey, 8 Dec. 1860.

7. See ibid., Captain General Francisco Serrano y Dominguez to Poey, 19 Feb. 1861, where the Cuban government grants 596 pesos for instruments. Publications of Poey's from this period include "Sur la neutralité de la force électro-magnétique de la terre et de l'atmosphère observée à la Havane durant les aurores boréales de 1859," *Annuaire de la Société météorologique de France* 9 (1861), 42–58; and *Relation historique et théorie des images photoélectriques de la foudre observée depuis l'an 360 de notre ère jusq'en 1860* (Paris, 1861). As the titles suggest, they are undisciplined collections of data. See Vilaseca Forne, "Matemáticas," pp. 201–2 for, the advisory educational role of the Economics Society.

8. PMN, Poey, "Legajo del observatorio," n.d.

prise organized into three classes. In the first class were the scientists in Paris who issued instructions, worked up the data, and prepared the eventual publications; in the second class were the French and foreign travelers who, for a price, worked overseas at the scientists' behest; and in the third class were the correspondents—hangers-on who could be charged with carrying out modest projects to collect data. The serious parts of the commission's work concerned natural history, public health, political economy, and ethnology. Direction of the committee on the physical sciences was confided not to a practicing physicist or chemist but to Marshal Jean-Baptiste Philibert, comte de Vaillant, one of Napoleon III's favorite ministers. The French voyagers set off during the latter part of 1864. They were nearly all young and unknown men who sought to use the expedition to propel themselves upward in Parisian society. It would not be inaccurate to say that none afterwards made a mark in the world of science.[9]

The brightest of the metropolitan savants saw the expedition as a golden opportunity to collect data that might be used in one or another of their ongoing research programs. Geologist Jean-Baptiste-Armand-Louis-Léonce Elie de Beaumont, for example, ordered his minions to make exacting descriptions of what they saw—in maps, charts, and samples; he sought support for his own global ideas about geological processes, which were then hotly disputed by Charles Lyell and Louis-Constant Prévost.

The French initiative intended to stimulate further research by Mexican counterparts. Early in 1864 a joint committee of French officers and Mexican scientists took shape to facilitate the tasks of the new scientific commission. The military took charge, Marshal Achille François Bazaine remarked, because of a shortage of French engineers, who would otherwise have been better qualified.[10] Instructions to the collectors of data on meteorology and geophysics specified the ephemeral nature of the enterprise:

> Permanent stations that can be set up by the commission will necessarily be few, and it is for the future Mexican government to see its way to developing the seeds sown by the generous initiative of France. We are carrying out an exploration; we cannot and we must not think of creating meteorological

9. Numa Broc, "Les grandes missions scientifiques françaises au XIXe siècle (Morée, Algérie, Mexique)," *Revue d'histoire des sciences* 34 (1981), 319–58; Lewis Pyenson, "Functionaries and Seekers in Latin America: Missionary Diffusion of the Exact Sciences, 1850–1930," *Quipu* 2 (1985), 387–420; Gary S. Dunbar, "'The Compass Follows the Flag': The French Scientific Mission to Mexico, 1864–1867," *Annals of the Association of American Geographers* 78 (1988), 229–40.

10. PAN, F17/2909/12, Bazaine to Vaillant, 24 Mar. 1864.

> observatories in a country so far away—especially since France herself has such a small number of them.[11]

The mission was one of inculcation. At best they would be perceived as paragons; at worst, as transient parasites.

Mexican men of learning carried out much of the serious collecting and measuring. In the exact sciences, however, the most talented Mexican observer, Francisco Díaz Covarrubias, declined to serve the new regime. In 1862 Díaz Covarrubias had persuaded the federal government to build an astronomical observatory in the Chapultepec garrison and to provide it with an Ertel meridian circle purchased twenty years previously for the national military college but never used; he abandoned his observatory when the French army gained the capital.[12] Even six months after Díaz Covarrubias departed, Maximilian's men hoped to enlist him as chief astronomer to aid in the preparation of maps; they settled on Francisco Jiménez (president of the Mexican commission) and José María Margui, who set up an English telescope at Mexico City.[13] Maximilian then appointed Matthew Fontaine Maury to Díaz Covarrubias's place. The organizer of the United States Naval Observatory, Maury had cast his lot with the Confederacy in the American Civil War. Following the South's defeat, he had gone to Maximilian with a scheme for settling Virginians in northern Mexico. Early in 1866 Director Maury left for London with £72,000 to purchase astronomical equipment for Mexico City; the settlement scheme fell through, and Maury decided not to return to Mexico.[14]

If Díaz Covarrubias could not be turned, a number of Mexican scientists nevertheless welcomed the French invasion as an occasion to demonstrate their capability for scientific work. Manuel Orozco y Berra, who was undersecretary of state for public works, vice-president of the commission in Mexico, and correspondent of the French commission,

11. Ibid., Program, n.d.

12. Marco Arturo Moreno Corral, "Algunos sucesos que dieron origen a la fundación definitiva del Observatorio astronómico nacional de México en 1878," *Quipu* 3 (1986), 299–309; Luis G. León, *Los progresos de la astronomía en México desde 1810 hasta 1910* (Mexico City, 1911), p. 6. Rámon Sánchez Flores, *Historia de la tecnología y la invención en México* (Mexico City, 1980), p. 369, writes that General Pedro García Conde created an astronomical observatory at the Chapultepec Military College in 1846; a French telescope arrived in 1847 and functioned for several months before being destroyed by the invading army of the United States.

13. PAN, F17/2911, Manuel Orozco y Berra to Minister of Public Instruction, 20 Dec. 1864.

14. Moreno Corral, "Sucesos," pp. 304–5. Moreno kindly informs me that the *Mexican Times*, a periodical of American secessionists in Mexico, accused Maury of embezzling the £72,000.

expressed concern to his French overlords that the task confided to him and his countrymen

> will be an element that may show abroad what we can do, something that shows our strength in the sciences; it will diminish somewhat the venom of the barbarians who slander us so unjustly albeit without contradiction; it will be for us to occupy ourselves with something serious when we waste so much time in useless charlatanry.

Orozco insisted that

> the French commission, arriving in Mexico, will have no problems with Mexican engineers, who are completely familiar with handling the most delicate precision instruments, with the most complicated calculation, and with resolving the most difficult problems of topography, geodesy, and astronomy.

Orozco certainly exhibited bravado in his report, yet in the absence of astronomers and geographers from France – none would come beyond those in the company of the invasion force – his words had a certain reassuring ring in Paris.[15] And, with the exception of Maury, only one French-speaking adept of the exact sciences worked directly with the scientist invaders: the Cuban Andrés Poey.

Poey, whose Havana Observatory was proving to be a great deal of work for little reward, proposed himself as a correspondent of the Mexican scientific commission. He wrote to astronomer Hervé-Auguste-Etienne-Albans Faye in February 1864 to report that he had been invited to set up a meteorological observatory at Lima and to ask for a letter of support and encouragement which could be shown to the Peruvian authorities; he had also contacted Mexican authorities about a similar project. This was precisely when Duruy breathed life into the commission, and it is at least possible to read in Poey's letter an urgent desire to be named director of meteorological research in Mexico. In replying to Poey, Faye was careful to specify that any meteorological director in Mexico would have to be a French citizen charged with organizing temporary stations and collaborating with Mexican observers; it would be up to the Mexicans to appoint a permanent director. To E. Lépissier, astronome-adjoint at the Paris Observatory, Faye was more direct: "If a real observatory is organized in Mexico, one that will certainly be the best situated in the entire world, I would like its direction to be given to a professional astronomer like you, just as the physical and meteorological part should be given to a real physicist."[16]

15. PAN, F17/2911, Orozco y Berra's report, n.d., edited by de Lemay.

16. Ibid., F17/2909/12, Poey to Faye, 6 Feb. 1864; Faye to Poey, 30 Mar. 1864; F17/2910, Faye to Lépissier, 12 May 1864.

Within months of the commission's formation, at any rate, Poey had received an appointment as corresponding member and had shipped out to join the expedition's organizers in Paris. To justify his leave of absence from Havana, he emphasized to the Cuban authorities that *"the power of science must form part of the science of power."* Meteorology served the interests of Spain's overseas empire, and to receive the latest word about telegraphical networks in meteorology, Poey had to travel to Europe. The trip would also allow him to standardize his observing techniques with those of the French. He left the observatory in charge of the first assistant, Ricardo Zenoz, and set off with two years of meteorological data, which he hoped to work up for publication.[17]

Poey informed his father from Madrid late in 1864 that Spain would not allow him to remain in charge of the Havana Observatory while he was in Europe and that the Cuban authorities wanted to suspend his salary. The French had provided him with a thousand francs per month, in his view a small sum. He traveled to Paris, where he fretted about Zenoz's intriguing to replace him and to confiscate his personal library and furniture. He obtained supporting letters from the Madrid Academy of Sciences to the effect that he was on leave from Havana for two years. He petitioned the Cuban governor to confirm his mission to Mexico. And he had his French patrons correspond on his behalf. Anatole Duruy, head of the cabinet in the ministry of public instruction and secretary of the Mexican scientific commission, informed Poey that he was unable to have his brother Victor, the minister, send a letter of support to the Cuban captain general because ministers were forbidden from communicating with foreign governments on their own initiative. So Duruy sent the letter to the ministry of foreign affairs for transmission through the French consul in Havana. Zenoz had support from the local Jesuits, and in July 1865 Poey—then commuting between Paris and Madrid—saw his salary diverted to his former assistant.[18]

While trying to retain control of his observatory, Poey was busy promoting his career as the premier Latin American meteorologist. He sought to create a Latin American meteorological network linked to Paris by telegraph. In September 1864, during his stay in Paris, he circulated a "Call to Hispano-American Nations," which consisted largely of testimonial letters from Faye and Urbain Jean Joseph Le Verrier. Faye's letter praised Poey for setting up the first meteorological obser-

17. PMN, Poey to Cuban Governor, 9 June 1864, emphasis in the original; Zenoz to Poey, 31 July 1864.

18. Ibid., Andrés Poey to Felipe Poey, 2 Nov. 1864, 4 Feb. 1865, 12 Mar. 1865; Poey to Governor and Captain General of Cuba, 10 Feb. 1865; Duruy to Andrés Poey, n.d.; Andrés Poey to Felipe Poey, 1 Sept. 1865 and 20 Oct. 1865.

vatory in Latin America and for proposing to coordinate meteorology in Mexico and Peru. Le Verrier, in a letter to Poey of 1860, likewise welcomed the possibility of New World data.[19] The various testimonials exacted a price. Late in 1864 Anatole Duruy named it: a royal Spanish order of merit. Duruy was soon to marry, and he wanted the decoration to impress his intended wife. Poey wrote some seven months later that Duruy had received entry into the American Order of Knights of Isabella the Catholic, nominated by both the president of Madrid's Academy of Sciences and Poey himself.[20]

It is one thing to haunt literary salons and ministerial antechambers; it is something else to prosecute science in the midst of a foreign war. In February 1866 Poey finally arrived in Mexico as the exploration commission's meteorologist. He wrote to his Parisian overseers that he looked forward to setting up a temporary observatory at Veracruz, site of the disembarcation of the French army, although he anticipated locating himself in Mexico City. He underlined the difficulty of combating "Mexican apathy and inertia and other obstacles inherent in the current state of the country"—a perception completely different from Orozco's soothing vision of industrious Mexican scientists.[21] Poey was quick to claim the mantle of Mexico's astronomers manqués. According to a French newspaper published in Mexico City, while the empire waited for Maury to organize his new observatory, Poey had begun to take meteorological observations in the Santa Clara barracks. Poey was no stranger to the Mexican milieu (at least he, unlike Maury, spoke Spanish fluently), and the newspaper anticipated the success of his work.[22]

France received Poey's initiative as something of a question mark. Edme-Hippolyte Marié-Davy, Le Verrier's colleague at the Paris Observatory, roundly criticized Poey's meteorological program. According to Marié-Davy, Poey demonstrated "inexperience with the laws of physics," a failing that undermined the high aims of the program, although the French savant admitted that perhaps "this is only the result of an inexperience with the French language." Marié-Davy was against including Poey's work in the commission's archives: "While taking into account the zeal of Mr. Poey and the services that he can render to the

19. Poey, *Appel aux nations hispane-américaines* (Paris, 1864); Faye's letter dated 30 Mar. 1864, Le Verrier's dated 16 June 1860.

20. PMN, Duruy to Poey, 14 Dec. 1864; Poey to Duruy, 6 July 1865.

21. PAN, F17/2909/12, Poey to Ministry of Public Instruction, 9 Mar. 1866. In PMN, Poey to Governor and Captain General of Cuba, 10 Feb. 1865, Poey signals his appointment as "head" of the expedition's "department of meteorology and geophysics."

22. PAN, F17/2912, articles from *L'ère nouvelle*, 1 Apr. 1866, and *Diario del imperio*, 2 Apr. 1866.

commission, I do not think that our travelers, all of them well educated, and even our correspondents may find his work really useful."[23]

If Poey's French fell short of Parisian norms, he had plenty of experience in promoting his own enterprises; when one project faltered, he touted another in its place. After six months in Mexico, however, Poey had reached something of a dead end. Only a generous critic would not have read failure in his activity. In letters to Paris he insisted on the Baconian nature of his work. He would establish catalogues of "the true, the only *facts* of science," and these he would find in Mexican libraries or collect from stations situated across the country. Poey sent back temperature and rainfall measurements as well as qualitative observations of the seasons and shooting stars. He wanted to stay on in Mexico, as he had to record data during all seasons of the year. He emphasized that his data had limited significance, for many of his instruments had arrived from Havana in an unusable state. Poey shifted his attention to philological inventory—reading through old meteorological accounts of Mexico; he proposed, in fact, to continue the research in Spain upon the expiration of his term with the French commission.[24]

Virtually nothing came from Poey's grand plans; volumes stemming from the expedition appeared up to 1915, but none bears Poey's name.[25] Early in 1867 the Havana Academy of Medical, Physical, and Natural Sciences (newly created in 1861) asked its fellow Felipe Poey to have his son Andrés vacate the rooms that had been placed at his disposal for the duration of the Mexican adventure.[26] Maximilian was also under pressure to retire, and his reign ended before a firing squad in June. By 1868 Poey was still claiming to direct the Havana Observatory; the Cuban administration wanted to know only that he had returned the meteorological instruments used to take measurements in Mexico.[27] Recognizing that the government would no longer keep him on the payroll, the observatory director sued to continue his functions without salary. The Cuban government declared an end to the meteorological experiment.[28] According to the government, Poey had failed to provide for his observatory during his European and Mexican travels, and he had

23. PAN, F17/2912, Marié-Davy to Vaillant, 7 Apr. 1866.

24. Ibid., F17/2909/12, Poey to Minister of Public Instruction, 9 Mar. 1866, for the quotation, emphasis in the original; 9 Sept. 1866; 20 Sept. 1866; 9 Nov. 1866.

25. The closest thing to a synthesis seems to be a prospectus and provisional report appearing in a Mexican newspaper, *La sociedad*, 3 Apr. 1866, in PAN, F17/2912.

26. PMN, F. Valdes y Aguirre to Felipe Poey, 15 Jan. 1867.

27. Ibid., Civil Higher Government to Poey, 6 Nov. 1868.

28. Ibid., Narciso de la Escosura to Poey, 5 June 1869; Poey to Governor and Captain General of Cuba, 2 Apr. 1869; Cuban Ministry of Justice to Poey, 18 Oct. 1869.

failed to demonstrate the worth of his travels for Cuban needs.[29] The observatory's budget of six thousand escudos was canceled; Poey received sixty escudos per year in recognition of a meteorological lecture series.[30]

Poey's demise was due to his principal rivals at Havana – the Jesuits – as much as to his personal failings. In 1854 the Jesuits had founded the Colegio de Belen at Havana, a secondary school with a royal patent. Four years later they were operating a meteorological observatory. The meteorological operation saw a succession of directors, one replacing another on a yearly basis, until 1870, when the young Spanish Jesuit Benito Viñes took charge. Viñes found his secular calling in the weather; he guided the fortunes of the observatory over the next twenty-three years.[31]

Poey retired to France, where he worked as a meteorologist and a convinced disciple of Auguste Comte. In a popular treatise on positivism published in Paris in 1876, Poey outlined how he had come to the new religion. He had, he claimed, been a secret Comtean since 1855, his "politico-religious circumstances" and an observatory directorship having prohibited public adherence to the doctrine. From 1871 to 1873 Poey had attended the lectures of Comte's designated successor, Pierre Laffitte. He spent the rest of his long life spreading the Comtean gospel.[32] In the end, Poey renounced *paix armée* and *pénétration pacifique*, the very policies he had willingly supported during Maximilian's reign in Mexico.[33]

Appropriation of Mexican Astronomy

During the years after the fall of Maximilian, the Mexican government promised to renovate the Chapultepec Observatory for Díaz Covarrubias. It managed only to place him at the head of an expedition to Japan for observing the transit of Venus in 1874. Two years later, the responsibility for renovating astronomy went to a former subordinate of Díaz Covarrubias's, Angel Anguiano.

In 1869 Díaz Covarrubias, then undersecretary of public works, had

29. Ibid., Cuban Ministry of Justice to Poey, 18 Oct. 1869. Articles 25 and 26 of the observatory regulations of 1860 specified both conditions (*Gaceta de Madrid*, 19 Dec. 1860).

30. PMN, Cuban Ministry of Justice to Poey, 18 Oct. 1869.

31. *Vregille*, pp. 38–46.

32. Poey, *Le positivisme* (Paris, 1876), pp. i – xxiii. Poey claimed to have made 100,000 meteorological observations during his nine months in Mexico in 1866 (p. ix).

33. Poey, *L'anarchie mondiale, sa psychologie morbide* (Paris, 1912), p. 124.

asked Anguiano to observe a partial solar eclipse. Anguiano, a twenty-nine-year-old engineering graduate of the Academia de San Carlos, was a roads inspector in Morelia with no experience in astronomy. Díaz Covarrubias trained the younger man, who was in a real sense his student. In 1876 the ministry of public works authorized Anguiano to begin organizing an observatory for astronomy, meteorology, and terrestrial magnetism; his definitive appointment as director came in 1880. Anguiano labored with insufficient means, finally receiving a 38-cm equatorial refractor and a 20-cm meridian circle on the occasion of the transit of Venus in 1882. At just this time, the government decided to move the observatory from Chapultepec to the grounds of the military college at nearby Tacubaya; the military college went to Chapultepec. Lodgings at Tacubaya being unsuitable, in 1884 the government finally allowed Anguiano to begin constructing new buildings.[34]

When Jean Jacques Anatole Bouquet de la Grye headed a mission of naval hydrographers to observe the 1882 transit of Venus at Puebla, Mexico, he sensed only admiration for France and its savants. French engineers were not conspicuous in the scientific and industrial life of Mexico, he admitted, but everyone read French books. "The Mexicans whom we fought twenty years ago were never for us real enemies. The French who remained in Mexico during and after the occupation were not threatened at all. The two peoples felt that one blood had united them, and after the fighting they shook hands."[35] Bouquet soon found a task for Mexican labor.

In 1887 Teodoro Quintana, an astronomer at the Tacubaya Observatory, used his institution's equatorial telescope to take a photograph of the moon. Director Anguiano forwarded the photograph to Bouquet de la Grye, who brought it to the attention of the Bureau of Longitudes, on whose council sat Admiral Mouchez. Impressed with the photograph, Mouchez invited Tacubaya to take part in his sky-chart project. Anguiano received instruments from Paris, and in 1891 he set Quintana and Guillermo Beltrán y Puga to begin the survey. The photographs turned out to be defective due to the poor configuration of the telescopical lens. Anguiano sent the lens with Lieutenant Colonel Quintana to Alvan Clark in the United States for refiguring.[36]

Tacubaya was still under construction in the 1890s. Anguiano in-

34. See, for the most part León, *Progresos;* see also Moreno Corral, "Sucesos," pp. 305–8. According to Leon, pp. 21–22, Anguiano retained responsibility for terrestrial magnetism and meteorology even though a separate and moribund Central Meteorological and Magnetical Observatory had been set up in 1877 at the National Palace.

35. PSG, colis 1 bis, 1452, Bouquet de la Grye, observations on the transit of Venus (1882); *Hydrogr.*, p. 313.

36. León, *Progresos*, p. 27.

The Tacubaya Observatory, Mexico, ca. 1900. For half a century it directed attention to taking photographs and measurements for the *carte du ciel*, a survey of the heavens supervised by Paris astronomers. *Seated, left to right:* Guillermo Beltrán y Puga, Angel Anguiano, Francisco Rodríguez Rey: *standing, left to right:* Abel Díaz, Manuel Moreno y Anda, Camilo González. (Courtesy of Mary Lea Shane Archives of the Lick Observatory, Santa Cruz, California)

formed Mouchez in 1891 that he had not forwarded photographs of the observatory grounds because only the astrophotographical installation had been completed. "The other instruments, such as the equatorial of 0.38 m, the meridian circle the description of which you have seen in our bulletin, the photo-heliograph, another small equatorial, and an altazimuth are provisionally installed."[37] Astronomy, for all the government's fine words, was a luxury in Mexico. In 1899 Anguiano resigned from Tacubaya to become director of the newly created Mexican geodesical commission.[38]

37. PO, "Tacubaya, 1891–1902," Anguiano to Mouchez, 24 Dec. 1891.

38. See Guillermo Beltrán y Puga, *Descripción del observatorio astronómico n. de Tacubaya* (Mexico City, 1893), pp. 43–44, for pride at possessing some of the world's largest telescopes. Leon, *Progresos*, p. 16.

Felipe Valle, who had worked for many years under Anguiano, then rose to become director. The new man conveyed his keen interest in finishing the sky-chart work to Maurice Loewy, director of the Paris Observatory.[39] Over the next decade he prosecuted his task with vigor. Valle died in 1910, several months before the outbreak of the Mexican Revolution. The next director, Valentín Gama, found the destinies of men more interesting than the disposition of stars. In 1914 he became rector of the National University and then secretary of agriculture and public works, ceding the observatory to Joaquín Gallo Monterrubio, a thirty-two-year-old engineer who had haunted the Tacubaya Observatory for more than a decade, had visited observatories in the United States and Europe, and had risen to become *astrónomo titular* and assistant director. Gallo continued on the sky chart through the revolution, suffering periods of inconstant funding. He took a doctorate at Northwestern University in 1928, taught and popularized astronomy, observed eclipses, and standardized Mexican clocks, but the sky chart constituted the principal scientific focus of his observatory. Gallo finished his commitment to the French astronomers in 1943, three years before he retired. His career remained outside the research front of twentieth-century astronomy, which, when it entered Mexico, did so at a new observatory built on a hill near the city of Puebla.[40]

The Refugee from French Despotism

Simon Schwartzman has emphasized that "France's drive to build a modern, efficient, and secular state after the Revolution was more politically meaningful to the Brazilian elite than the intellectual and cultural climate of England," where capitalist "economics took precedence over politics and sociology."[41] Brazilians were fascinated by scientism of the Fourierist and Saint-Simonian variety, and the 1840s and 1850s found graduates of the Ecole polytechnique at work on engineering projects along the Brazilian coast. They and other French émigrés spread the new religion of social progress through technological transformation.[42]

39. PO, "Tacubaya, 1891–1902," Valle to Loewy, 24 June 1899.

40. Marco Arturo Moreno Corral, "El catálogo fotográfico y la carta del cielo," *Anuario del observatorio astronómico* (Mexico City), 108 (1986), 172–79. In idem, ed., *Simposio de historia de la astronomía en México* (Mexico City, [1982]), see Joaquín Gallo Sarlat, "Entre eclipses y cometas: Reminiscencias de la vida de Joaquín Gallo," pp. 245–66; and Bart J. Bok, "Astronomía mexicana, 1930–1950," trans. Jana Benda, pp. 267–80.

41. Schwartzman, *Science and Higher Education in Brazil: An Historical Overview* (Washington, D.C., 1980) [Smithsonian Institution, Wilson Center, Latin American Program, Working Paper no. 9], p. 4.

42. A general survey appears in Maria Amelia M. Dantes, "Le positivisme et la science

Brazilians responded in kind; among Latin American countries around the middle of the nineteenth century, only Spanish Cuba supplied more students than Brazil to the Ecole centrale des arts et manufactures.[43] Brazil's first important astronomer was a French scientist who under other circumstances might have risen to great heights in his native land.

The astronomer, Emmanuel Liais, was born in Cherbourg in 1828. He completed the local collège in 1843 and then studied science independently. By the early 1850s he had published on diverse topics in the Cherbourg natural-science society (of which he was a founder) and in the *Comptes-Rendus* of the Paris Academy. He visited the Paris Observatory in 1852 and impressed its director Dominique François Jean Arago, who promised him the first vacant place in astronomy. Arago died late in 1853, before making the appointment. The directorship fell to Le Verrier, who, having to reconstitute Arago's operation, saw no reason not to place confidence in Liais. Le Verrier named Liais élève-astronome and then astronome-adjoint in 1854 and astronome titulaire in 1857. Liais collected membership in foreign scientific societies — the Pontifical Academy, the Royal Bohemian Society in Prague, and the Royal Society at Liège — and was inducted into the Legion of Honor. His own most vigorous champion, Liais seemed poised for a brilliant career.[44]

Le Verrier, jealous and misanthropic, stopped at nothing in transforming the observatory into a true state service. He placed Liais in charge of the scientific part of meteorology, no doubt recognizing that Liais's numerous publications, while testimonial to ardor and industry, lacked a sharp and critical edge. Liais wanted more — election to the Paris Academy in the astronomical section, for example. Le Verrier assimilated Liais's results for the greater glory of the observatory and its director. Liais persisted in publishing his own papers, and Le Verrier cut off his salary. After what Liais later described as "a violent scene"

au Brésil," in *Science and Empires*, ed. Patrick Petitjean, Catherine Jami, and Anne Marie Moulin (Dordrecht, 1992), pp. 165–72; one of the French engineers, Luis Léger Vauthier, is the subject of a long and rambling treatise by Gilberto Freyre, *Um engenheiro francês no Brasil*, 2 vols. (Rio de Janeiro, 1960).

43. In the classes of the Ecole centrale from 1841 to 1867, there were thirty-three Cubans, twenty-eight Brazilians, eighteen from the French Antilles, seven each from Argentina and Peru, and fewer from other Latin American countries (Châtenay-Malabry, archives of the Ecole centrale).

44. The principal source for Liais is Jacques Ancellin's exhaustive study, *Un homme de science du XIXe siècle: L'astronome Emmanuel Liais* (Coutances, 1985) [Cherbourg, Société nationale des sciences naturelles et mathématiques, Mémoires, 57].

between the two astronomers, Liais left the observatory grounds without the courtesy of a dismissal notice.[45]

Suspended without pay, Liais looked beyond Le Verrier's clutches. He asked the ministry of public instruction for a mission to Brazil, where he had family and facilities. The object of his mission concerned atmospheric measurements on his Atlantic passage and, secondarily, observations in Brazil to be taken during an upcoming solar eclipse. He requested a salary, since his post at the observatory had been discontinued "for purely administrative reasons."[46]

Liais's project received high marks from astronomer Hervé Faye, to whom it had been sent for an appraisal. Faye thought that if the ministry were to fund an eclipse expedition, then it had to give priority to sites in Peru previously identified by physicist Jean-Baptiste Biot as suitable for remeasuring the meridian arc defined in the eighteenth century by Pierre Bouguer and Charles-Marie de La Condamine. Faye concluded his recommendation with a plea in favor of funding Liais as a private scientist:

> It is true that one can object that the encouragement of the state must legally be directed to state institutions, but I protest in advance against such a doctrine, which essentially restricts the rights of the ministry and the initiatives of individuals. This novelty [the ministry's restrictive policy] – for I have no memory that such a strange thesis has ever been proposed – this novelty, I say, would not be favorable to the true interests of science. If it is permitted to carry on geodesy outside the confines of military establishments, astronomy outside the observatory, geology outside the *corps des mines*, etc., it must also be permitted to the government to grant its liberal support to ideas, projects, and enterprises. The sun and the government of the emperor exist for everyone.[47]

Faye's eloquence failed to produce the desired effect. Liais received no money for his voyage, but the ministry designated him an official representative of France in order to facilitate his obtaining support in Brazil. The ministry of the navy and colonies agreed to lend him a chronometer, a theodolite, and three magnetical compasses, worth in all some Fr 4,850. Liais then set off to practice astronomy in South America.[48]

Liais's voyage to Brazil came at an opportune time. In 1856 the tal-

45. See, generally, ibid. See also PAN, F17/2985A, "Emmanuel Liais," Liais to Ministry of Public Instruction, 26 June 1865, where Liais describes the circumstances of his departure; the dossier also contains a printed vita of 1858.

46. PAN, F17/2985A, Liais to Minister of Public Instruction, 9 Feb. 1858.

47. Ibid., Faye to Minister of Public Instruction, 27 Feb. 1858.

48. Ibid., First Bureau, Ministry of Public Instruction to Liais, 18 Mar. 1858; Ministry of the Navy and Colonies to Ministry of Public Instruction, 19 May 1858.

ented, French-acculturated mathematician Joaquím Gomes de Sousa had left the military school at Rio de Janeiro, the institution that administered the local observatory. The observatory itself, directed from 1845 to 1850 by Engénio Fernando Soulier de Sauve and afterwards by the former minister of war Antônio Manuel de Melo, was modestly equipped, but it lacked a serious commitment to research. The situation could be exploited by an enterprising expatriate.[49]

The French astronomer found a warm welcome from Pedro d'Alcantara, emperor of Brazil, Liais reported to the ministry of public instruction in 1858. The emperor placed two steamships at his disposal for observing the upcoming solar eclipse. "Here the sciences are held in higher regard than among the French public," Liais exulted, "and before my arrival one deplored the insouciance of Europeans with regard to studying the great physical questions lying at the base of eclipse observations." Liais reported that he had carried out the objectives of his eclipse expedition, verifying that the solar corona was the sun's atmosphere.[50]

Over the next months and years, Liais wrote a battery of letters to the educational authorities in France. He sought to dissuade them from thinking that he had been forced to leave France "under the cloud of disgrace"; he averred that Le Verrier's accusations had been delivered "not only against me, but also against more than twenty other persons who left the observatory under his administration." The ministry of public instruction considered Liais's request for amnesty, but nothing came of it except a letter dunning the expatriate for the return of the instruments that he had borrowed in 1858.[51]

Liais labored on astronomy, cartography, and natural history, all with money from Emperor Pedro II. Then he returned to Cherbourg to regain his health and flog his arrested career. In a letter of 1865, Liais underscored his attempts to spread French culture. He had served peace, he emphasized, advanced international relations, and "also struggled against the English and the Americans, who everywhere try

49. Abraão de Morais and Abraham Szulc, "A astronômia no Brasil," in *As ciências no Brasil*, ed. Fernando de Azevedo, 2 vols. (Rio de Janeiro, 1955), 1:8–161. In the same volume see F. M. de Oliveira Castro, "A matemática no Brasil," pp. 41–77. The history written by an early director of the Rio de Janeiro Observatory has recently been published: Henrique Morize, *Observatório astronômico: Um século de história (1827–1927)* (Rio de Janeiro, 1987) [Museu de astronômia e ciência, Coleção documentos de história da ciência, 1], here pp. 10–11. On one of Brazil's earliest accomplished mathematicians, see João Barcelar Portela, *Gomes de Souza e sua obra* (São Luis, Maranhão, 1975).

50. PAN, F17/2985A, Liais to Minister of Public Instruction, 6 Oct. 1858.

51. Ibid., 9 May 1859, for the quotations, and 7 June 1859, across which is written the word *amnistie;* Minister of the Navy and Colonies to Minister of Public Instruction, 14 Dec. 1861.

to substitute themselves for our nationals under the pretext of a practical superiority, which is really illusionary but which, a bit unfortunately, has become a prejudice."[52] After a decent interval, Liais made ready to present his candidacy for the Academy of Sciences in the geography and navigation section. The hopeless application met with its predestined rejection in the spring of 1866. Early in 1867, Liais set off for Brazil once more. There he observed the skies and worked as an engineer. His labors and self-promotion continued to resonate with the sensibilities of the Brazilian emperor. In 1870 he was called to direct the Rio de Janeiro Observatory, succeeding the late military director Antônio Joaquim Cruvelo d'Ávila.[53]

Liais made his acceptance conditional on a reorganization of the observatory. First of all, he wanted it separated from the military school. To provide the observatory with a committee of patronage, Liais had Emperor Pedro II create a longitudes commission. The commission's members, however, all senior military men or illustrious scientific figures, declined to be presided over by Liais, whom they considered their inferior. Liais hoped to break the administrative deadlock by receiving permission from the emperor to be addressed as "excellency," a title usually reserved for generals. After supervising the observatory for seven months, Liais returned to Paris with the object of purchasing instruments. He continued to style himself director of the observatory, but for three years the actual direction was carried out by the *visconde* de Prados, the only member of the longitudes commission to have recognized Liais's authority. Liais declined to return to Brazil until he received better terms from the emperor; only in November 1874 did Liais take up residence in Rio de Janeiro.[54]

The real reason for Liais's precipitous return to Paris, and the excuse for his extended stay there, is not difficult to discern. In January 1870, Le Verrier was dismissed as director of the Paris Observatory, following the resignation of his staff and their circulation of a *cahier de doléances;* Le Verrier's worst enemy, Charles-Eugène Delaunay, came in as his replacement. Liais would surely have been hoping for justice under the new regime. But the conditions offered, if there were any, could not compare with a directorship at Rio de Janeiro, and Liais, who had become comfortable dealing with imperial courts, found the Third Republic uninspiring. Delaunay died in a shipwreck at Cherbourg in 1872, and early the next year Le Verrier was rehabilitated as Delaunay's successor. From that point forward, Liais would have found no profes-

52. Ibid., Liais to Minister of Public Instruction, 26 June 1865.

53. On Liais in France and the Legion of Honor, see Ancellin, *Liais*, pp. 83–108; for Liais's call upon the death of Cruvelo d'Ávila, see p. 134.

54. Ibid., pp. 133–44; Morize, *Observatório*, pp. 66–70.

sional attraction in his native land. In 1874 there was additional personal motivation to flee overseas: death took away his wife, who had been his constant companion and assistant in France and Brazil.[55]

The director *in absentia* returned to take up his post at Rio de Janeiro late in 1874. His new equipment was adequate for serious work—a 25-cm equatorial from Dolland and a 37-cm azimuth constructed at the observatory—and the grounds had been refurbished.[56] He had as staff four astronomers, three student astronomers, and three maintenance men. One of his astronomers, Luis Cruls, acted as assistant director; another astronomer, Julião de Oliveira Lacaille, had spent three years training in Paris. Liais occupied his staff with routine meteorological and geographical tasks, and he suspended educating naval officers. Together with Cruls, he used the 25-cm equatorial to catalogue Southern Hemisphere double stars, and the two astronomers recalculated orbits of planets and asteroids. The epoch was auspicious for new undertakings in higher education: in 1874 the Escola central of Rio de Janeiro became an Escola politécnica, a reflection not of the French homonym controlled by the army but rather of the civilian Ecole centrale des arts et manufactures in Paris; French inspiration also produced, in the same year, an Escola de minas in Ouro Preto.[57]

It took only four years for Liais to come into conflict with three dissident staff, one of whom he considered to be incompetent. The leading dissident, astronomer Manoel Pereira Reis, was also a professor at the naval school and had strong support from Brazilian nationalists. The matter appeared in the daily press. By 1880 Pereira Reis had obtained authorization to erect a second observatory in Rio de Janeiro for the use of the navy, and Liais found his funding reduced, despite the vigorous support of his friend the minister of war. Liais's position weakened through professional infelicity. Astronomers from the United States had reevaluated Liais's coastal mapping and found it wanting; notably, his value for the longitude of Rio de Janeiro was off by more than 25 sec. The new value echoed Admiral Mouchez's questioning of Liais's latitude determinations during the 1860s.[58]

Late in 1880 one dispassionate observer found Liais in a precarious

55. *Astron.*, s.v. "Le Verrier"; Ancellin, *Liais*, pp. 129–32.

56. Acquisition and shipment of Liais's new equipment cost the Brazilian government 24 million reis (Morize, *Observatório*, p. 72). At the time, 1 million reis translated into about Fr 2,800–the monthly salary Liais demanded from the emperor in his negotiations from the middle of 1871 (Ancellin, *Liais*, p. 135).

57. De Morais and Szulc, "Astronômia," pp. 128–34. Lacaille followed courses at the Ecole des mines in Paris, although he did not take a diploma there (Margarida Rosa de Lima, *D. Pedro II e Gorceix: A fundação da Escola de minas de Ouro Preto* [São Paulo, 1977], p. 29).

58. Ancellin, *Liais*, pp. 142–57; Morize, *Observatório*, pp. 75–81.

state. At that time an independently wealthy Dutch physicist, Elie van Rijckevorsel, arrived in Brazil to construct a magnetical map of the country. Van Rijckevorsel's assistant Willem Rijklof Adriaan van Alphen described a meeting with Liais to Utrecht astronomer Jean Abraham Chrétien Oudemans. The director at Rio de Janeiro cut a poor figure with his bravado and self-promotion. He was little liked in Rio de Janeiro, Van Alphen continued, and enjoyed his position only because of the emperor's patronage. "He did not make a good impression on me at all, not as a scientist and not as a human being either." His observatory contained instruments that were largely unused. One very large telescope had not been mounted. "What a pity."[59]

Liais did not last long into 1881. He resigned and returned to France, where he put a colophon to his scientific work and made a career in provincial politics. His successor was Luis Cruls, a native of Belgium who, after receiving training at the University of Ghent, had become a military engineer. In 1871 Cruls signed on with the Brazilian geographical service and then found his way to the observatory at Rio de Janeiro. Adjunct astronomer in 1879, he was promoted to first astronomer early in 1881 and then shortly thereafter to interim director, a temporary appointment regularized in 1884.[60]

Cruls watched for the usual celestial events—planetary transits, comets, and eclipses—and he expanded the observatory's involvement in meteorology and geophysics. His observatory left much to be desired.[61] When Admiral Mouchez convened the first sky-chart meeting in Paris in 1887, Cruls was among the invited guests. He obtained a portion of the southern sky. The emperor promised Fr 40,000 for a Gautier telescope, and Cruls sat waiting for its delivery. But the astronomer had not secured funds to mount and house the new telescope, and the government decided to have its astronomers map the empire. The Brazilian legislative assembly came through with 30 million reis for the construction of a cupola and photographical darkroom. Cruls planned a campus for the new acquisitions. He did not anticipate that Pedro II would disappear into exile in Portugal late in 1889 and the military would proclaim a republic.[62]

Political revolutions do not challenge the authority of scientists so much as demand that the abstract knowledge under their command be

59. *ER*, p. 95.

60. Morize, *Observatório*, pp. 81–82.

61. PO, carton 10, "La Plata, 1886–1907"; François Beuf to Admiral Mouchez, 23 July 1887: "I saw Cruls briefly at Rio and visited his observatory. It is quite miserable; but Cruls is nice."

62. See Morize, *Observatório*, pp. 108–35, on the observatory during Cruls's regime.

turned to the material profit of the new regime. So it was in Brazil. Luis Cruls continued to direct the astronomical fortunes of the nation, but he had to educate engineers and surveyors and, in addition, devote much time to cartography. To discipline Brazil's master astronomer, the new military government transferred the observatory to the ministry of war.

Initially, Brazil's professional killers supported Cruls's commitment to the sky-chart work. In 1892, however, Cruls was seconded to the ministry of industry, rail, and public works, along with two astronomers and two mechanics, to survey the central Brazilian highlands for a new national capital. The depleted observatory, left in the hands of the assistant director, became a defensive battery during the civil war of 1893. For three years Cruls administered through a proxy, hoping that funding would come through for his telescopes and their installation. Finally, on the last day of 1896, the observatory passed from the ministry of war to the ministry of industry, and Cruls regained direct control of its activity. He took delivery of a Cooke photographical equatorial, the Gautier meridian circle, and the Gautier photographical equatorial. By 1900 he had made substantial progress toward mounting the latter, which was destined for measuring the primary geodesical question of the day—the variation of terrestrial latitude. The government would not pay to place the equatorial in working order, however; having cost the government two thousand pounds, it remained boxed. The sky-chart masters in Paris transferred Cruls's mandate to the Perth Observatory.[63] Cruls's health declined into the new century. In 1908 he died in Paris. It was left to his understudy and successor Henrique Morize to move the observatory to a new location and mount the telescopes.

The Envoy of Imperialist Designs

Cultural affinities may grow stronger in uncertain times, for in attempting to chart a new course we backsight along our experience. In cultural matters, Pedro II, emperor of Brazil, looked to Napoleon III, emperor of France, for inspiration and instruction. When the French empire collapsed and the German empire was proclaimed, Pedro saw no reason to change his imperial allegiance. In its push to modernize scientifically during the 1870s, the Brazilian empire imported models and manpower from France. Pedro II, whose scientific pretensions had been gratified by a position as foreign associate in the Paris Academy of

63. "Conferenza internazionale per la carta fotografica del cielo," *Memorie della Società degli spettroscopisti italiani* 29 (1900), p. 3 of a ten-page monographic reprint.

Sciences, asked France to design a new educational institution that could train men to exploit Brazil's mineral wealth.[64]

The Brazilian emperor visited Europe in 1871 and 1872. There he spoke with his academician colleague Auguste Daubrée about the best way to develop Brazil's resources. Daubrée suggested a project to map the geology of Brazil, with instruction to be provided by foreign professors or by Brazilians trained abroad. Pedro II implored Daubrée to set the scheme in motion on Brazilian soil, but Daubrée, recently named director of the Ecole des mines in Paris, clung to the pinnacle of his career. Daubrée, knowing that the Brazilian emperor would wait for him to designate a suitable candidate, looked over the recent graduates of France's finest finishing schools. Late in 1873 his gaze came to rest on Henri Gorceix.

The son of a small rural property owner, Gorceix lost his father at an early age. He attended lycée at Limoges on a scholarship, and entered the science section of the Ecole normale supérieure in 1863. Graduating in 1866 and placed as a teacher at the lycée in Angoulême, he implored Louis Pasteur, one of his professors at the Ecole normale, to find him a scientific position in the capital. The following year he became préparateur in geology and mineralogy at the Ecole normale. Two years in a junior capacity at his alma mater were enough to propel Gorceix abroad. When in 1869 a science section came to the French school in Athens (which until then had limited its activity to archaeology and history), Gorceix took charge of it. Gorceix carried out geological research and teaching in Greece and Asia Minor until 1874, when he returned to Paris to work up publications that would allow him to receive a university chair.[65] Two weeks in his native land brought a nomination from Daubrée for the Brazilian post.[66] Gorceix set off, as he wrote to his brother, on something of a temporary adventure: "I dreamed about a bit a glory, some to-do about me; I suffer the penalty of my pride."[67] Thirty-one-year-old Gorceix arrived in Rio de Janeiro with a mandate to organize geological instruction.

Gorceix's organizational focus reflected his training. His Brazilian school, far from offering practical training for miners, would produce mining engineers in the model of the écoles des mines at Paris and St-

64. The account here follows two fine studies: Rosa de Lima, *Pedro II e Gorceix;* and José Murilo de Carvalho, *A Escola de minas de Ouro Preto: O peso de glória* (São Paulo, 1978).

65. See Septime Gorceix, "Une grande oeuvre française au Brésil (avec des lettres inédites de l'empereur D. Pedro II)," *Revue de l'Amérique latine* 14 (1927), 304–15, for Gorceix's career before he went to Brazil.

66. Rosa de Lima, *Pedro II e Gorceix*, p. 269.

67. Gorceix, "Grande oeuvre," p. 307.

Etienne; the standards of general instruction would derive from the Ecole normale. The faculty of Brazil's Escola politécnica, just granted independence from the military, wanted no rival in the matter of producing generalist engineers. As a consequence, Gorceix's school went to Ouro Preto, far away from the imperial capital. According to Gorceix's original plan, classes were to be small, professors were to have modest salaries, and the state would not be obliged to hire the school's graduates. The school acquired a life of its own, evolving to set the pace for science education in Brazil. Gorceix's sequence of courses extended from an original two years to more than six years; the number of his students rose from four to eighty-seven. His handful of professors earned double the median salary of colleagues in engineering, law, and medicine at other institutions.[68] Among the teaching corps were Armand de Bovet, a graduate of the Ecole polytechnique, who stayed on until 1882; Arthur Thiré, a polytechnicien and ingénieur des mines, who taught from 1878 to 1887; and Paul Ferrand, also ingénieur des mines, who taught from 1882 to 1900.

Gorceix's graduating engineers found a depressed job market waiting for them when they left Ouro Preto. Industry preferred foreign-trained men, offering little in the way of salary inducements to Brazilians. Among other institutions of higher education—the polytechnical school, the national museum, and the Collegio Pedro II—none was favorably disposed to hire Gorceix's engineers. "From the beginning," Gorceix wrote to the emperor in 1879, "the school at Ouro Preto seems to have been ostracized!"[69] Geological surveys undertaken by the various governments of Brazil constituted a third employment sector. The imperial authorities found inspiration for mapping in the activity of Charles Frederick Hartt, a geologist from the United States who devoted much of his career to Brazil, and Orville Derby, another U.S. geologist, who found a position in the national museum at Rio de Janeiro and went on to direct the Geological and Geographical Commission of the Province of São Paulo. Derby's commission entailed creation of a meteorological service, complete with an observatory for making astronomical measurements.[70] Gorceix's students never re-

68. Murilo de Carvalho, *Escola*, pp. 28–61. The St-Etienne school produced the most famous chemist to work in Latin America during the nineteenth century, Jean-Baptiste Boussingault (Frederick William James McCosh, *Boussingault: Chemist and Agriculturist* [Dordrecht, 1984], p. 26).

69. Murilo de Carvalho, *Escola*, pp. 49–50.

70. The history of São Paulo's meteorological service and its organic relationship to Derby's commission is provided in "Esboço histórico," São Paulo, Universidade Instituto astronômico e geofísico, *Annaes* 1 (1928), 9–40.

ceived a mapping contract similar to the one obtained by the English-speakers.[71]

Entrepreneurial commitment would have been required for exploiting the Province of Minas Gerais (of which Ouro Preto was the capital until the end of the century). Enterprise of this kind was lacking in Gorceix, his staff, and his students. "We belong to the Latin races," Gorceix informed Pedro II in 1879 when he defended the French style of instruction. "We do not have the adventurous genius of the Saxons or the patience and tenacity of the Germans. We have other qualities of emulation, of intelligence. Let us use them, but let us not destroy them by following a program contrary to our genius."[72] Gorceix's genius led him to depend on the direct intervention of Pedro II to secure his school's budget. While Gorceix could bypass jealous competitors in obtaining funds, he was unable to generate demand for his graduates, whom he groomed to sit as senior managers in a corporate structure. The strategy, closely following the pattern of the grandes écoles in France, entailed teaching about natural phenomena irrelevant to mining engineers. When one student focused on geological exploration and metallurgy to the exclusion of physics, mathematics, and mechanics, he incurred Gorceix's displeasure.[73] The prestige of the French grandes écoles depended, however, more on the research carried out by their staff than on their restrictive admissions policy and advanced curriculum. And while the staff at Ouro Preto shined as teachers, they could point to little in the way of significant research. The school's periodical featured tedious and obscure reports describing minerals or the lay of the land — texts that could not command authority among colleagues.

Gorceix never ceased to dream of a career in France. He sent a stream of notes and communications to his former professor Achille-Ernest-Oscar-Joseph Delesse, who inserted them in French periodicals. Gorceix had no illusions about the importance of his publications; because of his pedagogical activity in Brazil, he wrote to Pedro II in 1878, he had to "give up all personal work that would have been able to call attention to me in France."[74] Delesse died in 1881, and Gorceix, then on leave in France, was proposed as his successor at the Ecole normale. Things did not go his way, and Gorceix returned to Ouro Preto

71. Rosa de Lima, *Pedro II e Gorceix*, pp. 131–33; Murilo de Carvalho, *Escola*, pp. 48–52. Gorceix implored Pedro II to map Brazil in the way that Ignacio Domeyko and Aimé Pissis were mapping Chile (Gorceix to Pedro II, 8 Feb. 1879, in Rosa de Lima, *Pedro II e Gorceix*, p. 174).

72. Gorceix to Pedro II, 21 June 1879, in Rosa de Lima, *Pedro II and Gorceix*, p. 177.

73. Gorceix to Pedro II, 21 June 1879, in ibid., p. 178.

74. Ibid., p. 71; Gorceix to Pedro II, 18 Feb. 1878, in ibid., p. 169.

in 1882.[75] In the following years he vainly intrigued to fill correspondent chairs in the Paris Academy vacated by John Lawrence Smith and Quintíno Sella; he succeeded in receiving a prize offered by the academy in memory of Delesse and endowed by Delesse's widow.[76]

Gorceix faced a student body increasingly resentful of the French professors who stood at the lectern.[77] He fell out with Arthur Thiré, whom he managed to expel from the school.[78] We need not wonder about Gorceix's devotion to Brazil: in 1885 he married a Brazilian and renegotiated a five-year contract as director.[79] It would be his last. The eclipse of Gorceix's school was written in the revolution of 1889 and the exile of Pedro II. Two years later, Gorceix left Ouro Preto. He stayed on for a while in São Paulo and then retired to Mont-sur-Vienne in Limousin, France.

Henri Gorceix and Emmanuel Liais died in provincial France, where they had roots and sentimental attachments. Andrés Poey, the Cuban scientist mercenary, spent much of his life in Paris. In their self-conceived missions of spreading French scientific influence, they served emperors in the New World as well as imperialist ministers in Paris. Their presence abroad was resented, and their missionary zeal produced little reaction among scientist colleagues at home. Entrepreneurial adventurers similar to French contemporaries in Chile and Argentina, they failed to execute scientific programs of interest to French patrons.

75. Rosa de Lima, *Pedro II e Gorceix*, p. 79; Murilo de Carvalho, *Escola*, p. 53.

76. Rosa de Lima, *Pedro II e Gorceix*, pp. 127–30. Apparently James Hall and Joseph Prestwich obtained the vacant seats.

77. Ibid., pp. 97–102; Gorceix to Pedro II, 29 Feb. 1884, where Gorceix cites the articles of a protestor who compared the school "to the colonies of England" (ibid., p. 212).

78. Ibid., pp. 125–26.

79. Gorceix to Pedro II, 20 Apr. 1884, 25 June 1885, 10 Dec. 1886, and 13 Jan. 1887, in ibid., pp. 214, 225, 230, 232. The marriage to Constance da Silva Guimarães took place in June 1885; Pedro II presided over the baptism of Gorceix's daughter, born in December 1886.

10

Quebec and Martinique: Reassertion of Metropolitan Authority

Patronizing French Canada

In 1920 the province of Quebec contained some 2 million French-speaking inhabitants, making it the largest French community beyond Europe. There was in Montreal a tradition of serious scientific research at English-language McGill University, where William Osler and Ernest Rutherford had begun their stellar careers.[1] No comparable tradition existed at French-language Laval University in Quebec City or at French-language medical and engineering schools in Montreal.[2] In the French sector, education at all levels remained under the control of several orders of the Catholic church. After fifty years of rapid expansion, Quebec's polyglot metropolis, Montreal, was also the commercial capital of Canada. It was the juncture where rail lines from the west met oceangoing ships. English-speaking financiers controlled the fortunes of the city and the destiny of the Canadian dominion. The Quai d'Orsay in Paris sent scientists into this inhospitable territory as agents of French culture. Their entry was facilitated by a special, bilaterally funded institution that had quarters in the newly founded University of Montreal.[3]

The University of Montreal emerged in 1918 when a branch of Laval

1. Stanley B. Frost, *McGill University: For the Advancement of Learning*, 2 vols. (Montreal, 1980–84).

2. Yakov M. Rabkin and J. Ann Lévi-Lloyd, "Technology and Two Cultures: One Hundred Years of Engineering Education in Montreal," *Minerva* 22 (1984), 67–95.

3. The nineteenth-century background to science in French Canada is surveyed in Clelia Pighetti, *Scienza e colonialismo nel Canada ottocentesco* (Florence, 1984), esp. pp. 165–202. The role of provincial politics early in the twentieth century is considered in Raymond Duchesne, *La science et le pouvoir au Québec* (Quebec City, 1978). The early years of the University of Montreal are sketched in Olivier Maurault, *L'Université de Montréal* (Montreal, 1952). The history of recent French influence in French Canada is provided in Luc Roussel, *Les relations culturelles du Québec avec la France, 1920–1965* (diss., Laval University, 1983).

University acquired a charter from the religious authorities in Rome. The new institution was presided over by a faction of the bourgeois and ecclesiastical elite of French Canada who sought to populate the liberal professions. By all accounts, secular scholarship and scientific research remained foreign to the desires of the founders. Far more important was to assemble a faculty who were devoutly Roman Catholic and practised in the art of lecturing. The pool of local talent being insufficient for these ends, university administrators looked abroad for staff. They did not think to fathom the French university system for candidates nor to inquire of the Belgian state universities, much less the University of Geneva. Only European Catholic institutions would do. For the post of biology professor, French Catholics identified Louis-Janvier Dalbis to the *Canadiens.*

At the time of his recruitment by the university rector, Dalbis sat as an undistinguished thirty-nine-year-old teacher at the Collège Stanislas in Paris. The latter, elite secondary school was closely affiliated with the Marianist order and claimed as a staff member the philosopher Jacques Maritain.[4] Dalbis was not in Maritain's league; although he had authored several teaching manuals, he seems never to have published an original scientific paper. When called to Montreal he – like nearly all his new colleagues – lacked advanced certification; in his case, he did not have the doctorat ès sciences. It took him five years to settle into his tasks in the new land. He received a salary of thirty-five hundred dollars and a travel bonus. Having the service of a demonstrator paid at around fifteen hundred dollars to relieve him of unpleasant chores, and without the least interest in himself advancing the frontiers of knowledge, Dalbis had time on his hands. He was active in setting up the Montreal Société de biologie and in founding a similar forum for other high-minded citizens, the Association canadienne-française pour l'avancement des sciences (ACFAS). Dalbis realized his principal contribution to scholarly life, however, through his role at the Institut scientifique franco-canadien (ISFC).[5]

As the ISFC was Dalbis's brainchild, he became its first president. The aim of the institute was to bring illustrious professors from France for short lecture tours in Montreal and the rest of Canada, to send French Canadian professors to the Sorbonne (the implication being that they would teach French Canadian studies there), and to organize special scientific "missions" of a nature mutually agreeable to French Canada and France. Lawyer Edouard Montpetit, secretary general of

4. Collège Stanislas, *Le Collège Stanislas (1804–1979)* (Paris, [1979]), p. 14.

5. Roussel, *Relations culturelles*, pp. 47–50, 82–84, 138–40, and passim; Robert Rumilly, *Le frère Marie-Victorin et son temps* (Montreal, 1949), pp. 89, 132.

the university and vice-president of the ISFC, emphasized that the new institute was a clearinghouse, for it had neither buildings nor professors of its own. From its foundation in 1926 until 1931 it found lodgings in the rooms of the University of Montreal; thereafter, it took refuge in the Ecole polytechnique, the school that served in lieu of the university's lacking engineering faculty.[6]

In Montpetit's eyes, the institute had been created to contribute directly to the technological development of French Canada. He wanted French scientists to bring "their methods, their ways of doing things," to Montreal. "We need French science, French methods," he exclaimed. The whole operation was "to serve the expansion of French culture." The problem with this formulation, however, was a lack of symmetry. French Canada had no illustrious scientists to send to France. Then, too, there was the matter of the propagation of the faith, a consideration that weighed heavily against inviting, say, Jewish or Marxist professors from France. It is typical of the situation that the medievalist and philosopher Etienne Gilson, a traveling propagandist for French culture overseas, came as the first French envoy, although one senses from remarks made in the opening session of the ISFC that this choice slighted the desideratum for visiting professors to lecture on the natural sciences.[7]

Dalbis had originally envisioned McGill's Department of Romance Languages as a party to his institute. McGill refused to contribute to this cause, but its administrative head, General Sir Arthur Currie, turned up at the ISFC's inauguration. Dalbis sent his distinguished visitors to the city's premier university. He wrote to McGill physicist Albert Norman Shaw: "The speakers of the Franco-Canadian Scientific Institute will always be very happy to go to McGill University and to speak there as often as possible." When the institute brought in Léon Brillouin for a short visit, McGill physicists were glad to have him talk. And another French physicist missionary, Georges Déjardin from Lyons, was elected an honorary life member of the McGill Physical Society. There was, of course, much showmanship involved in these exchanges between French and English institutions. Arthur Stewart Eve, professor of physics at McGill, was quick to invite Dalbis to Werner Heisenberg's lectures when Heisenberg passed through McGill in 1929.[8]

6. MUM, P12/A1, Edouard Montpetit, "L'Institut scientifique franco-canadien," n.d.

7. Ibid., for the first quotation; "Rapport sur l'Institut scientifique franco-canadien," n.d.; P12/E3, French Consul General in Montreal to Dalbis, 20 Mar. 1928, for the second quotation; P12/A1, document without title.

8. Ibid., P12/E1, McGill University to Dalbis, Apr. 1926; Dalbis to McGill University, 6 May 1926; Principal of McGill to Dalbis, 18 Jan. 1927; Dalbis to Shaw, 19 Mar. 1929;

Regarding the other French-language university in North America, Laval University, Dalbis was more circumspect. Its scientific pedigree was longer than the University of Montreal's, and it resented the expenditure of provincial monies for the ISFC at Montreal. Laval scientists seem to have concentrated their lobbying on ACFAS, an avowedly nationalistic pressure group that rejected French intrusion into the affairs of French Canada. ACFAS found a vocal supporter in Dalbis's principal rival at the University of Montreal, the botany professor and director of the botanical gardens, Brother Marie-Victorin. The two biologists fell out over the ISFC and over provincial politics.[9] Dalbis was obliged, at any rate, to inquire at Laval on behalf of his missionaries. Some made the trip north; others, like Brillouin and physicist Jean Cabannes, did not. As Abbé Philéas J. Filion, rector of Laval and a former professor of chemistry, demurred to Edouard Montpetit regarding Cabannes, "We don't believe that we can find an audience sufficiently well prepared for this sort of talk."[10]

The ISFC acted quickly to attract scientists of note late in the 1920s. In 1926 experimental physicist Aimé Cotton declined Dalbis's invitation, although he left open the possibility of a future engagement.[11] Dalbis angled unsuccessfully for mathematician Robert Fernand Bernard, vicomte de Montessus de Ballore (the brother of Chilean seismologist Fernand), who taught at the Catholic University of Lille.[12] Dalbis also invited Henri Villat from the Institute for Fluid Mechanics in Paris. Villat, an expert applied mathematician, was studying aerodynamics for the French air corps and could not come. He would in any event have received a mixed reception. Dalbis complained to the arch-

---

P12/E9, Arthur Stewart Eve to Déjardin, 18 Mar. 1932; P12/E4, Eve to Dalbis, 20 Mar. 1929.

9. See Madeleine Lavallée, *Marie-Victorin: Un itinéraire exceptionnel* (St-Lambert, Quebec, 1983), p. 225: Marie-Victorin called the ISFC "a real betrayal" of ACFAS. Laval University founded an Ecole de chimie in 1920 and promptly engaged six recent doctoral graduates from the University of Fribourg in Switzerland; among the new professors was Joseph Risi, who became French Canada's most prominent physical scientist during the interwar period (Danielle Ouellet, "Défricher un demi-siècle de sciences et de génie," *Contact: Le magazine de l'Université Laval*, Autumn 1987, 4 pp.).

10. MUM, P12/E5, Filion to Montpetit, 14 Sept. 1929.

11. Ibid., P12/E1, Cotton to Dalbis, 12 Dec. 1926.

12. Ibid., P12/E6, Montessus de Ballore to Dalbis, two letters of 9 Jan. 1930. Montessus's mooted tenure at the ISFC was apparently pegged to a "mission" to the United States that did not materialize. Ten years earlier, Montessus had presented himself to the Catholic hierarchy in Montreal as a candidate for a chair in mathematics at the University of Montreal; no action was taken on his application ([Illegible, in the Archdiocese of Montreal] to Robert, vicomte de Montessus de Ballore, 11 May 1920, kept in the private papers of the comtesse Lecointre, Le Mas, Persac, Lussac-les-Châteaux, France).

bishop of Montreal later that Villat's French lectures seemed "a bit too advanced" for the students of Quebec.[13]

Dalbis's efforts in exact sciences were first gratified by accident. In 1927 Léon Brillouin, the brilliant French theoretical physicist, went on a lecture tour of the United States, and he wrote to Dalbis that he wanted to spend about ten days visiting French and English universities in Canada. He was willing to speak on the new quantum physics, atomic structure, statistics, or telecommunications. Dalbis leapt at the opportunity. Brillouin came to Montreal and lectured at its two universities; he was flattered by Dalbis's attentions.[14]

Dalbis landed a larger catch in Jean Cabannes, a physicist from Montpellier. According to Cabannes's plans, during his ten-week stay in North America he was to give one lecture on the radiation of energy, meet three times with those doing research in laboratories, and give twelve lectures on the physics of radiation at the Ecole polytechnique. His talks at McGill and in the rest of Canada would be added to this charge. Dalbis wrote on Cabannes's behalf all across Canada, especially dunning the French consul in Vancouver. Although Cabannes was a young man, Dalbis wrote to the consul, he would "bring to Canadian universities the latest discoveries of French physics." For this reason, Dalbis asked for the consul's financial and moral support in sending Cabannes to the Pacific. Dalbis was delighted with Cabannes's triumphant tour, and he congratulated him on "the accomplishment of a mission as useful to Canada as to French scientific expansion."[15]

Cabannes's successful tour led in 1932 to the arrival of Georges Déjardin, a thirty-year-old professeur titulaire of physics at the University of Lyons who in 1929 had served as president of the examinations committee at the engineering school of the St-Joseph University in Beirut.[16] Déjardin, who was traveling in any case to Harvard and the Massachusetts Institute of Technology, spoke on the new physics and radiant energy.[17] He planned to provide some forty lectures of one hour each: "fifteen talks on radiation, the new physical theories, the optical properties of X-rays, visual and physical photometry; ten talks on photoelec-

13. MUM, P12/E6, Villat to Dalbis, 10 Sept. 1928 and 22 Dec. 1928; P12/E7, Dalbis to Paul-Napoléon Bruchési, 25 July 1930.

14. Ibid., P12/E2, Brillouin to Dalbis, 23 Dec. 1927 and 1 Apr. 1928.

15. Ibid., P12/E5, ISFC to unknown, 10 Aug. 1929; P12/E4, Dalbis to French Consul in Vancouver, 27 Sept. 1929; P12/E6, Dalbis to Cabannes, 14 Jan. 1930.

16. LO, Déjardin, "Fonctions, grades et titres, mandats divers et missions à l'étranger," accompanying Déjardin to Clinton Joseph Davisson, 13 Oct. 1941; Davisson sent a copy of the vita to William Hammond Wright on 28 Nov. 1941.

17. MUM, P12/E8, Déjardin to Dalbis, 6 Nov. 1931.

tricity, with an accent on photoelectric cells and their applications (phototelegraphy, television); five talks on thermionic phenomena (wireless valves, rectification of alternating current, etc.); three or four talks on research and several popular expositions." In the last presentations, he would go into radiation theory, industrial applications, and high-altitude atmospheric research. Déjardin wanted to lecture on the level of advanced instruction in France, and he asked Dalbis if that were appropriate.[18] As Dalbis did in Cabannes's case, he wrote to Canadian universities asking them to host Déjardin. Some claimed poverty; others declined to engage a physicist who could not speak English.[19] Dalbis had the French consulate in Vancouver contribute six thousand francs toward Déjardin's voyages west. The French physicist worked hard for his money: he gave twenty-seven talks in the United States and fifty-four in Canada.[20] And he alone among French visitors in exact sciences eventually published a paper in a French Canadian journal.[21]

All this activity cost a great deal of money. During the 1920s the government of Quebec Province contributed six thousand dollars annually, and the French ministry of foreign affairs gave the same amount in francs. A measure of the value that France placed on the Montreal enterprise is obtained from its continual renewal of the ISFC grant, even in the depths of the Depression, when it reduced funding of other "French works abroad" by 10 percent. The money from Quebec was placed in deposit, in dollars, at the Canadian National Bank in Paris; the French contribution went to the same bank in francs. For the daily functioning of the institute, the University of Montreal contributed about a thousand dollars in matériel and personnel, and occasional private donations ran to about five hundred dollars annually. Half the Quebec monies went for honoraria to the invited professors; the remainder financed travel inside Canada, as well as the salaries of the institute's administrators. Cabannes, for instance, received a thousand dollars from Quebec, Déjardin only five hundred, and Brillouin one hundred. The ISFC's agent in Paris, Etienne Gilson, dispensed francs

18. Ibid., 21 Oct. 1931.

19. On Toronto's claim of poverty, see ibid., P12/E9, Robert Falconer to Dalbis, 9 Feb. 1932. The University of Western Ontario raised linguistic considerations, although in the end it canceled Déjardin's talk because of an influenza epidemic (Sherwood Fox to Dalbis, 9 Feb. 1932 and 19 Mar. 1932).

20. Ibid., French Consul in Vancouver to Dalbis, 10 Mar. 1933; P12/E2, ISFC council minutes, n.d.

21. Déjardin, "L'ozone atmosphérique," *Revue trimestrielle canadienne* 21 (1935), 373–91, based on a lecture given at Cornell University in that year.

for transatlantic travel and, one thinks, some of the salary of the invitees. Gilson also received a handsome award for his services.[22]

When Dalbis — under pressure from Quebec nationalists — returned to France in 1932, he administered the ISFC's accounts. He was well paid, receiving two thousand dollars for his troubles in 1936–37.[23] With his repatriation, however, the train of physicist missionaries to Quebec stopped. The distinguished visitors left no mark on science in French Canada.

Earth Fire and Ocean Wind

Early in May 1902 Martinique's Mt. Pelée blew its top. The accompanying firestorm, mudslide, and tidal wave killed thirty thousand people and entirely obliterated the island's commercial hub, the city of St-Pierre. Sugar, the great eater of men, dominated life in the French colony. Warehouses and distilleries, where the wealth of sugar was measured in liters of rum, exploded under the rain of fire.[24] The Academy of Sciences in Paris named Alfred Lacroix, mineralogy professor at the Museum d'histoire naturelle, to report on the volcanic event. Along with naval hydrographer Maurice Rollet de l'Isle and mineralogist Jean Louis Giraud, Lacroix and his wife spent three months on Martinique; shortly after the team returned to Paris, a second eruption brought Lacroix back to Martinique, this time to set up a permanent observatory. Lacroix's letters appeared in the *Comptes-Rendus*, and he published a detailed, two-volume analysis in 1904. Even before the volumes appeared, Mt. Pelée propelled the forty-year-old mineralogist into the ranks of the academy that he had served so well.[25]

Lacroix brought the instruments he needed from Paris. These included seismometers, although not magnetical recording devices.[26] It is an irony of history that but for the lack of relatively modest support, geophysicists would have had unparalleled recordings of Mt. Pelée's first eruption. The story of meteorology and geophysics on Martinique

22. MUM, P12/E10, Dalbis to Athanase David (Secretary of the Province), 21 Nov. 1933, for the quotation; P12/A1, Edouard Montpetit, "L'Institut scientifique franco-canadien," n.d.; P12/1F2, "Etats financiers," 1926–27 to 1940.

23. Ibid., P12/1F2, "Etats financiers," 1926–27 to 1940.

24. Philippe Ariès, Charles Daney, and Emile Berté, *Catastrophe à la Martinique* (Paris, 1981). Rum production in the French Antilles benefited from the deadly phylloxera epidemic that had ravaged French vineyards — and curtailed brandy manufacture — over the last third of the nineteenth century (Hugh Barty-King and Anton Massel, *Rum, Yesterday and Today* [London, 1983], p. 131).

25. *DSB*, s.v. "Lacroix"; A. Lacroix, *La montagne Pelée et ses éruptions*, 2 vols. (Paris, 1904).

26. Lacroix, *Montagne Pelée*, 1, 90–91; 2, 478–81.

in the years before Lacroix's passage is especially instructive about metropolitan patronage of *fin-de-siècle* colonial science.

The Windward Islands are familiar with God's wrath, being visited by tropical storms of great intensity, in addition to volcanic and seismological disturbances. The physical environment called for systematic attention by French savants. It received less than it merited. By 1840 volunteers in both St-Pierre and its sister town Fort-de-France were transmitting observations to Paris.[27] The haphazard situation continued until 1891, when the local oligarchy organized a permanent meteorological observatory in its newly created agricultural-experiment station and laboratory at St-Pierre. The president of the Union of Sugar Manufacturers of Martinique, Dr. Auguste Guérin, apprised Eleuthère Mascart of the plan. The physics teacher at the local lycée, E. Manceau, would direct the station and would be happy to follow Mascart's instructions. Mascart's Central Meteorological Bureau had only to provide the requisite instruments.[28] Mascart replied that he welcomed the new development, especially since "we have not had until now any correspondent in Martinique." He would gladly offer instructions for taking observations, but before sending instruments Mascart needed to know how often Guérin planned to observe; "a single observation per day is quite insufficient and could not be published." Mascart wanted three readings daily. He also asked about the physical situation of the garden and whether the sugar manufacturers would be able to cover the cost of constructing a wooden shed for the thermometers.[29]

A hurricane ripped through Martinique several weeks later, killing seven hundred people in St-Pierre alone.[30] The disaster provided impetus to the meteorological project. In October 1891 Mascart sent Guérin a shipment of instruments for the new observatory: a psychrometer, two rain gauges, and six thermometers of various sorts.[31] Manceau promptly returned his initial recordings. In December Mascart sent a barometer and a warning about tropical depressions: "Please do not disregard anything, however insignificant it may appear. The study of these serious atmospheric perturbations cannot be done too care-

27. A. Romer, "Introduction," *Bulletin annuel du Service météorologique et de l'Observatoire géophysique, année 1932*, pp. 5–8, where Romer inaugurated his new periodical with a history of meteorological observation in the French Antilles.

28. PMN, Guérin to Mascart, 11 June 1891. Manceau's successor, however, referred to Manceau as a "chemistry teacher." (Gaston Landes to Mascart, 10 Apr. 1892). On Guérin, see Simone Chrétien and Robert Brousse, *La montagne Pelée se réveille: Comment se prépare une éruption cataclysmique* (Paris, 1988), p. 227.

29. PMN, Mascart to Guérin, July 1891, on which is written by apparent error "copied 10 July 1890."

30. Ariès, Daney, and Berté, *Catastrophe*, p. 112.

31. PMN, Mascart to Manceau, 26 Oct. 1891; separate list of the shipment.

fully."[32] But Manceau lasted only two months more. He fell ill, prepared to return to France, and confided the observatory to his physics-teacher colleague Gaston Landes, a licencié ès sciences naturelles who had been *chef de travaux* at the science faculty in Montpellier. Mascart asked the new man to take temperature readings to tenths of a degree. He also asked Landes to coordinate efforts with Captain Gerbault, head of the artillery on Guadeloupe, who was his correspondent there. Gerbault, an "excellent observer," had built a shed for the thermometers, and Landes would do well to follow his example.[33]

A string of colonial dependencies dotted the Caribbean, and storm warnings required international agreement. The French weathermen knocked on the door of the dominant meteorological power in the Caribbean, the United States, proposing to collaborate with the U.S. Signal Service. The initiative came from V. Lacroux, a pharmacist at the hospital in Fort-de-France, which had been the site of intermittent meteorological observations in the nineteenth century. Lacroux desired an intercolonial weather-prediction service to be centered at Fort-de-France, the traditional rival of St-Pierre on Martinique. As one of his first administrative acts Governor Noël Pardon used the budding meteorological rivalry to petition his ministry for an increase in the amount allotted both Lacroux and Landes's colleague at St-Pierre, the lycée's normalien naturalist Gustave-Etienne-Jules Saussine, for instruments—three thousand francs.[34] In fact the military pharmacists did receive funding to record basic meteorological data during the 1890s and into the early twentieth century.[35] The plan for a meteorological empire based in Martinique depended, however, on free use of telegraph lines with English, Spanish, and Danish islands, and the governments of Martinique and Guadeloupe were unable to provide es-

32. Ibid., Mascart to Manceau, 16 Dec. 1891.

33. Ibid., Manceau to Mascart, 24 Mar. 1892; *Annuaire de la Martinique*, 1900, p. 257; PMN, Mascart to Landes, 27 Apr. 1892. Meteorological data was recorded at Camp Jacob on Guadeloupe from 1891 to 1900 (Lacroix, *Montagne Pelée*, 2:450). On Gerbault, lieutenant colonel at the time of his death in Mt. Pelée's eruption, see Chrétien and Brousse, *Montagne Pelée*, p. 227.

34. SOM, Martinique, carton 30, dossier 272, note for the First Bureau of the Ministry of Commerce, Industry, and Colonies, 13 Mar. 1893; Lacroux to Governor General Martin, 16 July 1892; Governor General of Martinique to Minister of Colonies, 10 June 1893; Saussine, *La première année d'agriculture tropicale* (Paris, 1900), where Saussine provides an elementary exposition of agronomy. Cf. Chrétien and Brousse, *Montagne Pelée*, p. 235. Pardon replaced Dauphin Moracchini on 9 June 1893 (*Annuaire de la Martinique*, 1900, p. 55).

35. Their rainfall measurements from 1892 to 1903 appear in Lacroix, *Montagne Pelée*, 1:9.

timates of the number of messages that would have to be sent over the lines.[36]

The matter was then taken in hand by Léon Sully, a *propriétaire et photographe industriel* in St-Pierre, who sought to breathe new life into the moribund meteorological designs of the local sugar manufacturers.[37] Realizing that the hospital at Fort-de-France had been designated as a central observatory and that the local administrative council had promised four thousand francs toward its construction, he entered into an agreement with the Havana Observatory in 1895 whereby observations would be exchanged daily by telegraph.[38] In 1896 Sully petitioned the colonial government for a grant of three thousand francs; he received it the following year. With his grant, which he took to be the first of many annual disbursements, in hand, Sully introduced himself to Mascart as the architect of an international meteorological observatory. Sully's vision focused on prediction, and it extended to atmospheric electricity and "telluric currents." He counted on receiving an anemometer through the good will of Martinique's deputies in Paris. From Mascart he hoped to obtain the standard instruments of his trade. Sully claimed that the instruments already lent to Guérin "had never been used" and had deteriorated from abuse; he proposed to use them to standardize the instruments on neighboring islands once new equipment arrived at St-Pierre. In return, Sully promised to send Mascart readings every month.[39]

Mascart directed Guérin to pass the inactive instruments to Sully. He also wrote to Sully directly, offering moral support.[40] Sully replied in 1898, nearly a year later. He had received some of the instruments from Guérin, and these only at the end of 1897. Although his plans were being held up by lack of funds, he was transmitting weekly observations to the Hydrographic Office in Washington, D.C. – the United States

36. SOM, Martinique, carton 30, dossier 272, Third Bureau of the Ministry of Colonies to the Governor of Martinique, 25 Nov. 1893; Governor of Martinique to Third Bureau of Ministry of Colonies, 9 July 1894; Director General of Mail and Telegraph to Minister of Colonies, 14 Aug. 1894.

37. PMN, Sully to Mascart, 6 June 1897, where Sully gives his identification; Mascart to Sully, 24 July 1897, where Mascart says that he received no observations from the instruments that he had lent to Guérin in 1891.

38. Léon Sully, "Un observatoire météorologique à la Martinique," *Journal des colonies*, 9 Sept. 1895. At this time there were two meteorological observatories in Havana – the Jesuit one at the Colegio de Belen and a naval observatory created around 1890. The Jesuit director was Laurent Gangoiti (*Vregille*, pp. 44–47).

39. PMN, General Council of Martinique, session of July 1896, on which transcript Sully notes award of the grant on 22 May 1897; Sully to Mascart, 2 June 1897.

40. Ibid., Mascart to Guérin, 24 July 1897; Mascart to Sully, 24 July 1897.

being by then at war with Spain in the Caribbean. He would be glad to favor Mascart with the same information. Sully asked Mascart for more instruments, especially an electrometer to measure atmospheric electricity.[41] Mascart sent along thermometers and a psychrometer, but he urged Sully not to expand into atmospheric electricity until he could afford to purchase an instrument with photographical registration, all other models being unacceptably imprecise. He asked for details about the installation of Sully's equipment. Sully replied that the United States Weather Bureau had just judged his operation worthy of supplying Washington with observations twice daily. Mascart sought to discipline the amateur meteorologist by suggesting improvements to his equipment which would produce more reliable data.[42]

Professional weathermen are not alone in telling which way the wind blows, but their pronouncements carry authority. Early in 1898 Sully complained to Mascart about the arrival of A. Perrenod, a meteorologist who claimed to be a correspondent of Mascart's bureau and who sought to take charge of Martinique's weather service. Mascart reported that Perrenod had indeed asked for support; he would hold Perrenod's request in abeyance pending information from Martinique's governor about whether the colony could support a meteorological service. Sully then clarified his aims. He wanted to study tropical storms above all, and he hoped to be able to predict their arrival. To this end, he had just received a grant of two thousand francs from the navy for purchasing instruments. He could not provide all the data that Mascart desired, however, as his only assistant was his wife. Mascart suggested improvements to Sully's thermometer shed, which did not allow for sufficient circulation of air.[43]

Perrenod, for his part, conducted an international campaign to become Martinique's meteorological captain. He circulated a printed brochure to this effect, prefaced by an open letter to the island's governor and council. After reciting current ideas about hurricanes, Perrenod proposed the formation of a society to chart the weather. The society was to be formed from the nucleus of the twenty Martinique residents

41. Ibid., Sully to Mascart, 10 May 1898; J. E. Craig (Hydrographic Office, Washington, D.C.) to Sully, 22 Apr. 1898. The USS *Maine* sank at Havana on 15 Feb. 1898; on 25 Apr. 1898 the United States declared war on Spain, a declaration made retroactive to 21 Apr. 1898.

42. Ibid., Mascart to Sully, 5 July 1898; Sully to Mascart, 13 Aug. 1898; Mascart to Sully, 2 Nov. 1898.

43. Ibid., Sully to Mascart, 10 Feb. 1898, where Sully identifies Perrenod as Swiss; Mascart to Sully, 13 Mar. 1899; Sully to Mascart, 10 June 1899; Mascart to Sully, 30 June 1899.

who were members of the Astronomical Society of France; aided by government support, the society's members would record the weather. If an observatory were founded, it could be charged with determining the official time of day. In founding an observatory, the first step would be to send a trained man to Jamaica and perhaps also to Washington, there to receive instruments and instructions about their use. An annual grant of three thousand francs would be sufficient to guarantee the observatory's survival, as the operation would depend on voluntary observers at various locations. Perrenod proposed that astronomy be included in the observatory's mandate, and he referred to his own astronomical observations made there in 1897. The meteorological-astronomical observatory could be situated on a large hill two hundred meters above St-Pierre. There was a better site. "If the country had not been too impoverished by recent reversals," Perrenod wrote, "I would have asked for a larger sum to construct my station *on the summit or on the slopes of Mt. Pelée, at an altitude of more than 1300 meters.*"[44]

About Perrenod and the colony's meteorological-astronomical intentions, Mascart remained silent. Perrenod's campaign was unsuccessful, and Sully's efforts came to nothing. The illness of Sully's wife forced him to discontinue observations. Sully began again "without the aid of an administration systematically hostile to any private initiative." His grant of three thousand francs had been "reduced to ridiculous proportions" by the administration. He continued to send telegraphical observations to U.S. meteorologists at Havana; he had no time to transcribe observations, although he would lend his notebooks if Mascart wanted to publish the data.[45] Mascart gladly accepted Sully's offer. Sully sent along data for 1899–1900 and for the first half of 1901; Mascart summarized them and promised to publish the summary in his *Annales.*[46]

By April 1902 Sully was scrambling over Mt. Pelée and reporting on the creaks and groans of a volcano set to explode; on 7 May he reported on the avalanches of water and mud that had begun to flow toward the

44. See Perrenod, *Décembre 1898, un observatoire à la Martinique* (St-Pierre, 1898), p. 33, for quotation (emphasis in the original). A presentation copy sent to Willis L. Moore, chief of the U.S. Weather Bureau, by Perrenod on 16 May 1899 is in the library of the National Oceanic and Atmospheric Administration, Bethesda, Maryland. Perrenod had resided on Martinique for some time. He indicates (p. 16) that he reported to the Neuchâtel Observatory on a Martinique hurricane of 18 August 1891.

45. PMN, Sully to Mascart, 19 June 1901. After the Spanish-American War, the Americans took over the Spanish naval observatory at Havana and tried to displace the weather service run by the Jesuits at the Belen Observatory (*Vregille*, pp. 47–48).

46. PMN, Mascart to Sully, 20 July 1901; Sully to Mascart, 10 Sept. 1901; Mascart to Sully, 9 Dec. 1901.

sea.[47] Gaston Landes, in Lacroix's opinion "the most qualified naturalist at St-Pierre," provided detailed analyses during the weeks leading up to the first eruption.[48] There are no communications from Sully after the mountain blew up on 8 May; he perished during the holocaust. Sully's rival Perrenod does not figure in any of the scientific reporting from 1902. Landes was caught in a cloud of burning ash and died horribly by his home.[49]

Before leaving Martinique, Alfred Lacroix set up an observatory on Morne-des-Cadets, a promontory from which Mt. Pelée could be viewed in relative safety. He installed two Bosch horizontal seismometers and a Secchi warning seismoscope; these instruments remained in service until 1927.[50] The observatory continued under a series of part-time meteorologists until 1910, when it was taken in hand by A. Simon, science teacher at Fort-de-France. Simon ran the station – such was its designation – for the next eighteen years. He took regular observations and sent them to the Central Meteorological Bureau for publication.[51] France's seismological chief, Edmond Rothé, had little faith in Simon or his installation, which was "incapable of making good observations."[52] In 1926 Simon brought two Mainka seismometers to Fort-de-France. When he died in 1928, his understudy Boutin continued the observations. Mt. Pelée erupted again in 1929. Boutin and his assistants were able to take precise measurements, and they were assisted by Henri Arsandaux, a volcanologist sent out by Lacroix. Vulcanology and meteorology were then married in a geophysical observatory organized within Henry Hubert's colonial meteorological service and overseen by a committee comprising Lacroix, Charles Maurain, Rothé, and Arsandaux, in addition to Hubert.[53]

In 1930 Maurain chose André Romer to be the first director of Mar-

47. Lacroix, *Montagne Pelée*, 1:76–77, 172–73. According to Chrétien and Brousse, *Montagne Pelée*, p. 235, Sully was editor of the newspaper *Les Antilles* at the time of his death in the eruption.

48. Lacroix's opinion in his *Montagne Pelée*, 1:34. Landes's reports and telephone calls figure prominently in Lacroix's study (see, e.g., pp. 110, 180).

49. Ibid., pp. 283–86.

50. A. Romer, "La séismologie à la Martinique," *APGFOM*, 1936, pp. 39–41.

51. PMN, Alfred Angot to Simon, 8 Nov. 1920, where Angot remarks on the "great regularity" of Simon's observations. Recommended by Lacroix, Simon received thirty thousand francs from the Pasteur fund to renovate the observatory (E. Rothé, "Notice nécrologique, A. Simon [1868–1928]," *ASPG*, 1927, p. 130).

52. PO, carton 3, II-B–1, Comité français de géodésie et de géophysique, minutes of general meeting of 13 Apr. 1921, to which is appended the report of the section for seismology, cited here.

53. "Introduction," *Bulletin annuel du Service météorologique, année 1932*, where the early twentieth-century developments are sketched.

tinique's meteorological service and geophysical observatory at Fort-de-France.[54] Romer supervised construction of the observatory's buildings and installation of equipment, a good deal of it on Morne-des-Cadets. He surveyed weather, took magnetical readings, and kept watch on the active and ominous Mt. Pelée. Romer devoted "the major part of my time" to climbing up the volcano, "collecting gases and rocks, and analyzing them."[55] His staff published articles and short monographs on their own. Pierre Molard, a graduate of the Ecole des mines, arrived in 1935 as ingénieur-météorologiste-adjoint stagiaire; he was joined by a certain de Salles de Hys, former ingénieur-météorologiste-adjoint in French West Africa.[56] Serge Frolow arrived as ingénieur-météorologiste-adjoint in 1935, fresh from a meteorological tour of duty in French West Africa.[57] When Romer eventually moved up to administer the meteorological service of a grander colony, Frolow inherited the directorship at Fort-de-France.

The primitive state of research into terrestrial magnetism in 1930 suggests the task facing the new director. French hydrographers took intermittent magnetical readings in the nineteenth century, but beginning in 1905 the magnetical values for Martinique were provided by the Department of Terrestrial Magnetism of the Carnegie Institution of Washington. Romer surveyed the region around Fort-de-France (for which there were no accurate geographical maps!) and installed permanent recorders on Morne-des-Cadets. He hoped to predict a volcanic eruption by watching for fluctuations produced by changes in the magnetical permeability of lava in a slowly warming volcanic chimney. He was cheered in this prospect by a report from the geophysical observatory at Santiago de Chile that claimed to have predicted an earthquake twenty hours before it happened by recording such variations.[58]

By 1936 Romer had installed a twenty-ton Quervain-Piccard seismograph to help monitor the volcano, and around 1938 he began to take gravity measurements with a Holweck-Lejay seismometer.[59] In-

54. PIPG, note of 1930 directing candidates for the Martinique post to contact Maurain.

55. Ibid., Romer to Maurain, 30 Dec. 1933.

56. *APGFOM*, 1934, p. 159; 1935, pp. 96, 191.

57. See ibid., 1937, p. 192 for a note on Frolow's study of the climate of Niamey in French West Africa. Serge Frolow was the brother of the chief irrigation engineer in the French mandate of Syria, Vladimir Frolow, who received his engineering doctorate from the Sorbonne in 1934 (V. Frolow, *Le Niger moyen* [ing. diss., University of Paris, 1934]). See also *Congrès*, pp. 101–2, for the Frolow brothers.

58. A. Romer, "Le magnétisme terrestre à la Martinique," *APGFOM*, 1935, pp. 42–44.

59. See *APGFOM*, 1936, p. 190, for installation of the seismograph in 1936; Romer, "Séismologie," for details on the big seismograph; and PIPG, Romer to Maurain, 13 Aug. 1938, for the early Holweck-Lejay surveys.

stalling the seismometer, a sister to one operated by Rothé at Strasbourg, was a stroke of genius. The seismograph had been designed by Alfred de Quervain and August Anton Piccard and set up at three Swiss locations in the middle 1920s. The enormous, caged mass – essentially following on Emil Wiechert's astatic-pendulum innovation of the preceding generation – allowed for simultaneous registration of motion in three dimensions, and it improved on existing seismographical amplification by an order of magnitude.[60] Installed on Morne-des-Cadets, the seismographical leviathan surveyed both local earthquakes and those originating elsewhere in the New World.[61]

Between the two world wars, seismology was a theoretical science that used recordings of earthquakes to elucidate the deep structure of the earth's core as well as the shallow structure of the earth's crust. The former was of interest to astronomers and geodesists, the latter to petroleum geologists. Romer's big seismograph served to elucidate neither structure. Instead, the geophysicists on Morne-des-Cadets focused on microseismical agitation, the background vibration that apparently depended on wind and surf. After running the Quervain-Piccard for a number of months, the geophysicists on Martinique concluded that ocean waves in general and the surf in particular, which derived from Atlantic gales, produced no change in microseismical activity. The observatory's Galitzin seismograph, with a longer period, recorded microseismical agitation that showed no correlation with local meteorological conditions. Molard, the man in charge of the seismographs, believed that the Galitzin's variations corresponded to ocean perturbations at great distances from the observatory, and he appealed to similar findings published by the Jesuits at Tananarive and Zikawei. The Martinique seismologists hoped to sense distant ocean disturbances and predict the weather by modifying the Galitzins' photographical recording device to permit a direct reading (Galitzins were constructed to record automatically and hermetically for two days at a time) and by isolating the relatively light seismographs from vibration due to the wind.[62]

60. Robert Billwiller, "Alfred de Quervain," *Neue Zürcher Zeitung*, 30 Jan. 1927. Billwiller provides Piccard's initial, and I infer that the Piccard in question is August, who received a doctorate at Zurich under Pierre Weiss (Albert Einstein was co-director): Piccard, *Die Magnetisierungskoeffizienten des Wassers und des Sauerstoffs* (ing. diss., Federal Institute of Technology at Zurich, 1913). Piccard had taken a mechanical-engineering diploma at the institute in 1910. *DSB* gives Piccard's first name as "Auguste."

61. P. Molard, "Le séismographe de 20 tonnes de la Station du Morne-des-Cadets," *Bulletin annuel du Service météorologique et de physique du globe, année 1937*, pp. 73–87.

62. Molard, "L'agitation microséismique à la station du Morne-des-Cadets," *APGFOM*, 1937, pp. 157–58. On using microseismical activity to predict the weather generally, see J. Lacoste, "La séismologie dans les territoires d'outre-mer," ibid., 1936, pp.

Molard's proposal to have geophysics serve the needs of weathermen was entirely of a piece with Romer's hope that terrestrial magnetism would aid in predicting volcanic eruptions. Enlisting geophysical inventory for practical ends reflects the bicameral nature of Martinique's premier scientific institution, which owed its existence to earth fire and ocean wind. Such applications would have cheered the loose confederation of volcanologists, mineralogists, meteorologists, and geophysicists who surveyed the colonial observatory from Paris. There is, however, an air of unreality to these research programs. Molard's end could not be achieved with the means at his disposal, as it would have required a network of weather reporting over the North Atlantic Ocean. Romer's dream seems to have been based on unverified speculation—or wishful thinking. It was sufficient to propel him to the directorship of the Phu-liên Observatory in Indochina.[63]

In the 1920s French America drew renewed attention from physicists in the hexagone. The interest derived from indigenous needs. Volcanic eruptions and destructive hurricanes brought a geophysical observatory to Martinique. In French Canada, an expatriate biology teacher persuaded private and provincial agencies to collaborate with the Quai d'Orsay in having French savants lecture at a local, religious university. Each receiving population, desperately poor and generally unsophisticated, took instruction and discipline from the Catholic church. The cultural superiority of France was never in question. On these American shores, the civilizing mission in pure science relied on programs and precedents that had been set in place a generation earlier at other points of the compass.

33–34; idem, "A propos de travaux récents sur les ondes microséismiques," ibid., 1937, pp. 5–10; and Pierre Bernard, "Etude sur l'agitation microséismique et ses variations," *AIPG*, 19 (1941), 1–77, which is especially good on French contributions. A survey of the problem appears in Benjamin F. Howell, Jr., *An Introduction to Seismological Research: History and Development* (Cambridge, 1990), pp. 131–36.

63. Romer's successor Serge Frolow undertook a systematic investigation of pressure waves in the tropical atmosphere, and he appealed to data from many tropical observatories, but he made no attempt to bring microseismical activity into the picture (Frolow, *Variations simultanées de la pression dans la région tropicale* [Fort-de-France, 1941]).

11

Ecuador: Missions Religious, Military, and Diplomatic

Heritage of a Benevolent Theocracy

The instauration of exact sciences in middle nineteenth-century Ecuador may be traced directly to the fanatical energy and vision of Gabriel García Moreno. The young García Moreno chose to become certified in law in the 1840s, but his real love seems to have been science. In 1836, at age fifteen, he received a scholarship allowing him to study in Quito, at the university and (for sciences) at the recently secularized Colegio San Fernando. He also took private lessons under Sebastián Wisse, a French engineer contracted in 1844 to work and teach in Quito. Master and disciple together explored the Ecuadoran interior. García Moreno visited Europe with his brother in 1850 and then returned to Paris in 1855 to devote all his energies to the mathematical and experimental sciences. He attended lectures and laboratory courses for fifteen months. García Moreno's work attracted favorable attention from his French professors, and upon returning to Ecuador in 1856 he was elected a member of the Geological Society of France. Complementing his sympathy for secular learning and French administrative norms was deepening religious faith. No socialist revolutionary, he came to typify the conservative, nineteenth-century Catholic intellectual for whom science was the revelation of God's truth. As he wrote in a letter of 1856, he agreed with a proposition announced at the Sorbonne by Antoine Jérome Balard to the effect that "the degree of civilization of a people may be measured by the liters of sulfuric acid produced."[1]

1. García Moreno's scientific training and interests are admirably recounted by Francisco Miranda Ribadeneira in *La primera escuela politécnica del Ecuador* (Quito, 1972), pp. 1–11. My account often follows Miranda Ribadeneira. The controversial status of García Moreno in Ecuadoran historiography is considered by Germán Arciniegas, *Latin America: A Cultural History*, trans. Joan MacLean (New York, 1977), pp. 363–67. A 2,000-page edition of García Moreno's letters appeared in the 1950s: Wilfrido Loor M., *Cartas*

After his second stay in Paris, García Moreno became his country's leading scientific intellectual. In 1856 he was named rector of the University of Quito. The following year he drew up an ambitious plan to reform science instruction, at the center of which was an institute of technology. The plan received approval by the senate but became lost in the turmoil of politics and war that soon engulfed the country. García Moreno rose to become supreme commander of the army in 1859, and at this time he thought seriously about inviting France to establish a protectorate over the country.[2] As leader of the Ecuadoran conservative party, García Moreno was elected president in 1861. He ruled for four stormy years, during which he conducted an unsuccessful war against Colombia. He failed to bring about changes in education. In 1865 he resigned and went into eclipse. His successors declared war on Spain and, in 1868, saw their country visited by a monstrous earthquake. García Moreno engineered a revolution in 1869 and then became president once more in 1870. This time he would have no mistakes. He immediately turned Ecuador into a benevolent theocracy.

García Moreno's inspiration, it seems reasonable to assume, was the Jesuit empire in colonial Paraguay. In his view, the Jesuits alone could act against the decadence and squalor that afflicted his country. He accordingly placed them in charge of reorganizing education, especially science and technology. The Jesuits would disseminate the claims of natural law while at the same time promoting literacy and public morality. They were to make Ecuador safe for democracy. García Moreno quickly passed by Spain and France in his search for teachers best apprised of exact sciences and their applications. As natural knowledge was to build García Moreno's utopia, he handed the system of higher education over to the Jesuit hierarchy in Germany.[3]

"The university established in the capital of the State is converted into a polytechnical school," read García Moreno's decree of 1869, "destined exclusively to educate teachers of technology, civil engineers, architects, mechanical engineers, mining engineers, and science teachers."[4] Law and medicine were disaggregated into separate schools.[5] The

de Gabriel García Moreno, 4 vols. (Quito, 1954?–56); here see 2:3, 37, 39 (quotation), 49, 59, 79.

2. See Loor M., *Cartas*, 2:150, for García Moreno's idea in 1859 of a French protectorate "under conditions analogous to those existing between Canada and Great Britain."

3. On the background to García Moreno's call to the Jesuits, Wilfrido Loor M., *Los jesuitas en el Ecuador: Su ingreso y expulsión, 1850–1852* (Quito, 1959).

4. Miranda Ribadeneira, *Escuela*, p. 37.

5. In focusing on the dominant role of the German Jesuits in the exact sciences, it is well to record persistent overtures by García Moreno to French men of learning in law,

German Jesuits began to arrive in 1870. Juan Bautista Menten, professor of higher mathematics and experimental physics, had been an assistant of astronomer Friedrich Wilhelm August Argelander's at the University of Bonn and a coworker of Pietro Angelo Secchi's at Rome; he was preparing to direct the Jesuit observatory at Bombay when he received a call to become dean and director of García Moreno's new school. Teodoro Wolf, also a former student at the University of Bonn, whose imagination had long been fired by the work of the Paraguayan Jesuits, came as professor of geology and zoology. Luis Sodiro, a native of Italy who had studied at the University of Innsbruck, was the school's botanist. Other mathematicians were José Kolberg and Emilio Muellendorff. And chemistry and geology were remanded to Luis Dressel, who had been sent to Bonn along with Menten and Wolf for training to combat Charles Darwin's subversive ideas. These men were joined in 1872 by mathematicians Armando Wenzel and José Epping and by zoologist Cristian Boetzkes; in 1873 came mathematicians Eduardo Brugier and Alberto Claessen and chemist Luis Heiss.[6]

With physics laboratory equipment ordered from the firm of Munich professor Philipp Carl, Quito in short order assembled the finest school of exact sciences in Latin America, one with standards of teaching unsurpassed in the New World. Few if any of the staff had doctorates (Wolf received one *honoris causa* from Bonn in 1877), but into the twentieth century Jesuit scientists of distinction, like Oxford dons, did not necessarily suffer through this particular rite of passage. Their publications easily made up for the lack. And they were the least expensive science masters that García Moreno might have found – the total sum paid to the dozen Jesuit professors between 1870 and 1876 was only 60,400 pesos.[7]

As his staff reached critical mass, Menten pushed ahead with plans to erect a major observatory. He ordered a large refracting telescope from

---

medicine, literature, fine arts, and elementary instruction. See Victor M. Rendon, "La France et l'Equateur dans leurs relations politiques et littéraires," Paris, Academy of Moral and Political Sciences, *Séances et travaux de l'Académie: Compte-Rendu*, May–June 1931, pp. 458–88, on pp. 466–67; Rendon's observations merit caution, as in his assertion that García Moreno asked the French government to establish an astronomical observatory at Quito in 1865.

6. Miranda Ribadeneira, *Escuela*, pp. 48–51, 55, 60; Augusto N. Martínez, "Sesenta años de recuerdos: El doctor Teodoro Wolf," Universidad central del Ecuador, *Anales*, no. 287, Jan.–Mar. 1934, pp. 181–206 for a detailed appreciation, p. 180 for the anti-Darwinian mission of the three Jesuits at Bonn. It seems appropriate to retain the Spanish form of their given names.

7. Miranda Ribadeneira, *Escuela*, p. 62 for the instruments from Carl, pp. 365–74 for a list of the professors' publications, p. 301 for the payment.

Sigmund Merz in Munich, which ran to more than twenty thousand francs. He began to construct the observatory building.[8] All did not go well, however, in García Moreno's Jesuit-run paradise. Jealousies existed between the German Jesuits and those of Spain, the latter having traditionally enjoyed free license in Ecuador. Voices within the Jesuit establishment, including that of its commandant general, questioned the wisdom of directing *any* advanced scientific school – Quito being the first experiment of its kind for the order.[9] What was most serious, the German Jesuits at Quito suffered a crisis of faith. Difficulties in setting up his observatory and lack of confidence in his administrative abilities led Menten to leave the order in 1875 (he retained control of the fledgling astronomical institution).[10] He followed Wolf, who had parted ways in 1874 after the ecclesiastical hierarchy in Quito objected to his lectures on Darwinian evolution.[11]

The key to the entire enterprise, the nation's theocrat, might have been able to hold it together, but he fell to an assassin in 1875. With García Moreno's death, the dozen or so Jesuit scientists dispersed. Menten, Wolf, and Sodiro stayed in Ecuador. Wolf eventually published the first great geographical and geological treatise of the country. Menten, who never succeeded in placing the observatory in proper working order, became director of Ecuador's public works; during the regime of General Eloy Alfaro he took refuge in Colombia. Menten was hoping to return to Ecuador when death surprised him in 1900. The scholarly world has showered its accolades on Father Epping, who in 1881 revealed the subtle harmonies of ancient Babylonian astronomy.[12]

8. Ibid., pp. 317, 325.

9. Ibid., pp. 179–81, 191, 204, 206.

10. Menten describes his institution, including its 9-inch Merz refractor, in his *Historia y descripción del observatorio astronómico de Quito* (Quito, 1877). The equipment seems to have been adequate for significant observational work.

11. Miranda Ribadeneira, *Escuela*, pp. 231–32 and 245–55 on Menten, pp. 239–41 on Wolf; Martínez, "Sesenta años."

12. Menten's observatory went to Wilhelm Wickmann – a refugee from the Santiago de Chile Observatory – from 1887 to 1895. For Menten's death in exile, see Miranda Ribadeneira, *Escuela*, p. 331. On Epping, see A. Baumgartner, "Joseph Epping," *Zeitschrift für Assyriologie und verwandte Gebiete* 9 (1894), 427–33; Otto Neugebauer, *The Exact Sciences in Antiquity* (Providence, R.I., 1957), pp. 103–4; and idem, *A History of Ancient Mathematical Astronomy*, 3 vols. (New York, 1975), 1:348–49. On Epping's Babylonian lunar astronomy, see Johann Nepomuk Strassmaier and Epping, "Zur Entzifferung der astronomischen Tafeln der Chaldäer," *Stimmen der Zeit* [Maria-Laach], 1881, pp. 277–92. In 1877 Menten wrote that knowledge about Babylonian astronomy was limited to the nineteen-year lunar cycle permitting approximate calculation of lunar eclipses (Menten, *Historia*, p. 3). The Maria-Laach house organ was a preferred outlet for the polytechnic professor Dressel (Miranda Ribadeneira, *Escuela*, p. 51).

Academy and Army: Longitude Redux

It was on the wreckage of what might have become a German-inspired tradition of science that General Alfaro opened a chapter of French learning in Ecuador. Before he rose to govern the country, Alfaro had reorganized the Ecuadoran army following French advisers, and he had sent his son to be trained in the art of war at St-Cyr. He then made good use of a French desire to prosecute geophysics.

At the very end of the nineteenth century, French academicians and generals conceived an urge to measure the meridian arc at a point near the equator, the very task carried out in the eighteenth century by Pierre Bouguer and Charles-Marie de La Condamine. The resulting information made it possible to calculate the extent to which the figure of the globe departed from a perfect sphere—a determination used in a number of astronomical, cartographical, and geophysical investigations. Funding for such a long and expensive expedition followed initiatives by imperialist rivals. Only after Russia and Sweden mounted a similar campaign in Spitzbergen and after the United States Coast and Geodetic Survey expressed an interest in the Ecuadoran measurement did the French government agree to commit its resources.[13]

The French ministry of public instruction organized the departure of two officers in the geodesical service of the army; they left for Ecuador in 1899 with a budget of twenty thousand francs for a feasibility study. General Alfaro promised lavish support. Upon the duo's return to Paris, the form of the expedition took shape: the geodesical service would carry out the mission, but all scientific questions were to follow instructions from the Academy of Sciences. A law of 1900 placed the expense, amounting to some five hundred thousand francs, with the ministry of public instruction. Additional funding came from philanthropists Raphaël Bischoffsheim and Prince Roland Bonaparte,

13. See Georges F. Perrier, "La mission française de l'Equateur," *Revue économique française*, 1907, pp. 441–60, 489–506; idem, *République de l'Equateur: Aperçu général, relations entre l'Equateur et la France* (Paris, 1937), pp. 58–66; and the more extensive description in Perrier's *Figure de la terre: Les grandes opérations géodésiques* (Paris, 1908), pp. 338–444. The last monograph appeared as volume 2 of the *Revue de géographie annuelle.* Superficial discussion by two members of the expedition are Eugène Maurain, "Reconnaissance de l'arc du méridien de Quito," *La géographie* 2, pt. 2 (1900), 1–8; and Robert Bourgeois, "Opérations de la mission française chargée de la mesure d'un arc de méridien en Equateur," ibid. 5 (1902), 340–50. Deep background is given in Antonio Lafuente and José L. Peset, "La question de la figure de la terre," *Revue d'histoire des sciences* 37 (1984), 235–54; and Antonio Lafuente and Antonio Mazuecos, *Los caballeros del punto fijo* (Barcelona, 1987).

Monument erected by the French army during its early twentieth-century expedition in Ecuador. The expedition repeated the eighteenth-century longitude measurements of Charles-Marie de la Condamine and Pierre Bouguer. (From Georges Perrier, *La figure de la terre: Les grandes opérations géodésiques* [Paris, 1908])

to which was added Alfaro's commitment of eighty-three thousand francs.[14]

The most striking feature of the expedition relates to its *scientific* side. The Paris Academy would not send one of its own members as an observer. The distinguished mathematician Henri Poincaré spoke for his academician colleagues when he noted that the geodesical service was perfectly suited for the task at hand: "We are not sure of finding their equal; we are sure of not finding any better. We could not expect otherwise to have the cohesion, the familiarity of working together and applying the same methods, and above all the discipline that will permit the work to be carried out quickly and without hesitation."[15] Any ancillary scientific tasks, Poincaré implied, could be confided to career military men who knew how to follow orders. Poincaré and his colleagues

14. PAN, F17/13062, General Jean-Antonin-Léon Bassot to Minister of Public Instruction, 4 Oct. 1900 and 29 Aug. 1900; Bonaparte to Bassot, 10 Jan. 1900; Bischoffsheim to Bassot, 29 Jan. 1900; Joseph-Hippolyte Frandin to Théophile Delcassé, 27 Feb. 1901; Perrier, *Figure de la terre*, p. 355.

15. Poincaré, cited in Perrier, *Figure de la terre*, p. 346.

no doubt saw their factotum in the expedition's geodesist, Lieutenant Georges-Antoine-François-Jacques-Justin Perrier, who was the son of the late, first director of the army's geodesical service.[16] It is no reflection on Lieutenant Perrier, later to rise to the academy's heights, that the French scientific spirit in Ecuador during the time of the expedition found embodiment in the person of astronomer François Gonnessiat.

Gonnessiat, forty-four years old in 1900, had logged an unimpressive record as a researcher and teacher. A pupil of Charles André's, he found himself marooned at Lyons as an aide-astronome and *chargé de cours.* When Ecuador's General Alfaro sought to profit from France's military presence by calling a French astronomer to reorganize the observatory at Quito, the position fell to Gonnessiat. His tenure at Quito lasted until 1906, that is, for as long as the French army was traipsing around the Ecuadoran countryside.[17]

When Gonnessiat arrived in Quito, he found something akin to a museum. The observatory still had Menten's 9-inch Merz equatorial and a 6-inch Repsold meridian telescope, and around 1890 it had benefited from an input of more than 48,000 marks, but as Gonnessiat noted, "most of the instruments deteriorated before having been put to use."[18] Just as Gonnessiat began his tasks, General Alfaro ceded to General Léonidas Plaza Gutiérrez in a general election. Plaza was less of a Francophile and patron of science than his predecessor had been, and the new man promptly cut back on Gonnessiat's resources. The French astronomer protested to influential senators and met with Plaza, but he was pessimistic about the long-term outcome: "It is to fear that he [Plaza] considers geodesy and astronomy a bit superfluous in a country that lacks a great many things." As far as the operation of his observatory went, Gonnessiat felt saddled with a "quite mediocre and hardly conscientious" support staff. He hoped to be able to appoint a former French student–Jesuit as assistant astronomer, but he wondered if the government would allow it. To place himself in the best light with the local community, he emphasized, he lectured three times a week at the university, and he invited his fifteen pupils to carry out practical work in the observatory.[19] The university left much to be desired, and a good deal of the blame for this situation resided with the secondary schools: "The level of secondary instruction in science is so low that it is neces-

16. *DSB*, s.v. "Perrier."

17. *Pogg.*, s.v. "Gonnessiat."

18. PAN, F17/13062, Gonnessiat to Director [of the Paris Observatory?], 23 Dec. 1900.

19. Ibid., Gonnessiat to Bassot, 19 Nov. 1901.

sary to begin everything anew at the Faculty." In general, students bypassed science in favor of law and medicine.[20] Gonnessiat spoke to General Plaza about hiring a French assistant astronomer. It was impossible, he said, "to find a competent coworker among the young people of Ecuador; the Faculty of Sciences has been closed for five years and scientific instruction is nonexistent in the collèges." Plaza declined to make the French appointment.[21]

The French military expedition was also having difficulty adjusting to General Plaza's regime.[22] The organizational problems are clearly revealed in a report tabled by the expedition's leader in 1902. The leader, Robert Bourgeois, like Perrier later to be received into the Academy of Sciences, emphasized that Alfaro had failed to keep his promise of providing the mission with fifty thousand francs for the year 1901. Commandant Bourgeois identified the source of the problem as the French chargé d'affaires in Quito, Joseph-Hippolyte Frandin, who had systematically obstructed the work of the mission. Bourgeois traced the lack of cooperation to wounded pride: Frandin felt slighted at not having been asked to head up the military mission, and he coveted a senior rank in the Legion of Honor for his efforts on behalf of the geodesists. There were other sins. Bourgeois emphasized that Frandin had neglected to watch out for a German military contingent stalking the halls of the Ecuadoran ministry of war.[23]

The surviving French military missionaries returned to Paris with their data, which Perrier worked up in desultory fashion over the next thirty years. Gonnessiat, his unimpressive tenure as observatory director near an end, persuaded the local authorities to establish a French-staffed school of sciences. To this end, in 1905 four French teachers signed a five-year contract with the government at Quito. The new arrivals had barely unpacked their bags when a political revolution occurred. Three of the four men voided their contract and returned home, leaving only the chemist Gentey.[24]

Gonnessiat, too, returned to Paris with the military geodesists. In 1907 he succeeded Charles Trépied as director of the Algiers Observatory. Having promoted the civilizing mission of France in South America, he graduated to an analogous post in Algeria. The Quito director-

20. Ibid., Gonnessiat to Director [of the Paris Observatory?], 23 Dec. 1900.

21. Ibid., Gonnessiat to Bassot, 14 Jan. 1902.

22. Ibid., 19 Nov. 1901.

23. Ibid., Robert Bourgeois, "Compte-Rendu sommaire des travaux exécutés par la mission de l'Equateur en 1901" (Paris, 1902).

24. Perrier, *Figure de la terre*, p. 385; José Maria Vargas, *Historia de la cultura ecuatoriana* (Quito, 1965), pp. 394–95.

ship fell, by the beginning of 1906, to his junior colleague Joanny Philippe Lagrula, a thirty-six-year-old aide-astronome at Lyons who had recently completed a doctorate on lunar occultations of star groups. Lagrula returned to France in the middle of 1907 to become, in 1910, astronome-adjoint at the Nice Observatory.[25]

Here the standard histories of French science in Ecuador end. There was, nevertheless, an important sequel that reveals how Parisian mandarins sought to use exact sciences as a cover for achieving larger, geopolitical goals. The sequel centered on the secret mission of a newly naturalized French citizen, Milan Ratislav Štefánik, an astrophysicist who had previously represented France in Asia, Africa, Oceania, and Brazil. Štefánik's earlier career situates his activity in Ecuador, and to it we now turn.

The Slovak Agent of French Interests

The son of a Protestant pastor in Hungarian Slovakia, Štefánik studied exact sciences in Prague at the Institute of Technology and at the Charles University, as well as in Zurich. He took a doctorate from the Charles University in 1904 and then went to Paris with the hope of becoming an assistant at Pierre Jules César Janssen's astrophysical observatory in Meudon. By that time Janssen was an invalid, and the daily operation of his observatory fell to his wife and daughter, whom Štefánik charmed. Štefánik should have forsaken the lioness for a jealous lion, Janssen's lieutenant and successor Henri Alexandre Deslandres. Upon Janssen's death in 1907, Deslandres exacted revenge on both the widow Janssen and Štefánik, the latter of whom he called "an odd-ball, a grotesque man, an adventurer."[26]

First to the enmity of Deslandres. In 1910 he provided a concise apologia in a letter to William Wallace Campbell. Deslandres wrote to thank Campbell for hosting a splendid astrophysical congress on Mt. Hamilton and to congratulate him on his receipt of the Janssen medal of the Astronomical Society of France (Štefánik was proposed for the medal in 1907). For Deslandres, the medal was Janssen's positive legacy. "On other points," however, Deslandres advised Campbell that Janssen "was quite annoying." He bothered the astronomers under his command, produced no students, and "as Mr. Poincaré rightly said, he

25. *Astron.*, s.v. "Lagrula." Lagrula's appointment was effective 5 Aug. 1906, but his scientist son Jean-Louis was born in Quito on 9 Feb. 1906.

26. An enormous literature appeared about Štefánik during the interwar period. Short accounts may be found in Ludmila Mrkosová-Pajdušáková, *M. R. Štefánik, astronom* (Bratislava, 1970), pp. 60–61 for Deslandres and his opinion of Štefánik; and Josef Ihnat et al., eds., *General Milan R. Štefánik, 1880–1980: Historical Profile* (New York, 1981).

Scientist, spy, and soldier; Milan R. Štefánik at the Meudon astrophysical observatory, ca. 1905. (From Štepán Osuský and Bohdan Pavlů, eds., *Štefánik, kniha prvá: Spomienky a postrehy* [Prague, 1938])

halted the progress of physical astronomy in France for more than twenty years." Deslandres knew that the California astronomers were friends of Janssen's, as Janssen had taken George Ellery Hale's side in a priority controversy with Deslandres (about H and K spectra, which Deslandres claimed to have observed in 1891 and 1894). The new director at Meudon described how his predecessor, "beginning in 1904, was soft-witted, senile. In fact, the observatory was directed by his wife and his daughter, who took as their scientific adviser de La Baume Pluvinel, who is a serious man but merely an amateur. It was intolerable." Aymar, comte de La Baume Pluvinel, received "effective direction of the observatory" in 1906. At a conference of 1907 in Meudon, the Janssens heaped scorn on Deslandres, but Deslandres had his revenge: "After the death of Mr. Janssen, I undertook a lawsuit against Mrs. Janssen, which I won."[27]

Chased from Meudon by Deslandres, Štefánik became a roving astrophysical traveler in the service of those who found the new man at Meudon spiteful and unimpressive. He took observations on the new international observatory atop Mt. Blanc and went to Central Asia on expedition; in 1909 he traveled to the French Sahara with a view to

27. LO, Deslandres to Campbell, 21 Sept. 1910.

locating an astronomical observatory there.[28] Deslandres interfered with Štefánik's missions, forcing the Hungarian national to receive guarantees of safe passage from the Austro-Hungarian embassy in Paris.[29] Then Štefánik, who was living in Paris on charity and grants, received a mission from the Bureau of Longitudes to observe Halley's comet on Tahiti. Despite the initiative of Henri Poincaré, however, the French state would give no more than a hundred francs toward an expedition led by a foreign national. Štefánik borrowed instruments from Emile Schaer, astronome-adjoint at the Geneva Observatory (Schaer had outfitted the Mt. Blanc Observatory), and set off across America to Tahiti for the comet and nearby Vavau for an eclipse. He brought a 38-cm refractor and a 28-cm photographical equatorial, as well as five smaller telescopes.[30]

Štefánik arrived on Tahiti eager to establish a meteorological network with instruments supplied by Alfred Angot and to erect a permanent astronomical observatory.[31] Angot, in fact, had alerted the governor at Papeete about Štefánik's arrival. Štefánik's meteorological observations would "certainly provide great interest in the way of comparison" with the observations then taken at the Papeete hospital; it was in any case important for Štefánik to compare the hospital's instruments against the ones he brought directly from France.[32] Štefánik set up his telescopes on Mt. Faiere, above the administrative center Papeete, under an 8-m cupola. "The time has come for an observatory in the region of the equator," he wrote, "and Tahiti seems perfectly designed for it. What a pity if France did not seize the moment."[33]

The governor of French Oceania, Joseph François, apprised the

28. Štefánik to Henri Devouassoud, 23 June 1900, in *Zápisníky M. R. Štefánika*, ed. Josef Bartůšek and Jaroslav Boháč (Prague, 1935), p. 145; Mrkosová-Pajdušáková, *Štefánik*, pp. 94–95.

29. Štefánik to Felix Heyman, 1909, in *Zápisníky M. R. Štefánika*, pp. 162–63, where Štefánik refers to Deslandres's sabotage.

30. Mrkosová-Pajdušáková, *Štefánik*, pp. 105–20; Štefánik to Schaer, 27 Dec. 1909, 22 Jan. 1910, and 1910, in *Zápisníky M. R. Štefánika*, pp. 166–67, 169–70, 172–73. According to SOM, Fonds des missions, carton 79, Štefánik to Camille Guillaume Bigourdan, 22 Jan. 1913, a 40-cm Newtonian reflector was planned for but apparently not brought to Tahiti in 1910. On Schaer, see Henry C. King, *The History of the Telescope* (New York, 1955), pp. 419–21; and Georges Tiercy, "Un astronome artiste-opticien: Emile Schaer," *Archives des sciences physiques et naturelles* 14 (1932), 69–81, on p. 76, where Štefánik's 28-cm photographical refractor is listed as being with the Meudon Observatory.

31. PAN, F17/17286, Angot to Minister of Public Instruction, 24 Oct. 1910, where Angot declines to fund Štefánik but agrees to supply him with instruments.

32. PMN, Angot to Governor of French Oceania, 19 Feb. 1910.

33. Mrkosová-Pajdušáková, *Štefánik*, pp. 111, 134; Rostislav Rajchl, "Štefánik a Oceanie," *Říše hvězd* 18 (1937), 93–101; A. V. Novák, *Tahiti: Rajské ostrovy jižních moří* (Prague, 1923), photographs of Štefánik's observatory and residence facing pp. 16–17.

minister of colonies in August 1910 about Stefánik's progress. The Slovak astronomer had arrived bearing letters of recommendation from Senator Emile Chautemps, a former minister of colonies, and Henri Poincaré. François ordered his public works department to install Štefánik's equipment on Mt. Faiere; the cost came to two thousand francs. Štefánik, emboldened by François's cooperative attitude, asked the colonial administration to help finance his constructions, offering in return "not only his entirely unselfish efforts but also those of the instruments belonging to him." François was not hostile to the idea of having Štefánik organize a meteorological service, but he would not authorize local funding without obtaining special permission from Paris. The ministry of colonies asked the ministry of public instruction whether Štefánik's scheme could be carried on the educational budget there; public instruction offered only to allow Štefánik to keep the meteorological instruments that had been lent to him.[34] Štefánik spent nearly a year setting his house in order. Upon leaving, he confided his instruments to the keeper of Papeete's semaphore, for he hoped shortly to return.[35]

In Paris without a proper position, Štefánik started to plan for his return to paradise.[36] He became leader of the Paris Academy's mission to observe the Brazilian eclipse of 1912, for the academy provided forty thousand francs to cover his expenses and send him on to Tahiti. Štefánik observed the eclipse at Passa Quatro, Brazil, accompanied by the sculptor Jaromír Králíček, who had served as his assistant since late 1911. The two men then returned to Paris and made final plans for the South Seas.[37] Early in 1913 Štefánik emphasized to his patron the Bureau of Longitudes that his photographical equatorial and meridian

34. SOM, Fonds des missions, carton 79, François to Minister of Colonies, 10 Aug. 1910; note of 2e Direction, 4e Bureau, Ministry of Colonies, 14 Oct. 1910; Minister of Public Instruction to Minister of Colonies, 7 Nov. 1910.

35. R. Klima, "Milan Rastislav Štefánik à Tahiti," *Bulletin de la Société des études océaniennes* 8, no. 3 (1950), 85–88.

36. "M. R. Štefánik v rodinné korespondenci Jaromíra Králíčka," in *Štefánik. Kniha-druhá: Vzpominky, dokumenty a jiné přtspěvky*, ed. Josef Bartůšek (Prague, 1938), pp. 126–34, where Králíček's notes from late 1911 through early 1913 are reproduced. On 5 Dec. 1911 Králíček mentioned his tasks in the second Tahitian adventure. Štefánik intended to return to Tahiti in March and travel by way of Russia, China, and Japan; he hoped to bring a telescope worth three hundred thousand francs.

37. "Štefánik v rodinné korespondenci Králíčka," p. 128, note of 13 Apr. 1912, where the Paris Academy grants Štefánik forty thousand francs for the expedition to Brazil and Tahiti; 1 Mar. 1912, where Štefánik proposes to go to Tahiti directly from Brazil; 19 Sept. 1912, where Štefánik and Králíček will return to Europe before setting out for Tahiti. SOM, Fonds des missions, carton 79, Camille Guillaume Bigourdan, attestation of 24 Dec. 1912 for Štefánik's Tahiti mission, according to a note apparently presented to the ministry of colonies by Emile Chautemps.

telescopes remained at Papeete, along with his meteorological instruments, a darkroom, a library, and an embryonic meteorological network throughout the archipelago. Štefánik wanted to return there to catalogue southern nebulae and double stars. He thanked astronomer Camille Guillaume Bigourdan for recommending him to the ministry of colonies, to which he was applying for financial assistance.[38]

French colonial regimes, enjoying considerable autonomy, depended on the character and prejudices of their governors, who seldom held power for more than several years at a time. By 1913 a new governor had come to Tahiti, and he manifested no sympathy for Štefánik's plans. Governor Adrien Jules Bonhoure labeled Štefánik's earlier work on Tahiti as "personal studies, without practical utility for the colony, which in no way profited from them." Furthermore, Štefánik had not interacted well with the local movers and shakers. "He did not succeed in attracting the sympathy of the most responsible representatives of the population." As a consequence, "it would be quite difficult to obtain any sort of funding from the Administrative Council for this astronomer, who did not leave excellent memories in Tahiti." Despite these strictures, Bonhoure recommended that Štefánik be granted free transport for his instruments, be released from all customs duties, be provided with a rent-free site for the observatory, and be offered the services of the Tahitian public works.[39] The ministry of colonies came through with seven thousand francs and tried unsuccessfully to have the ministry of public instruction furnish additional support for Štefánik's endeavor.[40]

Benjamin Baillaud campaigned on Štefánik's behalf. He emphasized

38. SOM, Fonds des missions, carton 79, Štefánik to Bigourdan, 22 Jan. 1913.

39. Ibid., Bonhoure to Minister of Colonies, 17 Feb. 1913. Cf. Klima, "Štefánik à Tahiti," p. 86, where the native Tahitians "very much liked" Štefánik.

40. SOM, Fonds des missions, carton 79, "Subventions accordées à M. Štefánik," recording seven thousand francs from the ministry of colonies on 4 June 1913. PAN, F17/17286, Minister of Colonies to Minister of Public Instruction, 4 June 1913, where Štefánik, who is charged by the Bureau of Longitudes and the Central Meteorological Bureau to set up the Tahiti observatory, asks for a grant of seven thousand francs; colonies wanted public instruction to cover part of the cost, because his work "concerns science in general more perhaps than the colony itself." Public instruction declined to contribute to the cause. Angot to Minister of Public Instruction, 16 June 1913, declining to finance Štefánik's overseas activity; Benjamin Baillaud to Minister of Public Instruction, 25 June 1913, declining to support Štefánik on the ledger of the Bureau of Longitudes; Colonies to Public Instruction, 23 Oct. 1913, where colonies tries once more to pry loose money for Tahiti. SOM, Océanie C16, carton 104, Minister of Colonies to Governor of French Oceania, 31 July 1913, noting the ministry's grant of seven thousand francs and asking the governor to contribute to Štefánik's upkeep until early 1914, when "the astronomical service of your colony will be able to be constituted in a permanent fashion."

to the minister of colonies that Tahiti would be just the place for France's long-desired southern observatory. Baillaud had first thought of Madagascar for the observatory, but Elie Colin persuaded him that the physical climate was too unhealthy for regular night observations; Nouméa was too far from France, and its moral climate (the presence of convicts) too unappealing; French Guinea had cloudy skies. France had lost its close connections with two stations in South America, La Plata and Santiago de Chile. According to Baillaud, there were only two active observatories in the Southern Hemisphere – Charles Dillon Perrine's at Córdoba, and the British one at the Cape of Good Hope. Baillaud proposed that if the colony financed a Tahitian observatory, it would best be governed by the model of the Algiers Observatory; if metropolitan France financed it, then its administration would lie between those of Algiers and the observatories in the hexagone. Štefánik's practical, meteorological tasks received detailed discussion in Baillaud's petition, as did his plan to set up a wireless station. Baillaud estimated that the operating budget for the entire installation – including salaries for Štefánik, an assistant astronomer, other staff – would come to twenty thousand francs, plus the costs of transport, rent, and maintenance of the grounds.[41]

The Astrophysicist Spy

Late in 1911 Štefánik's activity for France brought him the two-thousand-franc Wilde prize from the Academy of Sciences, upon the recommendation of Henri Poincaré.[42] His profile, experience, availability, and new status (from 1912) as a naturalized French citizen recommended him to the French authorities as a secret agent.

As plans for the second Tahiti expedition advanced, Senator Emile Chautemps, in whose house Štefánik was a regular visitor, decided to send the astrophysicist to the South Seas through the Panama Canal. The arrangements fell to Anatole de Monzie, the undersecretary of the French merchant marine. De Monzie had been pondering the imminent opening of the American-controlled canal, anxious to secure a French-controlled coaling station on its Pacific side. Clipperton Island, a French dependency claimed by Mexico, was too far north of the main

41. SOM, Fonds des missions, carton 79, Baillaud to Minister of Colonies, 5 May 1913.

42. See ibid., Gaston Darboux to Minister of Colonies, 22 Jan. 1913, for the Wilde prize on Poincaré's recommendation; "Štefánik v rodinné korespondenci Králíčka," p. 128, note of 10 Jan. 1912, on receipt of the two-thousand-franc prize from the Paris Academy on 18 Dec. 1911.

route to French Oceania. Over the preceding decade, however, the Galápagos Islands had been much on the mind of French diplomats.[43] De Monzie recalled that one of the islands had become the property of a Basque expatriate, who had willed it to the small town of Barcus in the Basses-Pyrénées. Štefánik presented his project to de Monzie along the lines of erecting a radio transmitter on Tahiti. De Monzie identified the French-claimed Galápagos island as an essential relay station for Štefánik's project.[44] A French coaling station would naturally take shape next to the radio mast.

De Monzie and Štefánik hatched their plans in the course of the next week.[45] Štefánik would be granted a mission by the ministry of public instruction to reorganize the Quito Observatory. He would gain the confidence of the Ecuadorans by his scientific expertise, following which he would obtain a wireless concession for France. Notwithstanding its judgment against an observatory on Tahiti, the ministry of public instruction contributed two thousand francs toward Štefánik's proposed meteorological observatory and wireless installation in the Galápagos.[46] De Monzie presented Štefánik to the Société de géographie of Paris and persuaded its president, Etienne Gabriel Joseph, baron Hulot, to support a permanent meteorological station in the Galápagos.[47] Štefánik left for Ecuador in September 1913, stopping at Papeete only to survey the site of his future observatory.[48]

43. Wilhelm Clement, "Die Monroedoktrin und die deutsch-amerikanischen Beziehungen im Zeitalter des Imperialismus," *Jahrbuch für Amerikastudien* 1 (1956), 153–67; E. Taylor Parks and James Fred Rippy, "The Galapagos Islands: A Neglected Phase of American Strategic Diplomacy," *Pacific Historical Review* 9 (1940), 37–45. Around 1900, President Eloy Alfaro of Ecuador claimed that France offered him 100 million francs for a free port in the Galápagos.

44. Anatole de Monzie, *Destins hors série* (Paris, 1927), pp. 127–53.

45. SOM, Fonds des missions, carton 79, "Subventions accordées à M. Štefánik," recording five thousand francs from the undersecretary of the merchant marine to Štefánik in August 1913.

46. PAN, F17/17286, Štefánik to Minister of Public Instruction, 18 Aug. 1913; note of 29 Aug. 1913.

47. PSG, colis 76/4306, Štefánik to Hulot, 17 Aug. 1913, where Štefánik thanks Hulot for receiving him into the society and also for applying for a grant on behalf of his upcoming expedition to Ecuador and the Galápagos; Štefánik indicates that the letter was forwarded to him from the ministry of colonies, and he asks Hulot to send the grant to de Monzie for transmission to Ecuador. See Štefánik to Hulot, 20 Aug. 1913, for the "principal" task of setting up the Galápagos meteorological observatory.

48. According to Klima, "Štefánik à Tahiti," Štefánik's observatory disintegrated and his instruments "dispersed." Papeete was bombarded by the German cruisers *Scharnhorst* and *Gneisenau* in 1914, but the accounts do not mention Štefánik's observatory (Claude Farrère and Paul Chack, *Combats et batailles sur mer* [Paris, 1925], pp. 5–64; Emile Vedel, "La guerre à Tahiti [1914]," *L'illustration* [Paris], 24 May 1918, pp. 511–15, where a map locates the observatory on Mt. Faiere, some two hundred meters inland from the sema-

Arriving in Quito, Štefánik found the observatory almost in ruins. The roof leaked, and "in every corner there were spiders." He quickly entered into relations with the French ambassador, Henri Francastel, and the observatory's director, Luis G. Tufiño, who had served as Lagrula's assistant and had spent the years 1909–11 at the Paris Observatory. By the end of November, Štefánik had helped Tufiño place the Merz equatorial and the meridian circle in working order.[49] Tufiño's interest in Štefánik's visit is evident from a proposal he had circulated in 1911 for the regeneration of the Ecuadoran geographical service. Mapping the country would focus instruction in the Faculty of Sciences of the University of Quito; it would invest a military corps with prestige and honor. Tufiño volunteered to train the core of the service in his observatory, and to this end he asked the government for an increase in his budget, which was only half of what Gonnessiat had enjoyed. His project would be the salvation of the nation.[50] The savior needed foreign learning. He embraced his French cultural allies.

Early in December Štefánik and Tufiño planned for a meteorological observatory in the Galápagos that would complement "the work done in the observatories of Quito and Tahiti."[51] Then Štefánik sprang the wireless scheme on the unsuspecting Ecuadorans. They fell into his waiting arms. Ecuador promised the extraordinary sum of 3 million sucres (7.5 million francs) for setting up a wireless network. It would be headed by a French engineer, on contract for three or four years.[52]

Štefánik took to his role as a secret agent, setting down catty comments and geopolitical fantasies in a Slovak notebook. He sent coded telegrams to Chautemps in Paris, informing the senate of his progress.[53] He mused:

---

phore). In addition to several batteries on the mountain, the semaphore also had a cannon, which issued the first shots of the battle, intended to sound an alarm (Claude Farrère and Paul Chack, "L'épopée de Tahiti," *Revue des deux mondes* 23 (1924), 5–42, on p. 36.

49. Mrkosová-Pajdušáková, *Štefánik*, pp. 142–46, for quotation and work schedule; Vladimír Polívka, ed., *Zápisník Dr. M. R. Štefánika z Ecuadoru z r. 1913* (Banská Bystrica, Czechoslovakia, 1933), p. 67, for Tufiño's stay in Paris. See also *Astron.*, s.v. "Tufiño"; and Paul Stroobant et al., *Les observatoires astronomiques et les astronomes* (Brussels, 1907), p. 202, for Tufiño's assistantship under Lagrula.

50. Tufiño, *Servicio geográfico del ejército ecuatoriano y la única base práctica en los estudios de la Facultad de ciencias (proyecto)* (Quito, 1911). Gonnessiat received 16,000 sucres per year (p. 9).

51. *Zápisník Štefánika z Ecuadoru*, pp. 145–52, Tufiño and Štefánik to President of Ecuador, 6 Dec. 1913.

52. Ibid. The conversion of sucres to francs is from a note of Štefánik's.

53. See, e.g., ibid., p. 111, on 15 Dec. 1913 to Chautemps: "We hope to be able to do what you want. We should act carefully, without any responsibility or guarantee on your part. Speed up the decision as much as possible. Competitors are a worry."

> Situation: England will want to exercise its rights as soon as a foreign power occupies the Galápagos. Its interest is that the Galápagos remain under the jurisdiction of Ecuador. On the other hand, the United States wants the Galápagos to support Panama, and also, I think, because of the imperial utopia inaugurated by Roosevelt. If the United States does not succeed in purchasing them, it will succeed by other means. Ecuador will declare war for the Eastern Territories and will occupy the islands and sell them to the United States, or else the United States will send agents to the islands who will foment a revolution and acknowledge a republic, as in Panama.[54]

He continued several days later:

> It is in the interest of France that Ecuador be Francophile, which would put an end to the expansionism of the United States toward the south. Ecuador needs money and culture. France would be inclined to a discussion, because it wants new commercial possibilities, since investing in Germany would be suicide. But financial experts do not want to enter adventurously into this republic. And Russia, without consolidation of its debts, does not give enough guarantee. The Chinese government is pressuring French politicians about a certain financial arrangement. I came to explore the terrain. Scientifically, my trip is sufficiently motivated to be above all suspicion. And besides, our intentions are in reality friendly. It is a pity that my health and the scientific hopes do not let me carry through my intentions and my tasks. I think that he [Pierre Reinburg] is the better mediator for carrying through this arrangement between France and Ecuador. And thus, I could return as soon as possible. But I will not return empty-handed. I must have the wireless; and I will also leave a great scientific program for Quito.[55]

Štefánik's confidant Dr. Pierre Reinburg, an old hand in the Andes who had been charged with anthropological missions to Ecuador and Peru by the ministry of public instruction, acted as his translator during negotiations with Ecuador's ministers of state.[56] The Slovak secret agent did not trust the man on whom all his plans depended. Reinburg was, Štefánik noted privately, "a black, egoistical soul. He radiated confidence, but I didn't let myself be sucked in. Apparently he is able to help us, but he doesn't want to do anything that might threaten his position

54. Mrkosová-Pajdušáková, *Štefánik*, p. 144.

55. Ibid., pp. 142–43.

56. Reinburg, *Les accès dits éclamptiques* (med. diss., University of Paris, 1905). See PAN, F17/17284, for Reinburg's archaeological, anthropological, and geographical missions in 1908 and 1912. Reinburg provided the Central Meteorological Bureau in Paris with a copy of early geographical surveying in Chile undertaken by Albert Obrecht, Xavier Devaux, and Irénée Lagarde (see Obrecht et al., *Coordenadas jeográficas de algunas ciudades de Chile* [Santiago, 1890], extract from the *Anales* of the Santiago Observatory, in PMN, with Reinburg's stamp indicating his position as *chargé de mission* in Quito).

with the Ecuadoran government. I assured him that the wireless has nothing but a scientific character. If French colonies profit by it, that would be secondary."[57] The following day Štefánik continued his observations and ruminations in the wake of a dinner with Tufiño and Reinburg:

> Dr. R[einburg] came to us. He is very curious. He wants to know everything. But his friendliness always has a touch of arrogance. I spoke with him just about trivialities concerning Tahiti so that Tufiño could not speak in confidence about our plans. After dinner, Tufiño and I went for a walk, and I spoke with him about Dr. R. Apparently he is an egoist and small-time bandit. . . . R came on the scene successfully, founded a short-lived geographical society, and arranged for a monument to the [eighteenth-century?] French [geodesical?] mission in Ecuador. Ecuador gave him 6,000 sucres. But R is a bluffer. He plays the role of a confidence man. . . . R is thought to be a spy for Peru. Maybe not, but he is an opportunist.[58]

Štefánik worried that Reinburg had privileged access to Anatole de Monzie — whom Reinburg had known in collège — and he fretted that Reinburg had read his correspondence in the French embassy.[59] We are never laid so completely bare as when we speak in confidence about others. The immature, romantic, and frenetic adventurer was perfectly suited for the task of secret agent. He noted in his diary, alluding to a stomach condition: "All fixed up. . . . I did not like the diplomatic chore, and I feel gastric pains from time to time."[60]

The mission to Ecuador, serving to advance "French expansion," propelled Štefánik into the Legion of Honor, but not much else concrete resulted from it.[61] No powerful transmitter came to Tahiti, hence no relay station to the Galápagos and no French maritime intervention in Ecuador. Štefánik suspected that this would be the result. He noted privately while in Ecuador that contracts were "worth as much as in savage Fiji. . . . No signature is a guarantee."[62] Yet he corresponded with Tufiño, urging him "always to remain loyal to your second country, which you call France," for it was "the duty of France to make all necessary efforts to transform your beautiful country into a powerful, rich, independent, and cultured state."[63]

Tufiño published several issues of a monthly observatory bulletin

57. *Zápisníky M. R. Štefánika*, p. 258, Štefánik's notes for 1 Dec. 1913.
58. Ibid., Štefánik's notes for 2 Dec. 1913.
59. Ibid., p. 262, Štefánik's notes for 4 Dec. 1913.
60. Mrkosová-Pajdušáková, *Štefánik*, pp. 142–46.
61. Quotation in the citation for the Legion of Honor, in ibid., p. 147.
62. *Zápisníky M. R. Štefánika*, p. 272, Štefánik's notes for 11 Dec. 1913.
63. De Monzie, *Destins*, p. 146, citing a letter from Štefánik to Tufiño of 13 Sept. 1914.

and then fell silent.[64] His talent lay in ambitious, not to say pretentious, programs for future research. Late in 1914, for example, he elaborated a scheme for a central service to circulate meteorological data by telegraph; another plan had a radio receiver registering electrical activity associated with storms, but in formulating it Tufiño remained silent about Štefánik's French wireless contract.[65] Tufiño does not seem to have traveled to Paris for further training, and there is no indication that more French equipment found its way to Quito.[66] At the beginning of the war, German ships docked openly at Ecuadoran ports, and the Germans ran a wireless relay station in the Galápagos. Yet Anatole de Monzie asserted that during the war Štefánik's mission of cultural imperialism was "more useful than our tracts and warlike orders" in keeping Ecuador friendly to the French Republic.[67]

With Ecuador behind him, Štefánik still hoped to return to Tahiti. He received, however, a mission to work with General Louis-Hubert-Gonzalve Lyautey on Moroccan meteorology and geodesy. In Morocco on the eve of the outbreak of war, he learned about his induction into the Legion of Honor. Active military duty then claimed him. He elected to serve in the French air corps, where he applied his scientific talent to meteorology.[68] Štefánik was most useful to the French in organizing Czech and Slovak legions to fight under the tricolor, and recruiting for this cause took him to the United States and Russia. Wounded in Serbia, he convalesced in Italy and then climbed up the ranks to become brigadier general in 1918. In this capacity he went to Siberia to negotiate the European redeployment of Aleksandr Vasilyevich Kolčak's Czech-Slovak Legion. Having been a member of the Czech-Slovak Council in Paris, Štefánik became the first minister of war in the independent state of Czechoslovakia. Returning from Italy to take up his portfolio in 1919, he died when his plane apparently took hostile fire

64. See *Boletín mensual del Observatorio de Quito* 2 (1914), 1–3, for Tufiño's ambitious plans, which included participating in the astronomical *carte du ciel* project. The *Boletín* seems to have ended later that year with the combined numbers 3, 4, 5, and 6, for the months March through June.

65. Tufiño, "Tesis sobre meteorología agrícola," in *Proceedings of the Second Pan American Scientific Congress*, 11 vols. (Washington, D.C., 1917), 2:831–46, telegraph and radio on pp. 844–45.

66. Tufiño remained active in geophysics through the middle 1920s. See PIPG, Tufiño to Charles Maurain, 30 Apr. 1925, where he refers to having at Quito the first two volumes of the *Annales* of Maurain's geophysical institute.

67. Mrkosová-Pajdušáková, *Štefánik*, p. 149; de Monzie's global evaluation is on p. 151.

68. *Zápisníky M. R. Štefánika*, p. 296, where Štefánik explains how he learned about the Legion of Honor; Jules Rouch, "Historique du Service météorologique aux armées (mai 1915–février 1918)," *Revue maritime*, Feb. 1921, pp. 167–80, on p. 171.

from Slovak troops at the Bratislava airport.[69] His Tahitian observatory faded from memory. When, twenty-five years later, thoughts again turned toward endowing French Oceania with an astronomical observatory, Štefánik's pioneering work received no mention.[70]

Štefánik's involvement with Ecuador and Tahiti projects a dramatic image of French cultural imperialism. Late in the nineteenth century, Ecuador had become a field for interimperialist rivalries generally. The French ministries of war, the navy, the merchant marine, and public instruction funneled support to projects in pure learning designed to establish the cultural superiority of France and, that prestige won, to secure lucrative commercial advantages. The naturalized citizen Štefánik was an ideal agent of French interests abroad. A womanizing nomad with no close family ties and no prospects of a permanent scientific career in France, he was glad to render homage to his new land by traveling abroad. Like the other French missionaries in Ecuador—Robert Bourgeois, Georges Perrier, François Gonnessiat, Joanny Philippe Lagrula—Štefánik focused his attention on the administrative demands of higher authorities in Paris. The devotion catapulted him to the top of military and political institutions.

69. Milan S. Ďurica, "Milan R. Štefánik and His Tragic Death in the Light of Italian Military Documents," in Ihnat et al., *Štefánik*, pp. 3–43. See Vera Olivova, *The Doomed Democracy: Czechoslovakia in a Disrupted Europe, 1914–38* (Montreal, 1972), pp. 30, 62, 81, and passim, for Štefánik's political activity.

70. Jacques Ravet, "Les eléments météorologiques dans les établissements français de l'Océanie considérés au point de vue des observations astronomiques," *APGFOM*, 1937, pp. 58–60.

Conclusion: Rhetoric and Reality in French Cultural Imperialism

Ideology of Domination

From the summer of the July Monarchy to the autumn of the Third Republic, physicists and astronomers ranged across the globe as agents of French interests. Members of geodesical, hydrographical, and meteorological corps, complemented by career officers with scientific interests, rotated through postings in distant parts. Academic scientists – François Gonnessiat, Joanny Philippe Lagrula, Fernand Lataste, and Henri Gorceix – held a succession of positions overseas. War and disease led to geographical displacements among Jesuit scientists, for whom uncommon facility in foreign languages and cultures was an occupational hazard.

Administrators and scientists in the hexagone, variously secular and religious, attributed funds and supervised the manufacture of instruments. Men on the periphery realized that their career depended on responding to Parisian desiderata. Less would be gained by creating new research programs than by stewarding projects advanced by metropolitan mentors. French physicists and astronomers were functionaries more than researchers.[1]

1. I proposed the *functionary* ethos in previous publications: Lewis Pyenson, "Pure Learning and Political Economy: Science and European Expansion in the Age of Imperi-

French functionary scientists by no means restricted their activity to projects of lucrative exploitation. They pursued pure science, studying *general* questions in astronomy and physics that would return little immediate profit to metropolitan coffers. Pure science — in French, *science fondamentale* — has figured prominently in the preceding pages. There is nothing applied or practical about stellar astronomy. Mercantile fortunes are not made by determining intercontinental longitude differences to within a score of meters. Proposals for lavish funding based on arguments of lucrative gain — predicting the arrival of tropical storms by seismometry, orienting naval compasses by publishing global magnetical charts, locating mineral deposits by gravimetrical surveys — are travesties transparent to all but the most myopic politicians. What does register to local and more distant observers, however, are the claims of advancing disinterested knowledge.

Military authorities certainly supported projects in pure science because of the useful techniques or inventories they entailed. Astronomy produced maps; wireless communication — used by physicists for studying the movement of continent and ionosphere — was a crucial military technique of the twentieth century. Yet warriors also kept talented scientists on retainer, and for this reason they acquiesced to scientific projects of little or no military significance.[2]

Scientific *discourse* — what appeared in print — shows little variation across empire and sphere of influence. No flags draped a physical or astronomical paper issuing from the hand of civilian, warrior, or priest at Tamanrasset, Ifrane, Shanghai, Tananarive, Santiago de Chile, or Morne-des-Cadets. This invariability of form cannot result from disciplinary censure, for disciplines in the exact sciences have exercised little authority in modern France. Nor is it a function of military or religious regimen, even though many customs of the army and the church are observed by scientific institutions. The uniformity is a consequence of the civilizing mission, which depended on reinforcing an image of pursuing general principles and laws in the context of *French* traditions.

The past generation of scholarship has taught us that the exact sciences evolved differently from the biomedical sciences. For this reason,

alism," in *New Trends in the History of Science*, ed. R. P. W. Visser, H. J. M. Bos, L. C. Palm, and H. A. M. Snelders (Amsterdam, 1989), pp. 209–78; and "Functionaries and Seekers in Latin America: Missionary Diffusion of the Exact Sciences, 1850–1930," *Quipu* 2 (1985), 387–420.

2. Paul Forman, "Behind Quantum Electronics: National Security as Basis for Physical Research in the United States, 1940–1960," *Historical Studies in the Physical and Biological Sciences* 18 (1987), 149–229, p. 220 for the techniques and the retainer in another context.

we should expect that the pattern of French cultural imperialism in the present book may not extend to other fields of learning. French medicine, indeed, spread successfully through the efforts of the private Pasteur institutes, which focused on eradicating local scourges that menaced colonial overlords and plantation laborers. The pattern we have seen, furthermore, is special to modern France. Science in the German and Dutch empires exhibited no extreme centralizing and functionary characteristics;[3] universities and observatories set up by Americans at Arequipa, at San Luis (Argentina), at Beijing, at Beirut, and in South Africa received little in the way of instruction or funding from any government, whether metropolitan or peripheral. The French pattern in the exact sciences exemplifies the *mission civilisatrice* in practice. And in theory?

French overseas territory produced an ideologue who defined the career of *savant colonial.* He was Edouard de Martonne, a career officer in the colonial artillery and head of the geographical service of French West Africa. Commandant de Martonne focused on mapping the colony, using all the means at the disposal of a geographer: triangulation and cross-country treks, supplemented by astronomical and telegraphical determination of longitudes. He applied his talents to magnetical surveying. How this kind of inventory related to general laws and principles was not uppermost in his mind.[4]

In 1930 de Martonne published a book with the title *Le savant colonial* in a series directed by the director of the Ecole coloniale in Paris, Georges Hardy. De Martonne stated at the beginning that he wanted to sketch a new ideal type, one differing from the more familiar types of colonial administrator, colonial soldier, colonial physician, and colonial missionary. This was an ideal type of the future, instantiated as subject peoples increasingly received the "benefits of science" through "the intellectual *rayonnement* of the civilized nations" of the world. "*Science, placed in our colonial conquest, will up to a certain point counterbalance the mercantilism that is rampant everywhere.*" By observing scientific method, the indigene would be able to escape from his "*eternal childhood*" into

3. See *CIES* and *ER* generally.

4. De Martonne, *Rapport sur les travaux astronomiques et géodésiques exécutés en Afrique occidentale française avant la guerre, 1903–1914* (Paris, 1923) [reprinted from *L'Afrique française* 33 (1923)]; idem, "Le Service géographique de l'Afrique orientale française: Son fonctionnement, ses resultats (1922–1927)," *BC* 11 (1928), 321–51; idem, "Etat actuel de nos connaissances sur l'Afrique occidentale française: Magnétisme, avec une carte de la déclinaison au 1er janvier 1926," ibid. 8 (1925), 353–93, pp. 355–56, where de Martonne signals broader issues without, however, engaging them. Edouard de Martonne is not to be confused with Emmanuel-Louis-Eugène de Martonne, a geographer who rose to become a member of the Academy of Sciences and to direct the geographical institute of the Sorbonne.

"the progressive exercise of reason." Scientific manipulations, indeed, conferred "prestige" on French imperialists. Here, in de Martonne's opinion, was an unconscious moral factor acting in favor of the European colonialists. It followed that *"colonial savants are the agents of the propagation of French culture, just as colonists and functionaries are the agents of social and economic expansion."*[5]

Precisely what was colonial science? It was activity pursued "not by chance, but following a unified plan, executed according to a method," activity centered on "reasoned exploitation, accompanied by research into causes and the study of consequences." A meteorological service that only reported temperatures, pressures, and precipitation fulfilled no more than the simplest part of its mandate. The more difficult task was "to draw new indications about climatological questions from these facts, even to a certain extent to make predictions." Isolated observations, then, "were only the premises of knowing, which is based on integrating observations in time and in space, and on the criticism of results and the generalization of notions that have been collected. *Only the careful application of these methods confers the character of scientific work on the enterprise of establishments and services in the colonies.*" By paying attention to these criteria, colonial functionaries could transform themselves into "true colonial savants eminently useful to their country."[6]

The colonies were inhospitable to scientific adepts. Other places in Europe provided environments "where spiritual development is easier, construction of things is clearer, and where the intellectual worker benefits from a thousand imponderables that facilitate his task. The capitals are these privileged places, and great discoveries more frequently come from a well-equipped laboratory than from rural solitude, however rich in oxygen the latter might be." The colonies were not such "hothouses" of the mind. They sapped intellectual vigor. *"If one may work there, one may only rarely create there. Intellectual production is subject to too many distractions to permit long-term scientific enterprise."* Such obstacles served to underscore the extent to which science was predicated on "professional self-confidence," a character trait under threat from the dehumanizing effects of industrial society.[7]

Just as humanity in the industrial world could be degraded by the machine, so humanity beyond the industrial world could be—and to read de Martonne generally was—degraded. The indigene could do little with scientific concepts and scientific activity. "Facing abstract

5. De Martonne, *Le savant colonial* (Paris, 1930), pp. 21–22, 55, 32, emphasis (here and in the following) in the original.

6. Ibid., pp. 59–60.

7. Ibid., pp. 78, 91.

concepts . . . and a utility hardly apparent to him, the indigene will scarcely be motivated. Continuity fatigues him. His law is change without reason — at least, without apparent reason to us." Indigenes were unable to initiate action in a scientific context. The colonial scientist "*can only demand of them limited services, of a purely material sort and requiring the least possible initiative.*" It followed that the colonial scientist had as his task "raising the indigene from the sort of unconsciousness that pervades his life, prevents him from distinguishing cause from effect, and makes most of his reactions instinctive rather than reasoned."[8] And here the daily example of the colonial scientist was paramount.

That example, according to de Martonne, required living in a civilized fashion — drinking wine and eating steak — rather than adopting indigenous customs and fashions. This was the meaning of the mandates conferred on European powers by the League of Nations. At the same time, this image of civilized humanity, carefully cultivated by the colonial scientist, would serve to bind the colonies to their metropolitan master. It was wise to temper colonial restlessness and agitation by "the demonstration of our scientific superiority, alone able perhaps to enchain minds without contorting them. We must persuade these schoolchildren of civilization that the degree of emancipation grows naturally with the degree of culture, and that nothing may replace the maturity conferred by science." The colonial scientist was able to acquit himself of a number of responsibilities at once. He served his immediate superiors, he served the colony, he served France, and he served humanity at large.[9]

When we write about what we see, we describe who we are. De Martonne's observations are faithful to the character of scientists in the service of French cultural imperialism, but they are less successful as explanations of scientific activity abroad. Talented scientists did escape from administrative routine to seek generality in the colonies. More often than not, furthermore, their research depended on creative use of local talent, whether European subordinates or indigenes. Autochthonous energy powered the research of overseas savants from Tahiti to Brazil and China.

Ideology, in the sense of a skewed view of reality, does not end with ideas. It serves to channel human activity. Bound up with the French notion of the civilizing mission was, as de Martonne elaborated, a system of social control. The obvious part of that system — repressing conquered peoples and patronizing satellite civilizations — is of less interest here than the implications of the system for the life of French research-

8. Ibid., pp. 93–94, 109–10.
9. Ibid., pp. 106, 164 (for quotation), 151.

ers abroad. For in the matter of a scientific career, the civilizing mission gave rise to an antinomy.

Since original research served to enhance metropolitan prestige and radiate metropolitan norms, it had to be disciplined by metropolitan authority. The metropolis controlled promotions and rewards; it extended the promise of financing research and publishing results. Insofar as imperialist rivals and *indigènes sous tutelle* perceived research to be scientific, however, its end—publications—dared not appeal to authority. How, then, could pure science serve practical, political ends?

From our late twentieth-century perspective, there is an apparent resolution of the antinomy. Prestige would accrue from producing first-rate publications that in themselves were not obviously grounded in ideological prejudice or cultural myopia. The state would then reward its scientists for having devoted their life to an enterprise in which the state and the nation found no apparent place. Although direct pecuniary gain cannot be anticipated from the labors of scientists, pure science did and does result in practical spin-off. Astronomers in distant parts can calibrate the chronometers of naval vessels; physicists may repair broken radios and gyroscopes. Nevertheless, scientists—unlike mining engineers, livestock breeders, or medical healers—anticipate no monetary reward from their work. Lucrative practicality is not the main point of their activity.

French Jesuits—men who directly served the geopolitical and military interests of France—clearly perceived the importance of this kind of cultural imperialism in their own secular endeavor. The Zikawei Observatory was "not a commercial establishment," Louis Dumas recalled to his Chinese staff in 1940.[10] Its principal commodity was prestige. Pierre Lejay emphasized in 1938:

> There is no geophysical, astronomical, or radiotelegraphical congress where, in numerous commissions, we are not cited with sympathy and where, better, our cooperation is not requested.
>
> I have always thought that our influence in China could only be a corollary of this general recognition. Those Chinese scientists whom we can reach would only value us and would only respect us if, in the course of their studies at foreign universities or in their readings, they saw that we were well thought of by the experts.[11]

Spreading pure science abroad was essential to the French civilizing mission, just as it is central to the programs of today's cultural imperialists. In France, the question was never whether to pursue such a policy but rather how to implement it. The policy evolved, over a century,

10. SMB, Dumas to staff, 26 July 1940.
11. Ibid., Lejay, "Situation de l'observatoire," Oct. 1938.

from organizing scientific missions to establishing permanent research institutes. The policy's shortcomings derived not from the form of overseas expansion but from its restrictive spirit. The scientists *in partes infidelium* conducted themselves as apostolic ambassadors instead of as independent, creative thinkers.

Practicable Programs and the Hegemony of Language

Edouard de Martonne's vision extended across the entire tapestry of the French empire, from territories like Algeria, to colonies like Indochina, to protectorates like Morocco, to mandates like Lebanon. In these environments, civilian and military agents of the civilizing mission threaded their way through local administrative regimes. The imperialist vision was completely assimilated, furthermore, by Jesuit scientists, for whom all secular authority impeded progress toward the kingdom of God. The vision was manifest in French cultural missions to Latin cultures in the New World—in the Société scientifique du Chili, the Institut scientifique franco-canadien, and the star-chart satrapies.

In the late 1930s the ministry of foreign affairs conducted a general inquiry into French cultural imperialism in Latin America. The early twentieth century had seen a "crisis" in French influence, and the director of political and commercial affairs in the ministry asked ambassadors in Latin American countries for suggestions about restoring the French image. In the ambassadorial replies, one sees the signal role of French religious missions in promoting the French language. The best center of French instruction in Cuba was the Collège de la Salle, taught by French brothers; the Liceo Salvadoreño in San Salvador, run by the Marists, and the Instituto pedagógico de Varones in Managua, run by the Christian Brothers, had French directors; in Brazil, the religious schools, primarily under the Marists, were the "greatest source of French influence"; Argentina had more than fifty French religious colegios; the Bolivian ambassador asked for support of a number of French religious colegios, including one run by French-Canadian clerics. Many of the religious schools received funding from the ministry, whether in the form of outright grants or visiting instructors.[12] The

12. PAE, Amérique 1918–1940, dossiers généraux, 42, "Propagande générale: Enquête sur les moyens concrets de dévélopper l'influence française en Amérique latine": Cuban Ambassador to Ministry, 22 Apr. 1938; Ambassador to Central America (Guatemala) to Ministry, 1 June 1938; Brazilian Ambassador to Ministry, 27 June 1938; Argentine chargé d'affaires to Ministry, 6 Aug. 1938; Bolivian Ambassador to Ministry, 26 Aug. 1938. According to the ambassadorial reports, hundreds of thousands of francs went to support religious schools in Latin America annually.

Catholic church was the strongest overseas booster of the republic that had tried to eradicate religious instruction in the hexagone.

The general tenor of responses to the ministry's circular letter is one of pessimism and uncertainty. The ranking diplomatic officer in Bogotà wrote that the situation was "not only bad but very bad," French influence in Colombia being "null and our moral interests in complete decadence." Colombians were attracted to North American universities and seduced by German technicians; they saw the French as "a people in decadence," and they doubted that France was a great power. The reporter feared a popular, Mexican-style revolution in Colombia, which would only further compromise French interests. The Peruvian ambassador urged the Paris ministry to double to four the number of postgraduate scholarships offered in Lima and in this way attempt to counteract the fascist sympathies of the rector of the Universidad mayor de San Marcos. The Argentine chargé d'affaires emphasized that according to biochemist Bernardo Houssay, the German government had offered twelve postdoctoral scholarships to the Argentine Association for the Advancement of Science. There was reason to match the offer, since at that time France awarded only eight such scholarships.[13]

In the ministry's survey, science occupied a small part of French cultural interests in Latin America. When the Argentine chargé d'affaires proposed a dozen scholarships, he targeted only two of these for studies in basic sciences. Culture, according to French agents in Latin America, concerned literary and artistic enterprises, on the one hand, and practical matters, such as engineering and medicine, on the other hand. Poetry, beauty, public monuments, and personal health, all of which could be perceived directly by untrained minds, would radiate French values. The claims of pure science were diluted to the point of transparency. Prestige no longer derived from abstract research. French rhetoric remained much as it had been for fifty years, but French ambitions had changed. The Brazilian ambassador wrote to the ministry that French influence was "above all spiritual," hiding "no ambition of hegemony."[14] By the 1930s, however, France mounted no projects of cultural imperialism in exact sciences to compare with Admiral Mouchez's sky chart or Milan Štefánik's scientific mission to Ecuador. Energy in matters of scientific organization flowed away from Latin America to the French overseas empire.

Before the First World War, metropolitan academics (chemist Henry Le Chatelier and mathematician Paul Appell prominent among

13. Ibid., Bogotà Embassy to Ministry, 29 May 1938; Peruvian Ambassador to Ministry, 27 May 1938; Argentine Embassy to Ministry, 6 Aug. 1938.

14. Ibid., Brazilian Ambassador to Ministry, 27 June 1938.

them) formed an association to promote French learning in Latin America. Effort went into financing scholarships for Latin American students to study in France and visiting appointments or lecture tours for French professors in Latin America. Basing its appeal on race, on *Latinity*, the association became a respectable meeting place for French fascists and anti-Semites. In the years before 1940 it sent a score of French scientists to Brazil and Argentina (missionaries such as those sent to the Institut scientifique franco-canadien), but the ambitious plans for encouraging French instruction bore little fruit.[15]

According to association member Henri Goy, French schools were the principal means of spreading French influence in Latin America.[16] Neither the schools nor French learning sank roots. Unless people can see science and scholarship taking shape under their eyes, they may not have confidence in the grand phrases of occasional speakers. French observatory directors or institute chiefs, had they generated new wisdom, might have fanned Francophile sentiment, but such men and women did not turn up in Latin America after the beginning of the twentieth century. Centrifugal forces acting to shatter allegiance to an overseas imperial seat are counterbalanced only by the continual ministrations of a resident janissary.

A second volley of French cultural imperialism in Latin America – the military one – also failed to hit its target during the interwar period. The role of military missionaries in cultural imperialism is clear. With so much intellectual capital invested in its men under arms, France sent military men to carry out peaceful missions abroad. Scientific missionaries of this kind have been encountered in the preceding pages: Jules-Jean Feillet at the naval school in Santiago de Chile; Fernand, comte de Montessus de Ballore, in San Salvador; Georges Perrier and his colleagues remeasuring the Ecuadoran arc of longitude. Scientific gloss allowed military surveyors to act explicitly as agents of French interests, avoiding the charade played out by Milan Štefánik in wrangling a French wireless contract at Quito.

The gain anticipated from such initiatives is revealed in a report by

15. Ernest Martinenche, "L'action du 'Groupement' depuis sa fondation jusqu'à la fin de l'année 1913," *Bulletin de la Bibliothèque américaine (Amérique latine)* 4 (1913–14), 225–31. The *Bulletin* was the house organ of the Groupement des universités et grandes écoles de France pour les relations avec l'Amérique latine (see Patrick Petitjean, "Le Groupement des universités et grandes écoles de France pour les relations avec l'Amérique latine, et la création d'instituts à Rio, São Paulo et Buenos Aires [1907/1940]," in *Anais do Segundo Congresso latino-americano de história da ciência e da tecnologia, 30 de junho a 4 de julho, 1988*, ed. Ubiratan D'Ambrosio [São Paulo, 1989], pp. 428–42).

16. Goy, "Pour la culture française par l'école française," *Bulletin de l'Amérique latine* [successor to the *Bulletin de la Bibliothèque américaine*] 7 (1917–18), 115–22.

Camille Vincent, a captain in the French colonial army, about a mission carried out in South America on the eve of the First World War. In 1911 Bolivia asked the French Geographical Service to survey its frontiers. Vincent and an officer companion were seconded from the colonial army to the geographical corps and sent to delineate the Bolivian-Peruvian frontier. There they faced off against an English team hired by Peru to perform the same work, both teams being governed by a bilateral boundary commission. Vincent succeeded in persuading Bolivia to call on French expertise for further surveying. He urged the adoption of French wireless techniques for determining longitudes. "Only France, with its powerful military geographical service (metropolitan and colonial), would be able on occasion to lend its assistance to foreign states wanting to benefit from the great geographical progress in French science." Vincent noted that Uruguay had employed an officer in the French colonial army to survey the country, and this initiative had resulted in a mission of French officers to reorganize the Uruguayan army. Vincent hoped that once Bolivia had assimilated French science, it would invite in French military advisers.[17] Within twenty years German militarism rendered ineffective any attempt on the part of the French army to promote scientific work abroad: cultural imperialism is more subtle and slower to take effect than military adventurism.

Certainly the French version of cultural imperialism required time for indigenes to appreciate the foreigner in their midst who dined on steak and wine. What the representative of France generated, in terms of publications or other scientific monuments, was less important than how he comported himself. He served as a viaduct to the cultural life of Paris — relaying its fashions and passions to lesser peoples and extending the promise of introducing local talent to greater minds in the metropolis. For many cultural imperialists, synthesis was reserved for metropolitan institutions, an attitude sapping initiative and creativity on the periphery.

The desire to see the colonial scientist as Samson Agonistes — blind and shackled in distant obscurity, "there to labor in a common workhouse" — found expression by prominent scientists. In 1936 Alfred Lacroix, perpetual secretary of the Academy of Sciences, whose career rested on colonial vulcanology and mineralogy, could no longer conceive of building new centers of pure research in the French colonies. "There is no place in our colonies for major, new organizations devoted to purely scientific research. Local resources would not allow for their creation, and if a crisis of enthusiasm brought them into being today,

17. PSG, carton Vi-Vu, Vincent, 1028: Vincent, "Une mission géodésique française en Bolivie."

tomorrow their existence would remain uncertain." The centralized system of colonial service, in which functionaries changed colonial postings at regular intervals, militated against sustained research of a general nature at any particular location. Lacroix called for a formal, central authority to discipline colonial scientists, urging them to funnel observations and specimens to metropolitan savants. The new colonial council would place young scientists abroad, assigning them "a well-defined task." Then, "the research accomplished there, the study of the materials and discussion of the collected observations would be carried out in France." Lacroix called for a program to inventory the primitive world before it disappeared forever.[18] He provided a rationale for the overseas committee of the Conseil supérieur de la recherche scientifique, created in 1937.

Lacroix's vision resonated with the notions of Noël Bernard, director of the Pasteur institutes in Indochina. Successful colonial exploitation, Bernard observed, derived from inventorying the new land with a view to modifying its environment. The inventory had to be coordinated by a central agency in the metropolis that could set priorities essential to extending the power of the colonial regime. "The budget of our colonies does not support pure research, which sets out to discover the general laws of the physical world and of living things, and which must be independent of any obligation to provide an immediate return. Pure research is better situated in metropolitan laboratories, which provide the best conditions for specialization and work." Bernard would have botanists limit their colonial activity to exploitable resources: colonial entomologists had to study insect pests; colonial oceanographers had to apply their efforts to locating fish stocks. None of these scientists could be allowed to pursue questions of general interest: "These practical questions, which are connected to the progress of colonization, become an inescapable obligation for colonial scientists." By distinguishing themselves in the practical world, colonial scientists would transcend the world of research. They were to provide "inspiration, motivation, and counsel for increasingly ambitious work."[19]

Colonial science for Noël Bernard was precisely what Jean Perrin had in mind when he spoke about "directed research." All scientific institutions in the colonies had to be attached to major establishments in the metropolis, which would delegate students for overseas service. Men sent to the colonies had "to accept the obligation to carry out

18. Lacroix, "Pour une organisation des recherches scientifiques dans nos territoires d'outre-mer," *ACRACS* 13 (1937), 22–25; the text appears, slightly modified, in Lacroix's *Figures de savants*, 4 vols. (Paris, 1932–38), 4:224–29.

19. Bernard, "Au sujet de l'organisation des recherches scientifiques aux colonies," *ACRACS* 13 (1937), 12–17.

practical work in these distant countries." The metropolis would also establish "a division of labor between research done in the colonies and that – informed by colonial associates – which is carried out in metropolitan laboratories better equipped for the task. During sabbaticals in France, the associates work with their teachers to elaborate the projects for which they have collected the raw material during their overseas posting." Indigenes were to be selected and trained as laboratory assistants; no responsible positions could go even to local people with certification from institutions of higher learning. In this way, discipline could be maintained, standards ensured, and results made available to all the colonies.[20] Neither the Republic of Science nor the rights of humankind had a role in Bernard's notion of imperial centralization.

The minister of colonies in 1937, Marius Moutet, was less sure about the organization of imperial science. He opposed reducing science "to mere practical return, something that would risk making it a simple collection of useful recipes. Science is one and only merits the adjective 'colonial' in the immense field of its applications." Science was surely required for exploiting the French colonies, but its prosecution also constituted "a duty of our colonization, an example to give, a light to hold up to illuminate the path along which we are traveling." Moutet wanted to have the French civilizing mission become a habit of thought in the metropolis, and for that to happen,

> it is necessary to make it a category of mind. We need a kind of colonial thinking that is free, disinterested, unencumbered with either conformism or formalism. From such a movement of ideas, a colonial renascence must come about. In this way our colonization, which is not a colonization by force but a colonization of mind and of liberty, will keep its human face and its universal form. For these are its best part and its surest justification.[21]

Moutet's relatively enlightened vision enjoyed support from Fernand Blondel, a prominent engineer who compiled a geological bibliography of French overseas territory. Science had become an established feature of colonial life, he began in an essay about the colonial scientist. Because science was cultivated by scientists, the colonies had to respond to the material aspirations of their researchers. Blondel called for creation of "a scientific career properly colonial and entirely distinct from a metropolitan career." The time had passed when France could send metropolitan scientists on foreign missions to carry out research that would result in a call to a post in the hexagone; it was necessary to build permanent colonial institutions staffed by scientists who

20. See ibid., p. 15, for reference to Perrin and the long quotation.
21. Moutet, "Science et politique coloniales," ibid. 14 (1938), 17–25.

had no interest in a metropolitan posting. Permanence permitted a scientist to adapt physically and mentally to the colonies. Blondel envisaged a colonial career of fifteen years. This would allow for the appearance of a substantial body of work, after which the pensioned scientist could devote himself to a second career in France.[22]

Charles Jacob, a Sorbonne professor who had served as director of the Indochinese geological service and who rose to the Paris Academy's heights in 1931, provided an admirable illustration of Blondel's general scheme.[23] He supported Blondel's cry for an independent corps of colonial scientists. He wrote in 1937 about the urgent necessity "to create a cadre of colonial geologists, in which one would be able to follow a normal career, freeing colonial geologists from the constant worry of looking for another opening." Most colonial geologists were working toward a university diploma and dreamed of a university chair in France.[24]

"We must have a science for Africa, a colonial culture for colonials and for Frenchmen," Albert Charton, inspecteur général of education in French West Africa, emphasized in 1936. "It is an imperious necessity of the most urgent kind. Africa needs its own science just as it needs ports and roads; Africa needs it for its own exploitation, for the development of economic life, for the protection of agriculture and air navigation, for subterranean prospecting, for the multiplication of our resources and our riches, for our security and our *paix africaine*." Science lay at the center of France's civilizing mission. Through science France could extend her culture after military garrisons and governors general had departed. The "African peace" that France had brought to black civilizations was a humane intervention based on high scientific ideals. The resulting, hybrid Franco-African culture carried "the seeds of a renascence."[25]

Men who had achieved prominence in the scientific establishment debated the extent to which peripheral researchers would be tied to the metropolis. Ordinary scientists in French overseas territory simply wanted the opportunity to carry out original work. Georges Carle, an agricultural engineer with Moroccan experience, implored administrators to remember that science was open to all inquiring minds:

> Anyone able to establish a relationship between two phenomena or units that represent them observes a scientific fact. The latter is no longer the exclusive

22. Blondel, in ibid. 7 (1931), 80–92 (without title).
23. "Election à l'Académie des sciences," ibid., p. 93.
24. Jacob, report of the third section, Congrès de la recherche scientifique dans les territoires d'outre-mer, in ibid. 13 (1937), 97–102.
25. Charton, "Science, culture spirituelle et colonisation," ibid. 12 (1936), 177–81.

> domain of a few learned doctors who, encumbered with credentials, honor, and ermine, lecture *ex cathedra* about apparently certain truths that last for only a more or less limited time.

A scientific institute such as the one established in Morocco was a temple for maintaining international contacts with "the highest minds of the metropolis and all other countries." Science could not be placed under a particular branch of colonial administration. "To describe facts, to seek the connections between them, to inventory particular events—all these are acts carried out beyond any concern for time, money, or immediate utility." A scientific institution had to be above bureaucratic control so that it could "retain a freedom of thought and methods in the name of scientific truth, something no administrative organization would be able to do." Carle was concerned that the reorganization of the Sherifian Institute in the late 1930s would compromise free scientific inquiry, transforming colonial scientists into errand boys for scientific luminaries in the metropolis. Carle valued "the critical spirit, the spirit of observation, and the spirit of research" among colonial scientists. He hoped that colonial scientific institutions would promote the ultimate rights and liberties—those associated with "the search for Truth."[26]

The civilizing mission projected a reflection in the metropolis. Sébastien Charléty, rector of the Sorbonne and an old colonial hand in North Africa, sensed that France had been changed by its colonial experience. Oriental students were a fixture of the metropolitan universities, and Frenchmen no longer felt entirely at home in France. "The planet, formerly so far away, enters us, and some fear that it is submerging us." Charléty wondered whether the colonial phenomenon were not the central thread in history. France's historic mission was to administer her colonies in a manner above reproach. He urged the youth of France to forsake the routine of metropolitan administration for the magnificence and the danger of a colonial assignment, which rewarded initiative and entailed responsibility. He hinted at his own rise to metropolitan glory: "Who among our students dreams of being the great future leader after an imperial career?"[27]

The trick, a scientist knows, lies in formulating the right question. Charléty's question, with its appeal to personal gratification, goes to the heart of scientific inquiry in modern France. A German confronted

26. Carle, "La science au Maroc," ibid. 14 (1938), 97–103; for Carle's research, see idem, *Rapport sur la culture de coton au Maroc en 1924* (Paris, 1925).

27. Charléty, "L'université et l'expansion coloniale," *ACRACS* 8 (1932), 97–102. Charléty's elitist notion of education, not lacking in racist appeal, is summarized in Hans Zinsser's autobiography, *As I Remember Him: The Biography of R. S.* (Boston, 1940), pp. 197–99.

with something new demands to know whether it is right. A Netherlander questions whether it will hold water. But someone in France will want to know whether its external form is pleasing. The external manifestation of science is a printed text, and a French scientist consequently asks whether it appears in limpid, subtle prose; whether the text eschews tedious repetition and irrelevant digression; whether references are made to French precedent; whether, in short, the publication may represent French civilization. Can it decorate the table of the beef eaters and wine drinkers? Will the text transport its author to a council room or academy chamber where he rubs elbows with admirals and conquerors?

Language was the primary vehicle of French prestige and influence, Georges Hardy observed in 1929: "There is no longer any cultivated French person who does not believe in the particular virtues of the French language, in the seeds of clarity, of logic, and of humanity that it carries in the smallest folds of its syntax and its vocabulary."[28] The qualities of French were not sufficient, by themselves, to secure its preeminence. The language spread by regimented instruction. Albert Sarraut, minister of colonies, emphasized in 1923 that "instruction of the indigene, *generalized in the French language*, must have above all a practical and realistic character, whatever developments may follow in the sense of technology and culture." Sarraut cited with approval the observation of a former German consul general to the Belgian Congo about French colonial education: "Militarization and public instruction placed in the service of militarization are the means used for a vast and methodical Francization of the mass of indigenes, who are in fact destined, by a continuous and imperceptible reconciliation, to weld to the French mother country the parts of her colonial empire."[29] The French language also spread by learned inventory. Ambitious philological academies were prominent in Indochina, Madagascar, and Morocco.

The militant, patronizing vision of linguistic authority that underlay primary school and research institute led, paradoxically, to an attenuation of French influence in the world of science. Across the nineteenth and twentieth centuries, French scientific publications were increasingly irrelevant to foreigners not under the protective administration of French masters—the barbarians who dismissed the complicated conventions of a literary style removed from everyday speech. Physicists and astronomers especially became absorbed in complex and harmoni-

28. Hardy, "Problèmes d'enseignement colonial," *Revue de Paris* 36 (1929), 351–75, quotation on p. 371.

29. Sarraut, *La mise en valeur des colonies françaises* (Paris, 1923), pp. 97–98, emphasis in the original.

ous epiphenomena of form. By the end of the nineteenth century, science in France had shifted register; only men and women with special training in French could appreciate its counterpoint.[30]

Exact Sciences and the State Behemoth: France as a Modern Archetype

In the prosecution of the exact sciences, a close connection existed among priests, warriors, and colonial civil servants. The careers of fathers Elie Colin on Madagascar and Bonaventure Berloty in Lebanon provide examples of the symbiosis between scientific inventory and military pacification. Colin provided maps for Joseph Simon Gallieni's campaigns on Madagascar, and he supervised the colonial meteorological service; Berloty represented French interests in the Beqa'a Valley and provided weather reports for the French mandate in the Middle East. Colonial meteorologists, drawn in the nineteenth century from among military physicians and pharmacists, worked closely with military surveyors, naval hydrographers, and Jesuit geophysicists in the twentieth century. The affinity is characteristic: soldiers and sailors, traditional and conservative, viewed Catholic intellectuals with sympathy. When General Robert Bourgeois entered politics after rising to the Academy of Sciences, for example, he won election as a senator under the banner of the Catholic Parti d'action populaire in newly conquered Alsace.[31]

We may ask about the contribution of these researchers to French science as a whole. A rough index is obtained by using a classic survey of academic physics in 1900.[32] Physicists at Zikawei, Algiers, and various military schools — all of which are absent from the survey — would add about 20 percent to manpower estimates, a result having the French empire approach the British empire in personnel at institutions of higher learning. Overseas academic manpower continued to rise into the 1920s, with the creation of new university faculties at Hanoi and Beirut and with the expansion of institutions of higher learning in North Africa and Shanghai. Taking into account the publications of these and other physicists at Jesuit, military, and overseas locations —

30. Debate about the "decline" of French science, then, is misplaced. For a review of the declinist literature, see Robert Fox and George Weisz, "The Institutional Basis of French Science," in *The Organization of Science and Technology in France, 1808–1914*, ed. Fox and Weisz (Cambridge, 1980), pp. 1–28, on pp. 22–27.

31. *DBF*, s.v. "Bourgeois."

32. Paul Forman, John L. Heilbron, and Spencer Weart, *Physics circa 1900: Personnel, Funding, and Productivity of the Academic Establishments* (Princeton, 1975) [Historical Studies in the Physical Sciences, 5].

most of which appeared in house organs or as monographs – would also tend to equalize French and British scientific productivity.[33]

Whatever may be said about the heterogeneity of scientific establishments in France, one institution remained at the top of the heap – the Academy of Sciences. In the century after 1840 the academy drew its members to an appreciable extent from men who had made a career in the military or whose scientific work derived from overseas engagements. The exact sciences especially had a significant military component. The academy's structure reserved seats of honor for warriors.

During the nineteenth and twentieth centuries the Academy of Sciences had two divisions. The first division, *sciences mathématiques*, featured sections devoted to geometry, mechanics, astronomy, geography and navigation, and general physics; the second, *sciences physiques*, comprised sections of chemistry, mineralogy, botany, rural economy, anatomy and zoology, and medicine and surgery. Each of these eleven sections had six members. In addition, there was a division of ten *académiciens libres* and, in the twentieth century, six-member sections for nonresidents and for industrialist scientists.

The first division corresponds to what I have called "exact sciences." In this division, the section of geography and navigation was the domain of military and naval engineers and explorers; with the progressive mapping of the French empire, admirals and generals completely replaced civilian adventurers. The last nonmilitary explorer to sit in the section, Alfred Grandidier, was replaced by General Gustave Ferrié after the First World War. To the warriors' *chasse gardée*, which constituted one-sixth of the academy's exact-sciences division, must be added men admitted to the other sections or divisions as tribute to their overseas, naval, or military activity.[34] And as nearly all the latter academi-

33. Ibid., p. 116, where despite a poor showing in English and German bibliographies, France is considered to produce only one-sixth fewer papers than the British empire. French physicists are as a result held to be more productive than British ones (p. 119). This unreasonable conclusion (qualitative evidence suggests that physicists in the British empire published more, and more frequently, than their colleagues in the French empire) provides indirect evidence in support of my contention regarding the systematic underestimation of French manpower in the exact sciences.

34. Warriors and overseas activists outside the section of geography and navigation included: Admiral Mouchez in the first division's astronomy section; Jean-Baptiste-Joseph-Dieudonné Boussingault in rural economy; Auguste-Jean-Baptiste Chevalier in botany; Alexandre-John-Emile Yersin in medicine and surgery; Vice Admiral Abel Aubert du Petit-Thouars, Ferdinand-Marie de Lesseps, Vice Admiral Jean-Philippe-Ernest de Fauque de Jonquières, Marshal Ferdinand Foch, and Marshal Jean-Baptiste Philibert, comte de Vaillant, all as académiciens libres. Membership in the Academy of Sciences is given annually in the first number of the *CRAS*. My discussion is based on the years 1860, 1880, 1900, 1920, and 1939; it excludes foreign associates and correspondents. It is apparent that I argue for a revision of Robert Gilpin's evaluation that "a great cleavage has long

cians were observers of the physical world rather than speculators about ideal mathematical realms (the occasional physician and naturalist aside, the mathematician Vice Admiral Jean-Philippe-Ernest de Fauque de Jonquières is a notable exception),[35] warriors and imperialists exercised a strong voice — and sometimes a stentorian one — in the academy's deliberations about physics and astronomy.[36]

If priests, warriors, and colonizers played more than a cameo role in the drama of exact sciences in France, it is relevant to ask how that drama relates to the course of science generally. That is to say, are the Jesuits, the generals and admirals, and the colonial geophysicists nothing more than appurtenances of a scientific culture whose impact abroad diminishes as the French Revolution recedes in time? Is there wider meaning to the expanded picture of science in nineteenth-century and early twentieth-century France that is limned in the preceding pages?

The question calls for a comparatist response, and one is readily forthcoming. The three neglected dimensions of scientific activity in France find a reflection in the United States — a nation that, while certainly Francophile, is not usually identified as mirroring French scientific conventions. The United States has not figured in the preceding pages as a site of French imperialist designs, but American scientists often drew inspiration from French work and French models. The intent here is less to point to a transoceanic filiation than to suggest the extent to which French mandarins were the first to solve certain ineluctable problems of organizing modern scientific research.

During the nineteenth century, science in the United States reflected French norms of military involvement. The nation's premier scientific school in the first half of the century, the military academy at West Point, drew inspiration, syllabuses, texts, and teachers from the Ecole polytechnique. The U.S. army had its own corps of scientifically apprised explorers, the topographical engineers, whose exploits closely paralleled surveys by the French army in North Africa. Like their French homologues, American military physicians acted to funnel ecological observations from the Western domains to savants in the Eastern establishment. Beginning in 1830 the navy had a Depot of Charts and Instruments, which evolved into the United States Naval Observatory. Staffed by a naval corps of "professors of mathematics," the naval

existed between the academic world and the military establishment in France" (Gilpin, *France in the Age of the Scientific State* [Princeton, 1968], p. 258).

35. *DSB*, s.v. "de Jonquières."

36. The point is absent in a general survey by Maurice Crosland, "The French Academy of Sciences in the Nineteenth Century," *Minerva* 16 (1978), 73–104.

observatory and its cousin the Nautical Almanac Office became important sites for research in the exact sciences; the navy also had a hydrographical service and academy. To these military corps one may add paramilitary services, such as the Coast and Geodetic Survey and the various geological surveys. The civilian Weather Bureau emerged in the last third of the nineteenth century, in part from an army service. It is hardly surprising that America's first Nobel laureate, Albert Abraham Michelson, was trained as a professional killer at Annapolis.[37] When French astronomers sought American correspondents to help determine global longitudes on the eve of the First World War and again in the early 1920s, they naturally directed their inquiries to the navy.

Warriors have figured prominently in American science since the middle of the nineteenth century, when military operations absorbed around one-quarter of all federal funds allocated to scientific endeavors.[38] Viewed in this light, military penetration into research universities during the middle of the twentieth century reflects the desire of generals and admirals to reassert their authority as custodians of nature and interpreters of natural law.[39] The dominion of research universities

37. William H. Goetzmann, *Army Exploration in the American West, 1803–1863* (New Haven, 1959), p. 14, for West Point (the topographical corps comprised a total of seventy-two officers from 1838 to 1863, most of them West Point graduates); Edgar Erskine Hume, *Ornithologists of the United States Army Medical Corps* (Baltimore, 1942); Howard Plotkin, "Astronomers versus the Navy: The Revolt of American Astronomers over the Management of the United States Naval Observatory, 1877–1902," *Proceedings of the American Philosophical Society* 122 (1978), 386–99; T. G. Manning, *U.S. Coastal Survey vs. Naval Hydrographic Office: A Nineteenth-Century Rivalry in Science and Politics* (Tuscaloosa, 1988); *DSB*, s.v. "Abbe, Cleveland"; Steven J. Dick and LeRoy E. Doggett, eds., *Sky with Ocean Joined: Proceedings of the Sesquicentennial Symposia of the U.S. Naval Observatory, December 5 and 8, 1980* (Washington, D.C., 1983), esp. papers by Marc Rothenberg, Andreas R. Rechnitzer, Arthur L. Norberg, and Steven J. Dick; Donald R. Whitnah, *A History of the United States Weather Bureau* (Urbana, 1961); James Rodger Fleming, *Meteorology in America, 1800–1870* (Baltimore, 1990); A. Hunter Dupree, *Science in the Federal Government: A History of Policies and Activities to 1940* (Cambridge, Mass., 1957), pp. 184–94; Robert V. Bruce, *The Launching of Modern American Science, 1846–1876* (New York, 1987), pp. 201–14; Nathan Reingold, *Science in Nineteenth-Century America: A Documentary History* (Chicago, 1964), pp. 109–10, 135, 275–78. Reingold's book provides the single best treatment of these issues.

38. Nathan Reingold and Joel N. Bodansky, "The Sciences, 1850–1900: A North Atlantic Perspective," *Biological Bulletin* 168 (1985) [suppl., *The Naples Zoological Station and the Marine Biological Laboratory: One Hundred Years of Biology*], 44–61.

39. The recent literature on the American military-industrial complex and its connection with the academic world is largely silent about nineteenth-century antecedents and French precedents. No word is given to these issues in Everett Mendelsohn, Merritt Roe Smith, and Peter Weingart, "Science and the Military: Setting the Problem," in their edited volume *Science, Technology and the Military* (Dordrecht, 1988), pp. xi – xxix. For the army as a technological innovator, see Merritt Roe Smith, ed., *Military Enterprise and*

over science is indeed a short-term historical phenomenon, existing in Germany for about one hundred years and in the United States for half that span;[40] by the second third of the twentieth century major research installations had quit the groves of academia for the *campus martius*. To identify this commerce between learned professors and professional killers is not to assert that scientific *discourse* — what appeared in research publications — derived from military needs. It is to indicate the natural advantages of the military mind in carrying out many kinds of research project. The battlefield is an infernal laboratory. Athena is both goddess of war and avatar of wisdom.[41]

Junior officers may press gangs of draftees and indigenes to execute scientific projects; priests can act as their confessors and provide absolvement for greater or lesser sins. Priests, whose vision naturally transcends national boundaries and prejudices, are well suited by training and temperament for scientific research. A desire to uncover divine laws of nature runs through the fabric of American science. Cosmical and holistic apologies for scientific research, published in books currently shilled by Carl Sagan on the one hand and Lynn Margulis on the other hand, are genetically related to the sermons of orthodox naturalist divines who held forth a century ago at such elite, religious institutions as Swarthmore, Georgetown, and Princeton.

Little need be said about the religious side to the origins of higher education in the United States. We may nevertheless recall the progressive secularization of religious institutions over the course of the late nineteenth and twentieth centuries. Within living memory, test acts in one form or another flourished at America's finest colleges and universities. How secularization occurred is a complicated issue, but there can be no doubt that it took place under pressure from and through ex-

---

Technological Change: Perspectives on the American Experience (Cambridge, Mass., 1985), especially the historiographical remarks by the editor (pp. 1–37).

40. A perceptive account of the rise of the American research university appears in Nathan Reingold, "Graduate School and Doctoral Degree: European Models and American Realities," in *Scientific Colonialism: A Cross-Cultural Comparison*, ed. Reingold and Marc Rothenberg (Washington, D.C., 1987), pp. 129–49. The research university rose dramatically between 1880 and 1900, just at the time that the military and naval slice of federal science funding fell from 27 percent to 7 percent (Reingold and Bodansky, "Sciences," pp. 53–54).

41. I do not dispute the contention of Bernhelm Booss and Jens Høyrup, who see little connection between the discourse of mathematical science and military practice in nineteenth-century France (Booss and Høyrup, *Von Mathematik und Krieg* [Marburg, 1984], appearing in preliminary form as *Tekst* no. 64 [1983] of the Institut for Studiet af Matematik og Fysik samt deres Funktioner i Undervisning, Forskning og Anvendelser, Roskilde; Høyrup has also circulated an English translation).

amples set by a federal government that, like the Third Republic, was not given to encouraging religious institutions.[42]

Finally, it is apposite to invoke the figure of science in the American empire, whose decline is now heralded in learned treatises and on the silver screen. At the beginning of the twentieth century, the empire stretched from Cuba through Central America and the Caribbean to Panama, up to Alaska and across the Pacific to Hawaii and the Philippines. American scientists were active in all these domains. Furthermore, they lectured and published at American institutions from Shanghai to Beirut; they controlled observatories and national surveys in South America and South Africa. In these operations, many of which produced significant research, the Americans responded to metropolitan directives. The metropolis – whether in the form of the U.S. navy, Harvard University, the Carnegie Institution, or the Rockefeller Foundation – consecrated research programs and synthesized results.[43] Here we find yet another French solution to problems generated in a foreign cultural setting.

42. For general, comparative remarks, see Fritz Ringer, *Education and Society in Modern Europe* (Bloomington, 1979), pp. 248–59.

43. See *CIES*, pp. 127, 196–205, 271–72, for American-run institutions in Western Samoa, Argentina, and Shanghai; and *ER*, pp. 65–66, for American astronomers in South Africa.

Glossary of French Terms

Adjudant. Junior-grade officer.

Adjudant-astronome. Executive astronomer.

Aide-astronome. Assistant astronomer.

Aide-météorologiste. Assistant meteorologist.

Aide-physicien. Assistant physicist.

Agrégation. State examination for admission to teaching posts at *lycées* and universities.

Agrégé. One who has passed the *agrégation.*

Anciens élèves. Literally, "former pupils," usually meaning graduates.

Arrêté. Decree.

Arrêté viziriel. State ministerial decree in Morocco.

Assistant. Permanent, junior position at a state research institution.

Astronome-adjoint. Associate astronomer.

Astronome titulaire. Principal astronomer.

Baccalauréat. Secondary-school diploma.

Bachelier ès sciences. Bachelor of sciences.

Brevet d'ingénieur civil. Civil-engineering certificate.

Calculateur. Calculator.

Capitaine de frégate. Naval commander.

Capitaine de vaisseau. Naval captain.

Carte du ciel. Photographical map of the heavens, begun late in the nineteenth century.

Certificat. One of several university certificates required for a French *licence.*

Chargé de cours. Assistant lecturer.

Chargé de(s) travaux. Demonstrator.

Chef de service. Manager.

Collaborateur indigène. Native associate.

Collège. Secondary school, sometimes municipal or religious.

Conseil. Council.

Correspondant. Associate, nonresident member of an academy or learned society.

Corps des mines. State service of mining engineers.

Cumul de fonctions. Literally, "plurality of offices."

Diplômé. University graduate.

Diplôme d'études. University diploma.

Doctorat d'université. So-called university doctorate, established in 1898 for foreign students and not carrying the privileges of a usual state doctorate.

Doctorat ès lettres. State doctorate in humanities.

Doctorat ès sciences. State doctorate in sciences.

Ecole d'arts et métiers. Polytechnic or trade school.

Elève-astronome. Apprentice astronomer.

Faculté. Institution certified by the French state to provide instruction in higher education.

La France d'Outre-Mer. French Overseas Territory.

Grandes écoles. A handful of institutions of higher learning at the pinnacle of the French academic system, originally outside the university structure.

Hexagone. The vaguely six-sided polygon of the French Republic bounded by mountains and water.

Ingénieur. Literally, "engineer"; with qualifications, designating a rank in a state technical or scientific service.

Ingénieur-adjoint. Assistant engineer.

Ingénieur-agronome. Agronomist.

Ingénieur des mines. Mining engineer.

Ingénieur des ponts et chaussées. Civil engineer in government service.

Ingénieur-en-chef. Principal engineer.

Ingénieur-géophysicien. Geophysical engineer.

Ingénieur-météorologiste. Meteorological engineer.

Inspecteur général. Usually the chief civil servant in a state service.

Juvénat. Stage in the career of a Jesuit devoted to literary studies, following the *noviciat* and preceding the *régence.*

Lauréat. Recipient of a prize or grant.

Licence. Usual undergraduate diploma, originally a teaching certificate.

Licencié ès sciences. Bachelor of sciences.

Lieutenant de vaisseau. Naval lieutenant commander.

Liste d'aptitudes. Waiting list for state academic positions.

Lycée. Classical state secondary school.

Maître de conférences. University lecturer.

Mathématiques générales. First-year university mathematics course.

Médecin principal. Senior physician.

Membre-adjoint. Associate member.

Météorologiste-adjoint. Associate meteorologist.

Mutilé de guerre. Disabled veteran.

Navale. The Ecole navale at Brest.

Normalien. Graduate of the Ecole normale supérieure.

Noviciat. Period of probation at the beginning of a Jesuit's career.

P.C.N. Short course of study in physics, chemistry, and natural sciences for medical students, instituted at French universities in 1893.

Pépinière. Literally, "nursery."

Pharmacien major. Officer apothecary.

Physicien. Physicist.

Physicien-adjoint. Associate physicist.

Polytechnicien. Graduate of the Ecole polytechnique.

Préparateur. Laboratory assistant or demonstrator.

Professeur sans chaire. Literally, "professor without a chair," a position established in 1920 for senior university staff.

Professeur titulaire. Full professor.

Régence. Stage in the career of a Jesuit devoted to teaching, following the *juvénat* and preceding philosophical studies.

Résident. Ranking French authority in a protectorate or mandate.

Scholastique. Jesuit not yet ordained as a priest.

Sciences appliquées. Applied sciences.

Sciences fondamentales. Pure sciences.

Sous-directeur. Deputy director.

Stagiaire. Trainee.

Surveillant général. Vice principal.

Supérieurs majeurs. Jesuit provincial superiors.

Suppléant. Literally, "substitute" or locum tenens.

Titres. Diplomas and academic credentials.

Tutelle. Trusteeship or protective administration.

Note on Sources

The jewel of French libraries, the Bibliothèque nationale, is a source of information about everything, and the exact sciences in French spheres of influence find a place in it. Nevertheless, the library's holdings in science grow thin as the twentieth century progresses, its indexing system is occasionally confusing, and items requested are not infrequently unavailable. Restrictions on the number of volumes that may be consulted by a reader discourage systematic study of periodical runs and scientific monographs. Here is still the best collection of French doctoral dissertations, where much information may be obtained about French researchers overseas.

Among Parisian libraries with more accessible collections, one may note those at the Institut Henri Poincaré, the Faculté des sciences in Place Jussieu, the Institut de physique du globe, and above all the central office of Météorologie nationale on avenue Rapp. Beyond France, no location for printed material surpasses the libraries in greater Washington, D.C. In addition to collections at the Library of Congress, unusually important holdings are found in the libraries of the Smithsonian Institution, the Naval Observatory, and the National Oceanic and Atmospheric Administration (which has absorbed the collections of the Weather Bureau), as well as in the History of Medicine Division of the National Library of Medicine in Bethesda.

The published record of French Jesuit scientific activity merits special bibliographical study. A beginning for the Université l'Aurore in Zikawei is found in *Bibliotheca Missionum*, vol. 14, *Chinesische Missionsliteratur, 1910–1950*, pt. 1, ed. Johannes Rommerskirchen and Nikolaus Kowalsky (Rome, 1959), pp. 609–21. For the nineteenth and twentieth centuries, publications are available in large science libraries. Epistolary ephemera have found their way into many large Jesuit collections, for example, the one at Les Fontaines in Chantilly. Much information may be culled from various published mission, seminary, and academic reports, notably *Relations de Chine*, *Lettres de Jersey*, *Maduré-Madagascar*, and the *Bulletin de l'Université l'Aurore*, as well as the more general *Jesuit Science Bulletin* and the *Revue des questions scientifiques*.

No single location is strong in material relating to French scientific activity in the Americas, although the Latin American collection of the University of Texas at Austin offers a useful and accessible starting point. Much richer are the library of the Sociedad Científica Argentina, founded in 1872, and the Asociación Biblioteca José Babini, both in Buenos Aires.

If the record of French scientific expansion overseas may be retrieved only indirectly from published sources, archives are more obliging. In the papers of the ministry of public instruction, kept at the central deposit of the Archives nationales in Paris, may be found much material about observatories, expeditions, and expeditionaries, as well as information about astronomy and geophysics in Algeria. The historian of science does well to begin with *Inventaire des papiers de la division des sciences et lettres du Ministère de l'instruction publique et des services qui en sont issus (sous-série F17)*, ed. Marie-Elisabeth Antoine and Suzanne Olivier, 2 vols. (Paris, 1975–81). Indispensable for questions relating to overseas expansion are the archives of the Société de géographie, kept in the Bibliothèque nationale and admirably indexed by Alfred Fierro in *Inventaire des manuscrits de la Société de géographie* (Paris, 1984).

The archives of the Ministère des affaires étrangères, located at the Quai d'Orsay, contain much about French overseas initiatives in foreign parts, especially Asia and Latin America. Political and commercial correspondence is indexed chronologically by country; broader thematic dossiers are available, such as an inquiry of the late 1930s into French influence in Latin America (Amérique 1918–1940, dossiers généraux, 42) or an investigation of French professors in China during the 1920s (Asie 1918–1940, Chine 482). These archives are complemented by material kept in the Archives nationales, Section d'Outre-Mer, at Aix-en-Provence. The latter (which have absorbed the collection previously housed at the rue Oudinot in Paris) contain what remains of French colonial archival correspondence. The overseas archives are well-indexed, but there are notable lacunae in the Indochinese and Algerian material especially. A third major source of information is the military archives (army, navy, and air force) housed at the Château de Vincennes, just outside Paris. Hospitality ranges from spartan and grudging (the army) to efficient and tranquil (the navy) to relaxed (the air force). The archives of the naval Etat-Major contain material of most interest to the historian of exact sciences, as the navy had, in addition to its colonial responsibilities, a major role in overseas astronomical and hydrographical expeditions. In general, naval dossiers are more detailed and more informative than army and air force ones. There are numerous published guides to the material in these three places. Area introductions may be found in Archives nationales, *Guide des sources de l'histoire de l'Amérique latine et des Antilles dans les archives françaises* (Paris, 1984); and Commission française du Guide des sources de l'histoire des nations, *Sources de l'histoire de l'Asie et de l'Océanie dans les archives et bibliothèques françaises*, vol. 1, *Archives* (Munich, 1981).

French Jesuit archives are extensive and accessible. Of greatest latitude is the incoming material contained at the Cantilliaceusis Domus Sessionum et Exercitiorum, Les Fontaines, Chantilly. Here are found the papers of the former Jesuit Province of Paris, which governed Jesuit scientific institutions in China. The archives of the former Jesuit Province of Toulouse, which governed Jesuit science on Madagascar, are available at the Tolosana Residentia Maior, Cordis Iesu, in the rue des Fleurs, Toulouse. Some material concerning scientific institutions in Lebanon is contained in the Curtismonasteriensis Schola Apostolica Sancti at Cormontreuil (Marne). As a rule, Jesuit archives have large working libraries.

Scientific institutions in France contain much unpublished material of great importance. The Paris Observatory has recently made available its holdings for the nineteenth and early twentieth centuries; many files deal with overseas themes. The Institut de physique du globe holds the papers of its first director, Charles Maurain; these contain a great deal of correspondence with colleagues in overseas spheres of influence. The Institut de physique du globe has been unable to locate the archives of its dependency on Martinique, and I have been unsuccessful in corresponding with the Antilles. The library of the central office of Météorologie nationale, on avenue Rapp, has some unprocessed correspondence of the directors of the Bureau central météorologique and the Office national météorologique. I was not granted access to the archives of the Bureau des longitudes.

Relatively little professional correspondence from French overseas physicists and astronomers has found its way into archives beyond France. Occasionally, relevant material may be found in the Lick Observatory Archives (McHenry Library, University of California at Santa Cruz), the Hussey Family Papers (Michigan Historical Collection, Bentley Historical Library, Ann Arbor), and the Brera Observatory Archives (Milan). The papers of Fernand, comte de Montessus de Ballore, are held privately in Santiago de Chile, and Ernesto Sábato's papers are at his home, near Buenos Aires. I have not received permission to see either collection. The papers of Milan Ratislav Štefánik, tended with great care during the years before 1939, are apparently in Bratislava.

Archival research offers challenges to scholars over the coming generation. The files of the Zikawei Observatory, some twenty-five hundred items of correspondence from the years 1879–1952 located at the Shanghai Meteorological Bureau, are currently being processed. North African and Madagascan holdings have yet to be explored. Much material remains to be studied in Chile, Argentina, Brazil, and Mexico. I hope that it is seen before the paper disintegrates.

Acknowledgments

I am grateful to staff at the archives on which this book is based. I have received special help from librarians in greater Washington, D.C. – at the Library of Congress, the United States Naval Observatory, the National Oceanic and Atmospheric Administration, the Smithsonian Institution, and the History of Medicine Division of the National Library of Medicine. Librarians at the University of Michigan, the University of Toronto, and Princeton University have tolerated my cometary passage. Women and men at the libraries of McGill University, whose good will extends beyond the bounds of collegiality, have often come to my aid. Friendly assistance has been provided by the staff of the Bibliothèque de théologie, Collège Jean-de-Brébeuf, in Montreal. I owe a continuing debt to the interlibrary-loan staff at Vanier Library, Concordia University, Montreal.

My research has been supported by the Social Sciences and Humanities Research Council in Ottawa, the Izaak Walton Killam Program at the Canada Council of Arts, the Herbert C. Pollock Fund at the Dudley Observatory in Schenectady, and the Faculté des arts et sciences, Université de Montréal. I am grateful to the Macdonald-Stewart Foundation and its director Dr. Liliane Stewart, the Killam Program and its director Dr. Mel MacLeod, and the Dudley Observatory and its administrator Professor Ralph A. Alpher for helping to see the present volume through press.

The ideas elaborated in the preceding pages were aired at a number of colloquia over the past decade. For insight on these occasions I especially thank Professor Natalie Zemon Davis, Père Joseph Dehergne, S.J., Professor Dr. med. Wolfgang U. Eckart, Professor Roderick W. Home, Mme Catherine Jami, Professor Kathryn Olesko, Dr. Eduardo L. Ortíz, Mr. Patrick Petitjean, Mr. Roshdi Rashed, Professor Martin J. S. Rudwick, Dr. Juan José Saldaña, Dr. José Manuel Sánchez Ron, Professor Douglas Skopp, and Professor dr. Robert P. W. Visser. I have been helped in various ways by Ms. Joanne Allen, Professor Dr. Máximo Barón, Dr. Robert J. Brugger, Mr. Jacques Damiens, Ms. Luba zum Felde, Mr. George Franko, Professor Marcelo Montserrat, Dr. Terry Shinn, Mr. Philippe Véron, Mr. Xu Weimin, and Dr. Zhou Zuyi. My greatest debt is to my wife and children.

Index

About the Author

Lewis Pyenson, professeur titulaire in history at Université de Montréal, was born and raised in southern New Jersey. He is an advisory editor for the journals *Historical Studies in the Physical and Biological Sciences*, *History of Science*, and *Tractrix*. He is honorary curator of scientific instruments at the David M. Stewart Museum; he sits as a curator of the Osler Library for the History of Medicine at McGill University; and he is a member of the Toms River Schools Hall of Fame. From 1989 to 1991 he was a Killam Senior Fellow of the Canada Council of Arts. Among his books are *Neohumanism and the Persistence of Pure Mathematics in Wilhelmian Germany* (1983), *The Young Einstein* (1985), *Cultural Imperialism and Exact Sciences* (1985), and *Empire of Reason* (1989). He lives with his wife, Susan Sheets-Pyenson, and their three children on the western tip of the island of Montreal.